Lecture Notes in Computer Science

Edited by G. Goos, J. Hartmanis, and J. van Leeu

T0253622

Springer
Berlin
Heidelberg
New York
Hong Kong
London
Milan
Paris
Tokyo

Harrie de Swart   Ewa Orłowska
Gunther Schmidt   Marc Roubens (Eds.)

Theory and Applications of Relational Structures as Knowledge Instruments

COST Action 274, TARSKI
Revised Papers

 Springer

Series Editors

Gerhard Goos, Karlsruhe University, Germany
Juris Hartmanis, Cornell University, NY, USA
Jan van Leeuwen, Utrecht University, The Netherlands

Volume Editors

Harrie de Swart
Tilburg University, Faculty of Philosophy
P.O. Box 90153, 5000 LE Tilburg, The Netherlands
E-mail: H.C.M.deSwart@uvt.nl

Ewa Orłowska
National Institute of Telecommunications
ul. Szachowa 1, 04-894 Warsaw, Poland
E-mail: orlowska@itl.waw.pl

Gunther Schmidt
Universität der Bundeswehr München
Fakultät für Informatik, Institut für Softwaretechnologie
85577 Neubiberg, Germany
E-mail: Schmidt@informatik.unibw-muenchen.de

Marc Roubens
University of Liège, Department of Mathematics
Sart Tilman Building, 14 Grande Traverse, B37, 4000 Liège 1, Belgium
E-mail: M.Roubens@ulg.ac.be

Cataloging-in-Publication Data applied for

A catalog record for this book is available from the Library of Congress.

Bibliographic information published by Die Deutsche Bibliothek
Die Deutsche Bibliothek lists this publication in the Deutsche Nationalbibliografie;
detailed bibliographic data is available in the Internet at <http://dnb.ddb.de>.

CR Subject Classification (1998): I.1, I.2, F.4, H.2.8

ISSN 0302-9743
ISBN 3-540-20780-5 Springer-Verlag Berlin Heidelberg New York

Springer-Verlag is a part of Springer Science+Business Media

springeronline.com

© Springer-Verlag Berlin Heidelberg 2003
Printed in Germany

Typesetting: Camera-ready by author, data conversion by Olgun Computergrafik
Printed on acid-free paper   SPIN: 10976067   06/3142   5 4 3 2 1 0

Preface

Relational structures abound in the daily environment: relational databases, data mining, scaling procedures, preference relations, etc. Reasoning about and with relations has a long-standing European tradition. Today, there are strong European research groups in the theoretical as well as the applied branches.

European research in the field may be divided into three broad areas:
1. Algebraic Logic: algebras of relations, relational semantics, and algebras and logics derived from information systems.
2. Computational Aspects of Automated Relational Reasoning: decidability and complexity of algorithms, network satisfaction.
3. Applications: Linguistics, Psychology, Economics, etc.

While there is a wealth of theoretical knowledge to be used, there has been little interaction between basic and applied research in the field. For this reason, a European Concerted Research Action has been implemented, designated as COST Action 274: TARSKI (Theory and Applications of Relational Structures as Knowledge Instruments).

The main objective of this book is to advance the understanding of relational structures and the use of relational methods in applicable object domains. There are the following sub-objectives:
1. to study the semantical and syntactical aspects of relational structures arising from 'real world' situations;
2. to investigate automated inference for relational systems, and, where possible or feasible, develop deductive systems which can be implemented into industrial applications, such as diagnostic systems;
3. to develop non-invasive scaling methods for predicting relational data; and
4. to make software for dealing with relational systems commonly available.

We are confident that the present book will further the understanding of interdisciplinary issues involving relational reasoning. The study and possible integration of different approaches to the same problem, which may have arisen at different locations, will be of practical value to the developers of information systems.

The first five papers concern the *mechanization of relational reasoning*. This group of mechanization papers starts with a comparative report on two already existing systems by Rudolf Berghammer, Gunther Schmidt, and Michael Winter. The GUHA article by Petr Hájek, Martin Holeňa, and Jan Rauch refers to the well-developed system in Prague which derives information relations from information systems and is therefore some sort of a program for relational data mining. While there have been extensive studies in automated reasoning for propositional logics, Renate Schmidt and Ullrich Hustadt give a respective overview for modal and description logic reasoning systems. An attempt to develop a formal basis for theory extraction from relational data guided by some ontology is undertaken by Gunther Schmidt. Pasquale Caianiello, Stefania Costantini, and Eugenio Omodeo focus on definitional extensions applied to relational formalisms as a way of overcoming expressive limitations of logical formalisms.

The next three papers concern the field of *relational scaling and preferences*. Kim Cao-Van and Bernard De Baets discuss how a proper definition of a ranking can be introduced into the framework of supervised learning. Agnieszka Rusinowska gives an overview of axiomatic and strategic approaches to bargaining problems. Harrie de Swart et al. give an overview of the four major categories of voting procedures and their flaws.

The last four papers deal with the *algebraic and logical foundations of real world relations*. Wojciech Buszkowski presents relational representability results for the classes of algebras related to the Lambek syntactic calculus. Ivo Düntsch and Günther Gediga study modal-like approximation operators determined by binary relations and present their applications to practical problems that require a qualitative data analysis. Ivo Düntsch, Ewa Orłowska and Anna Radzikowska introduce and study a class of weak relation algebras based on not necessarily distributive lattices. Ingrid Rewitzky developed a relational model of programming languages whose commands may involve both angelic and demonic non-determinism.

Referees

Ricardo Caferra	Roger Maddux	Dimiter Vakarelov
Jules Desharnais	Ewa Orłowska	Hui Wang
Sašo Džeroski	Irina Perfilieva	Michael Winter
Marcelo Frias	Marc Roubens	
Günther Gediga	Gunther Schmidt	
Wendy MacCaull	Harrie de Swart	

Acknowledgements

We owe much to the referees mentioned above and are most grateful to them. Jozef Pijnenburg was instrumental in editing this book because of his highly appreciated expertise in LATEX. The cooperation of many authors in this book was supported by COST action 274, TARSKI, and is gratefully acknowledged.

Editors

Prof. Harrie de Swart, Chair, Tilburg University, The Netherlands
Prof. Ewa Orłowska, Institute for Telecommunications, Warsaw, Poland
Prof. Gunther Schmidt, Universität der Bundeswehr, München, Germany
Prof. Marc Roubens, Université de Liège and Faculté Polytechnique de Mons, Belgium

Table of Contents

RELVIEW and RATH –
Two Systems for Dealing with Relations

Rudolf Berghammer[1,*], Gunther Schmidt[2,*], and Michael Winter[3,*]

[1] Institut für Informatik und Praktische Mathematik
Christian-Albrechts-Universität zu Kiel
24098 Kiel, Germany
[2] Fakultät für Informatik
Universität der Bundeswehr München
85577 Neubiberg, Germany
[3] Computer Science Department
Brock University
St. Catharines, Ontario, Canada, L2S 3A1

Abstract. In this paper we present two systems for dealing with relations, the RELVIEW and the RATH system. After a short introduction to both systems we exhibit their usual domain of application by presenting some typical examples.

1 Introduction

In the area of logical reasoning, people began soon to look for subsets easier to handle than, for example, full predicate logic. This attempt resulted not least in relational reasoning. Already as early as 1915, Leopold Löwenheim postulated that one should resort to reasoning with relations in the "Gebietekalkul", and should "Schröderize" all of mathematics. This approach is certainly burdened with a loss in expressiveness. Nevertheless, such a loss has been accepted in the past by many scientists, as everything looks much simpler and it does not deteriorate expressiveness too much.

When working with relations today, one usually asks for additional computer aid. Three systems with quite different approaches have been proposed from our groups the last years. First, there may be just a specialized support in formula manipulation as in RALF (see [7, 8]), amended even by some automated features. A second approach is completely "on the model side" as with RELVIEW. Here, instead of working with binary predicates that may result in *true* or *false*, one works with Boolean matrices. This is a paradigm shift allowing to incorporate techniques known from linear algebra. In the RELVIEW system this has been elaborated in great detail to the extent that now something is available which might be compared to a "numerics package" – this time however for relational algebra. Thirdly, one may remain on the syntactic side, still avoiding to work

* Co-operation for this paper was supported by European COST Action 274 "Theory and Applications of Relational Structures as Knowledge Instruments" (TARSKI).

H. de Swart et al. (Eds.): TARSKI, LNCS 2929, pp. 1–16, 2003.

in a model. This means concentrating solely on the algebraic rules valid in the relational fragment. This characterizes the RATH approach. Logical reasoning is facilitated since the RATH system offers precise type control. Negation, e.g., need not be avoided, as due to the type restriction no unacceptably large result will show up. RATH also works if some of the rules of relation algebra are abandoned focusing on Dedekind categories, division allegories, etc. All the common aspects are handled simultaneously.

Considered in the context of the newly founded COST action 274: TARSKI, all systems seem extremely well-suited to fostering mechanization. Given the observation that many people keep inventing ideas to cope with relational structures arising around real-world phenomena, there is always the task to study whether these ideas are really helpful – whether they really work. The systems offer detailed computer help in different directions. Here, we exhibit in which way they may be used. Since RALF is currently not maintained, we concentrate on RELVIEW and RATH.

There is however a lot of work going on to further mechanize any form of work with relations. On the one hand side, a successor of RALF is currently under construction. On the other hand side, methods to decompose relations with respect to different criteria have been developed recently [14].

2   Relation-Algebraic Preliminaries

In this section, we briefly introduce the basic concepts of relation algebra, some special relations, and some relation-algebraic constructions. For more details concerning the algebraic theory of relations, see e.g., [4, 13].

Given non-empty sets X and Y, the set of all (set-theoretic or concrete) relations with domain X and range Y is denoted by $[X \leftrightarrow Y]$ and we write $R : X \leftrightarrow Y$ instead of $R \in [X \leftrightarrow Y]$. If X and Y are finite and of cardinality m and n, respectively, then we may consider R as a Boolean matrix with m rows and n columns. This matrix interpretation is well-suited for many purposes. Therefore, in this paper we frequently will use matrix concepts and notations also for relations. Especially, we will speak of rows and columns, and we will denote membership by R_{xy} instead of $(x, y) \in R$.

We assume the reader to be familiar with the basic operations on relations, viz. R^{T} (transposition), $\overline{R}$ (negation), $R \cup S$ (union), $R \cap S$ (intersection), RS (composition), $R \subseteq S$ (inclusion, subrelation test), and the special relations O (empty relation), L (universal relation), and I (identity relation). With the set-theoretic operations $\overline{}, \cup, \cap, \subseteq$ and the constants O, L such relations, respectively Boolean, matrices form a complete Boolean lattice. Further well-known laws for operations on relations are, for instance:

$$R^{\mathrm{T}^{\mathrm{T}}} = R \qquad Q(R \cap S) \subseteq QR \cap QS \qquad (RS)^{\mathrm{T}} = S^{\mathrm{T}} R^{\mathrm{T}}$$

The theoretical framework for such laws to hold is that of a *relation algebra*. First, such an algebraic structure is a category. I.e., there is a class of objects; for every pair A, B of objects there is a class $\mathcal{R}_{AB}$ of morphisms, and for all triples

$\mathcal{R}_{AB}$, $\mathcal{R}_{BC}$, $\mathcal{R}_{AC}$ there is a composition from $\mathcal{R}_{AB} \times \mathcal{R}_{BC}$ to $\mathcal{R}_{AC}$ such that associativity holds and for all $\mathcal{R}_{AB}$ there exists precisely one left identity from $\mathcal{R}_{AA}$ and one right identity from $\mathcal{R}_{BB}$. The morphisms are called (abstract) relations and for their composition and the identity relations we use here the same notation as for concrete relations. However, this category is extended by a transposition operation mapping relations from $\mathcal{R}_{AB}$ to $\mathcal{R}_{BA}$, where we use again the notation of the concrete case. Furthermore, the following properties are demanded to hold:

1. Every class $\mathcal{R}_{AB}$ is a complete Boolean lattice with the usual operations $\overline{}$, $\cup$, $\cap$, the ordering $\subseteq$, and the least (empty) relation O and greatest (universal) relation L.
2. For all relations $Q \in \mathcal{R}_{AB}$, $R \in \mathcal{R}_{BC}$, and $S \in \mathcal{R}_{AC}$ the following so-called *Schröder equivalences* hold:

$$Q^{\mathrm{T}}\overline{S} \subseteq \overline{R} \iff QR \subseteq S \iff \overline{S}R^{\mathrm{T}} \subseteq \overline{Q} \tag{1}$$

Often, in particular within the RELVIEW system, the following so-called *Tarski rule* is required as a further axiom; it is strongly connected to a generalization of the notion of simplicity known from universal algebra:

$$\mathsf{L}R\mathsf{L} = \mathsf{L} \iff R \neq \mathsf{O} \tag{2}$$

Note that for $R \in \mathcal{R}_{BC}$ in the equality of (2) there occur three – possible different – universal relations, viz. from $\mathcal{R}_{AB}$ and $\mathcal{R}_{CD}$ on the left-hand side and from $\mathcal{R}_{AD}$ on the right-hand, which all are denoted by the same symbol.

Let R be a (concrete or abstract) relation. Then R is called *univalent* (or *functional* respectively a *partial mapping*) if $R^{\mathrm{T}}R \subseteq \mathsf{I}$, and *total* if $R\mathsf{L} = \mathsf{L}$. As usual, a *mapping* is a univalent and total relation. Relation R is called *injective* if R^{T} is univalent and *surjective* if R^{T} is total. A *bijective* relation is an injective and surjective relation.

Now, let R in addition be *homogeneous*, i.e., a relation for which the specific product RR exists. (In the abstract case this is equivalent to $R \in \mathcal{R}_{AA}$ and in the concrete case this is equivalent to $R : X \leftrightarrow X$.) Then R is called *reflexive* if $\mathsf{I} \subseteq R$, *transitive* if $RR \subseteq R$, and *antisymmetric* if $R \cap R^{\mathrm{T}} \subseteq \mathsf{I}$. A *partial order* is a reflexive, antisymmetric, and transitive relation. The *transitive closure* of R is defined as $R^{+} = \bigcup_{i>0} R^{i}$, where $R^{0} = \mathsf{I}$ and $R^{i+1} = RR^{i}$ for all $i \in \mathbb{N}$. Using R^{+}, the *reflexive-transitive closure* R^{*} of R may be defined through $R^{*} = \mathsf{I} \cup R^{+}$. If $R^{+} \subseteq \overline{\mathsf{I}}$, then R is said to be *acyclic*.

A relation v with $v = v\mathsf{L}$ is called a *(row-) vector*. In the case of a concrete relation $v : X \leftrightarrow Y$ this condition means that an element from X is either in relation to none of the elements or to all elements of Y. Hence, v equals a Cartesian product $X' \times Y$, where X' is a subset of X. As for a concrete vector the range is without relevance, we consider in the following frequently vectors $v : X \leftrightarrow \mathbf{1}$ with a singleton set $\mathbf{1} = \{\bot\}$ as range and write then v_x instead of $v_{x\bot}$, i.e., suppress the second index. Such a vector v may be considered as a Boolean matrix with exactly one column, i.e., as a Boolean column vector. It describes the set $X' = \{x \in X \mid v_x\}$.

Sets may also be described via embedding mappings. Given an injective mapping $\imath : X' \leftrightarrow X$, we may regard X' as a subset of X. Then the vector $\imath^\mathrm{T}\mathsf{L} : X \leftrightarrow \mathbf{1}$ describes X' in the above sense. A transition in the other direction, i.e., the construction of an injective mapping $\mathsf{inj}(v) : X' \leftrightarrow X$ from a given non-empty vector $v : X \leftrightarrow \mathbf{1}$ describing X' in such a way that $\mathsf{inj}(v)_{yx}$ if and only if $y = x$, is also possible. Using matrix terminology, one only has to remove from the identity matrix those rows which don't correspond to an element of X'. We call $\mathsf{inj}(v)$ the *injective mapping generated by* v. A relation-algebraic axiomatization of this construction can be found in [2].

The *left residual* of S over R is defined by $S \mathbin{/} R = \overline{\overline{S}R^\mathrm{T}}$ and the *right residual* of S over R is defined by $R \mathbin{\backslash} S = \overline{R^\mathrm{T}\overline{S}}$. One also considers relations which share properties of left and right residuals simultaneously, viz. *symmetric quotients*. This construction is defined by $\mathsf{syq}(R, S) = (R \mathbin{\backslash} S) \cap (R^\mathrm{T} \mathbin{/} S^\mathrm{T})$. In the case of concrete relations we have that $(S \mathbin{/} R)_{xy}$ if and only if R_{yz} implies S_{xz} for all z, that $(R \mathbin{\backslash} S)_{xy}$ if and only if R_{zx} implies S_{zy} for all z, and that $\mathsf{syq}(R, S)_{xy}$ if and only if R_{zx} is equivalent to S_{zy} for all z.

3   A Short Introduction to the Systems

In this section, we want to give an impression of two computer systems, called RELVIEW and RATH. Applications will be presented in Section 4. More details and advanced applications can e.g., be found in [2, 3, 9, 10].

3.1   The RELVIEW-System

RELVIEW is an interactive and graphic-oriented computer system for calculating with relations and relational programming. In it all data are represented as relations which the system visualizes in two different ways. First, for homogeneous relations it offers a representation as directed graphs, including sophisticated algorithms for drawing them nicely. Alternatively, arbitrary relations may be depicted as Boolean matrices. This second representation is very useful for visually editing and also for discovering various structural properties that are not evident from a representation of relations as directed graphs. Because RELVIEW computations frequently use very large relations, for instance, membership, inclusion, and size comparison on powersets, the system uses a very efficient implementation of relations via reduced ordered binary decision diagrams. See [10] for its detailed description.

The RELVIEW system can manage as many relations simultaneously as memory allows and the user can manipulate and analyse them by pre-defined operations, tests and user-defined relational functions and relational programs. The pre-defined operations on relations include e.g., ^, -, |, &, and * for transposition, negation, union, intersection, and composition; the relational tests include e.g., incl, eq, and empty for testing inclusion, equality, and emptiness of relations. All that can be accessed through command buttons and simple mouse-clicks.

But the usual way is to use the pre-defined operations and tests to construct relational functions and relational programs.

A declaration of a relational function in the programming language of the RELVIEW system is done as usual in mathematics. Hence, it has the form $f(R_1, \ldots, R_n) = E$, where f is the name of the function, the R_i, $1 \leq i \leq n$, are the formal parameters (standing for relations), and E is a relation-algebraic expression over the relations of the workspace of the RELVIEW system that can additionally contain the formal parameters R_i.

As a simple example, the following unary relational function hasse computes the so-called Hasse diagram $R \cap \overline{RR^+}$ of an acyclic relation R:

```
hasse(R) = R & -(R * trans(R)).
```

In this declaration, a call of the pre-defined operation trans yields the transitive closure of its argument.

A relational program in RELVIEW essentially is a while-program based on the datatype of relations. Such a program has many similarities with a function procedure in languages like Pascal or Modula-2. It starts with a head line containing the name of the program and the list of formal parameters. Then the declarations of the local domains, functions, and variables follow. The last part of a program is its body, a sequence of statements which are separated by semicolons and terminated by the RETURN-clause.

We give again a simple example. If $g = (X, R)$ is a directed graph with the set of arcs given by the relation $R : X \leftrightarrow X$ and $s : X \leftrightarrow 1$ is a vector describing a subset X' of X, then the vector $(R^*)^\mathsf{T} s : X \leftrightarrow 1$ describes the set of those vertices which are reachable from some vertex of X'. Without using the reflexive-transitive closure, the latter vector may be computed by the following relational program: .

```
reach(R,s)
    DECL u, v
    BEG  u = s;
         v = R^ * u & -u;
         WHILE -empty(v) DO
            u = u | v;
            v = R^ * u & -u OD
         RETURN u
    END.
```

RELVIEW can be used to solve many different tasks. First, it assists the formulation and the proof of relation-algebraic theorems. In this field, the system can help to construct examples which support the validity of a theorem or to find – via random generated relations – counter-examples to disprove the considered relation-algebraic property. Relational program development is a second very important application of RELVIEW. Whereas relation algebra in combination with a programming logic (e.g., the Hoare calculus) gives a formal basis for ensuring correctness of the derived relational programs, RELVIEW supports many validation tasks for the development of a relational algorithm. For example, it can be applied to check the formal relational problem specification against the informal fixed requirements. Experimenting with relations and relation-algebraic propositions, the system may also help to find loop invariants or other decisive properties necessary for a correctness proof of a relational program. As a third application, the execution of a relational program or a piece of it by means of RELVIEW in the course of a program derivation can reveal alternative development steps and possibilities for optimization.

3.2   The RATH-System

The RATH-System presents a library of Haskell modules that allows to explore relation algebras and several weaker structures such as categories, allegories, distributive allegories, division allegories (see e.g., [5]) and Dedekind categories (see e.g., [12]) by providing tools to construct and test such structures. These modules constitute a common framework for calculational work with all the structures mentioned. It takes into account that they share concepts and properties so as to be able to, for instance, introduce the idea of division only once for division allegories and to directly reuse it for the more specific Dedekind allegories as well as for relation algebras.

For example, in RATH a parameterized data structure Cat obj mor representing categories with objects given by the type obj and morphisms given by the type mor may be defined as follows:

```
data Cat obj mor = Cat
   {cat_isObj   :: obj -> Bool
   ,cat_isMor   :: obj -> obj -> mor -> Bool
   ,cat_objects :: [obj]
   ,cat_homset  :: obj -> obj -> [mor]
   ,cat_source  :: mor -> obj
   ,cat_target  :: mor -> obj
   ,cat_idmor   :: obj -> mor
   ,cat_comp    :: mor -> mor -> mor}
```

Here cat_objects yields the list of objects and cat_homset yields for a pair of objects the list of morphisms. Source and target of a morphism are computed via cat_source and cat_target. Composition of morphisms is given by cat_comp and cat_idmor applied to an object yields the corresponding identity morphism. Finally, cat_isObj and cat_isMor test whether an element from obj respectively mor is indeed an element of the category. This is necessary, as one might be using the datatypes Int or String to denote the objects.

As an application of the above data structure, we want to implement the one-object category of truth values. It may be defined as an element of the data structure of categories as follows:

```
catB :: Cat () Bool
catB = Cat
  {cat_isObj   = const True
  ,cat_isMor   = const $ const $ const True
  ,cat_objects = [()]
  ,cat_homset  = const $ const [False, True]
  ,cat_source  = const ()
  ,cat_target  = const ()
  ,cat_idmor   = const True
  ,cat_comp    = (&&)}
```

There is exactly one object () of type () and two morphisms True and False of type Bool. Composition is given by intersection such that True becomes the identity. Now, the comprehensive test mechanism of RATH could be used to verify that catB is indeed a category. An execution of

performAll acat_TEST catB

will apply all pre-defined tests for categories on this structure. Of course, it is also possible to perform other tests on such a structure in order to find models with a specific property by using the underlying language Haskell. Later on, we will demonstrate this approach by an example.

A next step in the hierarchy of structures could be the representation of allegories. Following [5], an allegory is a category with some extra structure, in particular an intersection, a transposition operation, and an inclusion test. Within RATH, a corresponding parameterized data structure looks as follows:

```
data All obj mor = All
  {all_cat    :: Cat obj mor
  ,all_transp :: mor -> mor
  ,all_meet   :: mor -> mor -> mor
  ,all_incl   :: mor -> mor -> Bool}
```

Similarly, also other relational categories, including relation algebras, can be introduced. In doing so, it is also possible to compute different relation-algebraic expressions. Usually, such an execution is not as efficient as in RELVIEW. But in contrast with RELVIEW, the RATH system provides the possibility to switch to nonstandard relation algebras, i.e., to exchange the underlying model of the relational category in question.

RATH also provides means to construct new algebras from given ones as product algebras, subalgebras and matrix algebras. Last but not least, it is possible to generate a specific relational category by defining the corresponding operations on the set of atoms and taking the complex algebra over this atom structure (see [11]). This reduces the size of the algebra and the complexity of the operations. Using the representation of relation algebras by their atom structure, a wide variety of such algebras was generated with the system. For the moment, this variety contains 4527 different integral relation algebras (see again [11]). Besides these algebras, there are several examples of relation algebras included, which model quite simple everyday situations such as compass directions, interval interdependency, spatial information with "mereology", etc. Note that the number of available algebras is much larger since one may apply the constructions mentioned above to those integral algebras, too.

The sources of the whole RATH-System constitute executable Haskell code. A first account on RATH is given in [9]. This report contains examples of non-standard relation algebras – in particular algebras which, considered from the classical viewpoint, fail to correspond to our imagination of relations as sets of pairs. A very well-known non-standard relation algebra goes back to R. McKenzie and is described in detail in [13].

4 Applications

This section is devoted to some applications of RELVIEW and RATH. First, we concentrate on RELVIEW and show how to solve problems on concrete relations with its help. In the second subsection, we then use RATH to verify that a relation-algebraic proof of a property obviously holding for concrete relations requires the Tarski rule (2) as an additional axiom and that a well-known property of direct products and disjoint unions requires representability.

4.1 Computing Cut Completions and Concept Lattices

Let (X, R) be a partially ordered set, i.e., $R : X \leftrightarrow X$ be a partial order relation. Furthermore, assume $\varepsilon : X \leftrightarrow 2^X$ to be the membership relation between X and its powerset 2^X. This means that ε_{xs} if and only $x \in s$. For $s \in 2^X$, let $\mathrm{Ma}_R(s)$ denote its upper bounds wrt. R and $\mathrm{Mi}_R(s)$ denote its lower bounds wrt. R. Then $c \in 2^X$ is called a *(Dedekind) cut* of (X, R) if

$$c = \mathrm{Mi}_R(\mathrm{Ma}_R(c)), \tag{3}$$

i.e., if the first-order formula

$$\forall x : x \in c \leftrightarrow x \in \mathrm{Mi}_R(\mathrm{Ma}_R(c)) \tag{4}$$

holds. Obviously, formula (4) is equivalent to the formula

$$\exists s : \forall x : (x \in c \leftrightarrow x \in \mathrm{Mi}_R(\mathrm{Ma}_R(s))) \wedge c = s. \tag{5}$$

It is well-known that for an element $x \in X$ the set

$$(x) := \mathrm{Mi}_R(\mathrm{Ma}_R(\{x\})) = \{y \in X \mid R_{yx}\}$$

is a cut, called the *principal cut* generated by x. Now, let $\mathcal{C}$ denote the set of cuts of (X, R). Then $(\mathcal{C}, \subseteq)$ is a complete lattice, called the *cut completion* of (X, R), and the function mapping x to the principal cut (x) is an injective order homomorphism.

For a relation-algebraic construction of the cut completion of (X, R), we start with the definition that $y \in X$ is a lower bound of $s \in 2^X$ if and only if for all z from $z \in s$ it follows R_{yz}. Then we describe s by a vector $v : X \leftrightarrow 1$ and use the property of left residuals given at the end of Section 2. We obtain that the set $\mathrm{Mi}_R(s)$ is described by the vector $\mathrm{mi}(R, v) = R / v^{\mathsf{T}}$. Transposing the relation R yields $\mathrm{ma}(R, v) = R^{\mathsf{T}} / v^{\mathsf{T}}$ as the vector describing $\mathrm{Ma}_R(s)$. In the language of RELVIEW, hence, we obtain the following two relational functions mi and ma for computing lower and upper bounds:

$$\mathrm{mi(R,v)} = \mathrm{R} / \mathrm{v\hat{}}. \qquad\qquad \mathrm{ma(R,v)} = \mathrm{R\hat{}} / \mathrm{v\hat{}}.$$

If the second argument of these functions is not a vector but an arbitrary relation, then obviously they compute lower and upper bounds column-wise.

Using relation algebra and the formulae (4) and (5) – for the characterization of cuts (5) is more suited since it immediately leads to a symmetric quotient construction $\mathrm{syq}(\varepsilon, \ldots)_{cs}$ – in combination with the three relational functions mi, ma, and syq, we obtain the vector $\mathrm{cutvector}(R) : 2^X \leftrightarrow 1$ describing the elements of 2^X which are cuts, i.e., the set $\mathcal{C}$, as follows:

$$\mathrm{cutvector}(R) = (\mathrm{syq}(\varepsilon, \mathrm{mi}(R, \mathrm{ma}(R, \varepsilon)))) \cap \mathsf{I})\mathsf{L}$$

This relational specification may immediately be transformed into a relational program in the language of RELVIEW. The result is:

```
cutvector(R)
  DECL M, O, I
  BEG  M = epsi(On1(R));
       O = On1(M^);
       I = I(O * O^)
       RETURN dom(syq(M,mi(R,ma(R,M)))) & I)
  END.
```

In this program On1, epsi, I, and dom are pre-defined operations. The call On1(S) yields the universal vector with one column and the same row number as S, the call epsi(v) yields the membership relation with the cardinality of the base set given by the row number of the vector v, the call I(S) yields the identity relation with the same dimension as S, and a call dom(S) computes the composition of S with a one-column universal vector.

Let $v : 2^X \leftrightarrow 1$ abbreviate the vector $\mathrm{cutvector}(R)$ and $\mathrm{inj}(v) : \mathcal{C} \leftrightarrow 2^X$ be the injective mapping generated by v. Furthermore, define the relation $C : X \leftrightarrow \mathcal{C}$ by $C = \varepsilon \, \mathrm{inj}(v)^{\mathsf{T}}$. Then a little reflection shows for all $x \in X$ and $c \in \mathcal{C}$ the equivalence of $x \in c$ and C_{xc}. This means that the columns of C describe the cuts of (X, R). Cuts are ordered by inclusion. Using the property of right residuals given at the end of Section 2, we get for all $c, d \in \mathcal{C}$ the equivalence of $c \subseteq d$ and $(C \setminus C)_{cd}$. Hence, the ordering on $\mathcal{C}$ equals the right residual $C \setminus C$. If we formulate the procedure just described in the language of RELVIEW, we arrive at the following relational program:

```
cutcompletion(R)
  DECL v, C
  BEG  v = cutvector(R);
       C = epsi(On1(R)) * inj(v)^
       RETURN C \ C
  END.
```

Now, let us turn to concept analysis. Here one deals with *(formal) contexts* which are triples (G, M, I) consisting of a set G of objects, a set M of attributes, and an *incidence relation* $I : G \leftrightarrow M$. A *(formal) concept* is a pair (a, b), where $a \in 2^G$, $b \in 2^M$, $a' = b$, and $b' = a$. Here the sets a' and b' are defined as follows:

$$a' = \{y \in M \mid \forall x \in a : I_{xy}\} \qquad b' = \{x \in G \mid \forall y \in b : I_{xy}\} \qquad (6)$$

If (a, b) and (c, d) are two concepts, then (a, b) is defined to be *less general or equal* than (c, d), denoted by $(a, b) \leq (c, d)$, if $a \subseteq c$ or, equivalently, $b \supseteq d$. With this relation the set $\mathcal{K}$ of all concepts constitutes a complete lattice, in [6] called *(formal) concept lattice*. Sometimes, e.g., in [1], also the term *Galois lattice* is used. It is obvious that the concept lattice $(\mathcal{K}, \leq)$ is isomorphic to $(\mathcal{K}_G, \subseteq)$, with the carrier set defined as $\mathcal{K}_G = \{a \in 2^G \mid \exists b \in 2^M : (a, b) \in \mathcal{K}\}$, and also to $(\mathcal{K}_M, \supseteq)$, with the carrier set defined as $\mathcal{K}_M = \{b \in 2^M \mid \exists a \in 2^G : (a, b) \in \mathcal{K}\}$.

In the following, we concentrate on the computation of $(\mathcal{K}_G, \subseteq)$. Fundamental for this is the simple fact that the equations of (6) generalize the notions of upper bounds and lower bounds from partial order relations to arbitrary relations. Using also the notations $\mathrm{Ma}_I(a)$ and $\mathrm{Mi}_I(b)$ instead of a' and b', we, furthermore, get for $a \in 2^G$ the equivalence of $a \in \mathcal{K}_G$ and

$$a = \mathrm{Mi}_I(\mathrm{Ma}_I(a)). \qquad (7)$$

Property (7) is exactly the defining equation (3) for a set to be a cut. Hence, the lattice $(\mathcal{K}_G, \subseteq)$ generalizes the construction of a cut completion from a partial order relation to an arbitrary (incidence) relation. As a consequence, the relational program cutcompletion can also be used to compute for a context (G, M, I) the ordering of the lattice $(\mathcal{K}_G, \subseteq)$.

We have tested this approach with many examples. The following table shows the execution times (in seconds) for contexts of the specific form $(G, G, \bar{\mathsf{I}})$. They constitute the worst case since they lead to 2^n concepts, with n being the cardinality of G. The tests have been carried out on a Sun Fire-280R workstation running Solaris 7 at 750 MHz and with 8 GByte main memory.

n	10	11	12	13	14	15	16	17	18	19	20
cutvector	0.01	0.02	0.43	0.96	2.16	4.64	10.0	28.1	54.2	118	257
cutcompl.	0.19	0.22	0.74	1.89	5.84	13.3	38.9	89.1	204	400	1127

On a modern PC we obtained even better results. E.g., for $n = 19$ instead of 400 only 258 seconds are needed to compute the concept lattice of $(G, G, \bar{\mathsf{I}})$.

In many applications of context analysis, experts learn from contexts by carefully inspecting their concept lattices. Therefore, these lattices have modest size because otherwise they are hard to analyze visually. This is ideal for applying RELVIEW, especially since in such a case the system not only allows the fast computation of the ordering of the lattice but also its nice drawing as a graph and its further interactive graphical and relation-algebraic manipulation. To a certain extent RELVIEW can also be used for larger examples. But this requires more sophisticated relational programs which avoid the use of a membership relation. We have developed such a RELVIEW-program. Starting with the incidence relation it stepwise generates the ordering of $(\mathcal{K}_G, \subseteq)$ by gradually inserting missing least upper bounds and greatest lower bounds. Its detailed description, however, is out of the scope of this paper.

4.2 Investigating Properties of Abstract Relations

As mentioned in Section 3.1, the RELVIEW system may be used to construct counter-examples of a relation-algebraic property in question. Since RELVIEW uses just one specific model, the relation algebra of concrete relations between finite sets, it may fail. Using RATH we are able to switch to some abstract relation algebra providing the required counter-example. In this section we want to demonstrate this approach.

In any relational category one may prove that the following formula (8) is valid if one of the universal relations on the left hand side is homogeneous:

$$LL = L \tag{8}$$

A relational category such that the equation (8) holds in general, i.e., it is valid for all objects A, B, C and universal relations from $\mathcal{R}_{AB}, \mathcal{R}_{BC}$ and $\mathcal{R}_{AC}$, respectively, is called a uniform one. Obviously, the Tarski rule (2) implies uniformity. One may ask if there exists a non-uniform relation algebra. An example was given in [15]. This algebra has two objects and at most 4 relations in the corresponding sets of morphisms.

As mentioned in Section 3.2, it is possible to define the operations on relations just on the underlying set of atoms. Therefore, we define the following data structures and lists of elements in Haskell:

```
data Obj = A | B deriving (Eq, Ord, Show)
objseq = [A, B]

data A2 = At1 | At2 deriving (Eq, Ord, Show, Read)
atoms :: Obj -> Obj -> [A2]
atoms A A = [At1, At2]
atoms _ _ = [At1]
```

Consequently, we will have 4 relations between A and A and 2 relations otherwise. Notice, that the deriving-clause generates an equality, a linear ordering, a show respectively a read function for Obj respectively A2. Transposition and composition of atoms are defined as follows:

```
transpTab :: Obj -> Obj -> A2 -> A2
transpTab _ _ x  = x

atComp :: Obj -> Obj -> Obj -> A2 -> A2 -> [A2]
atComp A A A At1 At1 = [At1]
atComp A A A At2 At1 = []
atComp A A A At1 At2 = []
atComp A A A At2 At2 = [At2]
atComp A A B At1 At1 = [At1]
atComp A A B At2 At1 = []
atComp B A A At1 At1 = [At1]
atComp B A A At1 At2 = []
atComp B A B At1 At1 = [At1]
atComp A B A At1 At1 = [At1]
atComp _ _ _ At1 At1 = [At1]
```

The operations and data structures are converted into a category respectively an allegory by the following declarations:

```
aCat_NUW :: ACat Obj A2
aCat_NUW = ac where
  ac = ACat
  {acat_isObj  = const True
  ,acat_isAtom = (\ s t a -> a 'elem' atoms s t)
  ,acat_objects = objseq
  ,acat_atomset = atoms
  ,acat_idmor  = acat_idmor_defaultM ac
  ,acat_comp   = atComp}

aAll_NUW :: AAll Obj A2
aAll_NUW = AAll
  {aall_acat = aCat_NUW
  ,aall_converse = transpTab}
```

Finally, the corresponding relation algebra is given as the complex algebra over the allegory aAll_NUW. This is done by the following declaration:

```
ra_NUW :: RA Obj (SetMor Obj A2)
ra_NUW = atomsetRA aAll_NUW
```

Again, the comprehensive test mechanism of RATH can be used to verify that ra_NUW is indeed a relation algebra. Furthermore, the pre-defined test for uniformity ded_uniform_TEST may be applied to this structure. An execution of

```
performAll ded_uniform_TEST (ra_ded ra_NUW)
```

shows the following test result:

```
=== Test Start ===
non-uniform
 Objects:
  A
  B
  A
 Morphisms:
  SetMor ({At1},A,B)
  SetMor ({At1},B,A)
  SetMor ({At1, At2},A,A)
  SetMor ({At1},A,A)
=== Test End   ===
```

This confirms that ra_NUW is not uniform. Furthermore, it gives us a counterexample for the validity of equation (8). The composition of the greatest relation SetMor({At1},A,B) from A to B with its transposed SetMor({At1},B,A) yields SetMor({At1},A,A), which, however, is not equal to the greatest relation SetMor({At1, At2},A,A) from A to A.

If we denote the cartesian product and the disjoint union of two sets A and B by $A \times B$ and $A + B$, respectively, we have the well-known isomorphism

$$2^A \times 2^B \cong 2^{A+B}. \tag{9}$$

Within the theory of relations there are abstract counter-parts of cartesian products, disjoint unions and powersets, called the *(relational) product*, the *(relational) sum* and the *(relational) power*. One may ask whether (9) is valid in all relational categories. In [15] it is shown that this is true if all required objects exist. Furthermore, it is shown that every relation algebra may be embedded into an algebra with relational sums and powers. On the other hand, the existence of products implies representability, i.e., the algebra may be embedded into the relation algebra of concrete relations. Since there are non-representable algebras, a relation algebra exists with an object 2^{A+B}, which is not isomorphic to the product of 2^A and 2^B. The proof sketched so far is non-constructive. We may use RATH to develop a concrete example with the required property. This was done by using the matrix algebra over the non-representable algebra of R. McKenzie. A detailed description of this model and its implementation in RATH is, however, out of the scope of this paper.

In both examples given in this sub-section the relation algebra in question was basically known. The RATH system was used to verify the corresponding properties. On the other hand, it is possible to find a specific model just by testing the required property for all relation algebras available within RATH.

5 Conclusion

In this paper we have described the two computer systems RELVIEW and RATH for dealing with relations and have exhibited their usual domain of applications by presenting some typical examples.

The current investigations based on RELVIEW and RATH are manifold. To give two examples for RELVIEW, we presently use the system to solve tasks of systems architecture and re-engineering and for computing permanents of specific matrices appearing in physics. Concerning RATH, a work in progress is, for example, the inclusion of the theory of Goguen categories (see [16] for details) into the system. This kind of a relational category constitutes a convenient theory for dealing with so-called $\mathcal{L}$-fuzzy relations. Our next aim is to get a prototype of an $\mathcal{L}$-fuzzy controller using RATH.

RELVIEW is a system for set-theoretic relations, i.e., works within the standard model of relational algebra, whereas RATH is geared towards working with non-standard models. In this sense, the systems complement each other. But there is also some overlap in their functionality. E.g., the RATH system provides some possibilities for exploration and programming with concrete relations. There is, of course, the drawback that the naive Haskell list implementation of relations is not particularly efficient. For the future we plan to use the foreign-function interface of Haskell to connect RATH with the efficient implementation of relations in the kernel of RELVIEW. In the last months the latter has been isolated from the entire system and collected in a package called KURE (Kiel University Relation Package). See http://www.informatik.uni-kiel.de/~kure.

Much of the field of TARSKI can be circumscribed by mentioning how often words like *vague, rough, fuzzy, qualitative, uncertain* are used in presenting real world phenomena. Among these, we identified the handling of geographic information in GIS (Geographic Information Systems), dealing with lots of other vaguely defined spatial objects, their region connection etc., and the methods of automatic reasoning on spatial properties. Other activities are devoted to the difficulties of banking and investment corporations in decision making when a multitude of possibly divergent criteria must be taken into account. People work on the design of databases and information systems for large companies in industry including questions of information analysis, knowledge representation document management, and how to organize flexible querying. A diversity of intelligent systems for industry, such as data mining, work-flow design, software development with relational methods, including the demonic specification approach is studied. Times, locations, and events are handled with computer aid, employing temporal and other modal logics. Uncertainty is handled in common-sense reasoning, rough, vague, or approximate logical consequences (in human computer studies, psychology, e.g.); logical formalizations together with inference in general, proof systems from the logical side and automated reasoning as afterwards applied in artificial intelligence. A comprehensive bibliography may be found on the RelMiCS-homepage:

http://www.relmics.org

In any case, the two relational systems RATH and RELVIEW are intended to improve mechanization for any method proposed. They are thus really central, as it is easier to estimate usefulness of a method when it can be shown with computer aid that it really works.

While the examples in this paper are meant to demonstrate the systems in rather theoretically oriented examples, researchers are encouraged to try out the systems. With the help of the inventors, they seem capable to solve even intricate problems. Both systems are available via the Internet:

RELVIEW: http://www.informatik.uni-kiel.de/~progsys/relview.shtml
RATH: http://ist.unibw-muenchen.de/relmics/tools/RATH/

Acknowledgements

We are grateful to Barbara Leoniuk and Ulf Milanese who greatly contributed to the RELVIEW system and to Wolfram Kahl and Eric Offermann who did the same for the RATH system.

References

1. Barbut M., Monjardet B.: Ordre et classification: Algébre et combinatoire. Hachette (1970)
2. Behnke R., Berghammer R., Schneider P.: Machine support of relational computations. The Kiel RELVIEW system. Bericht Nr. 9711, Institut für Informatik und Praktische Mathematik, Universität Kiel (1997)
3. Behnke R., Berghammer R., Meyer E., Schneider P.: RELVIEW – A system for calculation with relations and relational programming. In: Astesiano E. (ed.): Proc. Conf. "Fundamental Approaches to Software Engineering (FASE '98)", LNCS 1382, Springer, 318-321 (1998)
4. Brink C., Kahl W., Schmidt G. (eds.): Relational Methods in Computer Science, Advances in Computing Science, Springer (1997)
5. Freyd P., Scedrov A.: Categories, Allegories. North-Holland (1990)
6. Ganter B., Wille R.: Formal concept analysis: Mathematical foundations. Springer (1999)
7. Hattensperger C., Berghammer R., Schmidt G.: RALF – A relation-algebraic formula manipulation system and proof checker. Notes to a system demonstration. In: Nivat M., Rattray C., Rus T., Scollo G. (eds.): Proc. 3rd Internat. Conf. "Algebraic Methodology and Software Technology (AMAST '93)", Workshops in Computing, Springer, 405-406 (1994)
8. Hattensperger C.: Rechnergestütztes Beweisen in heterogenen Relationenalgebren. Dissertation, Fakultät für Informatik, Universität der Bundeswehr München (1997)
9. Kahl W., Schmidt G.: Exploring (finite) relation algebras using tools written in Haskell. Report Nr. 2000-02, Fakultät für Informatik, Universität der Bundeswehr München (2000)
10. Leoniuk B.: ROBDD-basierte Implementierung von Relationen und relationalen Operationen mit Anwendungen. Dissertation, Institut für Informatik und Praktische Mathematik, Universität Kiel (2001)
11. Offermann E.: Konstruktion relationaler Kategorien. Dissertation, Fakultät für Informatik, Universität der Bundeswehr München (2003)
12. Olivier J.P., Serrato D.: Catégories de Dedekind. Morphismes dans les Catégories de Schröder. C.R. Acad. Sci. Paris 290, 939-941 (1980)

13. Schmidt G., Ströhlein T.: Relationen und Graphen. Springer (1989); English version: Relations and Graphs. Discrete Mathematics for Computer Scientists, EATCS Monographs on Theoret. Comput. Sci., Springer (1993)
14. Schmidt G.: Decomposing relations – Data analysis techniques for Boolean matrices. Report Nr. 2002-09, Fakultät für Informatik, Universität der Bundeswehr München (2002)
15. Winter M.: Strukturtheorie heterogener Relationenalgebren mit Anwendung auf Nichtdeterminismus in Programmiersprachen. Dissertation, Fakultät für Informatik, Universität der Bundeswehr München (1998)
16. Winter M.: A new algebraic approach to $\mathcal{L}$-fuzzy relations convenient to study crispness. Information Sciences 139, 233-252 (2001)

The GUHA Method
and Foundations of (Relational) Data Mining

Petr Hájek[1], Martin Holeňa[1], and Jan Rauch[2]

[1] Institute of Computer Science, Academy of Sciences, Prague
[2] University of Economics, Prague

Abstract. The GUHA method of automatic generation of hypotheses and its underlying logical and statistical theory is surveyed. Links to the theory of information relations and to relational data mining are discussed. Logical foundations present an original approach to finite model theory with generalized quantifiers.

1 Introduction

The aim of the present paper is, first, to give a survey of the present state of the GUHA method of automatic generation of hypotheses and its underlying theoretical foundation, general enough to serve as foundations of a broad class of data mining methods; second, to relate the GUHA approach to the theory of "information relations derived from information systems" as presented in Demri and Orlowska(2002) as well as to the notion of "relational data mining" presented in Džeroski and Lavrač(2001).

GUHA stands for General Unary Hypotheses Automaton and its origins go to 1966 when the first (English) paper Hájek et al.(1966a) on GUHA was published. That paper formulates the principle of the method as to let the computer generate and evaluate all hypotheses that may be interesting from the point of view of the given data and the studied problem. Moreover, the paper formulates a notion of a hypothesis almost identical with the notion of an *association rule* introduced by Agrawal et. al. during the advent of interest in *data mining* in nineties of XX century (see e.g. Agrawal et al.(1996)). Much earlier GUHA and its theory underwent intensive development, whose milestones are the monograph Hájek, Havránek(1978a) (now freely accessible on web see Hájek, Havránek(1978b)), two special issues of the International Journal of Man-Machine studies Hájek(1978), Hájek(1981), the Czech book Hájek et al.(1983) and the paper Hájek et al.(1995); these works contain references to several other papers on GUHA.

Regrettably, GUHA has remained unknown in the mainstream of *data mining* and *knowledge discovery from databases;* from our side, comparison was made in Hájek and Holeňa(2003), in a series of papers Rauch(1978) till Rauch(2002a) and in Hájek(2001a), Hájek(2003).

The second domain with which comparison is possible and desirable, is the theory of *information relations derived from information systems* as presented in Demri and Orlowska(2002) and more generally to data analysis inspired by

H. de Swart et al. (Eds.): TARSKI, LNCS 2929, pp. 17–37, 2003.

the notion of rough sets (see also Düntsch and Gediga(2000)). Needless to say, the COST Action 274 TARSKI is an ideal platform for this. A first step was tried in Hájek(2001b).

Third, the book Džeroski and Lavrač(2001) on *relational data mining* is also highly relevant; also here comparison seems well possible. (This has been stated by Hájek on the TARSKI meeting in Aquilla and is mentioned in his Czech note Hájek(2002).)

Needless to say, these comparisons are intended for both-side profit; the mentioned domains can learn from GUHA as well as GUHA can learn from them.

The paper is structured as follows: we discuss the assumptions on underlying *data*, sketch the *logic* used (note that the most important feature is use of *generalized quantifiers* and *finite structures.*). Particular attention is paid to *associational quantifiers* used to express various kinds of association (dependence) between properties (attributes) of objects under investigation. *Computational complexity* of our logic is discussed, relation to *statistical hypothesis testing*, and possibility of testing *fuzzy hypotheses*. We also comment on existing *implementations*. Some particular meeting points with information relations and with relational data mining are presented. In particular, we offer a certain duality between our notion of an observational quantifier and the notion of information relation of Orlowska et. al. The paper ends by a survey of further research plans in the frame of TARSKI.

Acknowledgement

Support of COST Action 274 (TARSKI) is acknowledged. Thanks are due to Mrs. D. Harmancová for technical help.

2 Data

Data concern a finite set M of *objects* and a finite set $\{P_1, \ldots, P_n\}$ of their *attributes* (properties); mathematically this can be represented in several equivalent ways. First, as $\langle M, f_1, \ldots, f_n \rangle$ where f_j is a function assigning to each object $u \in M$ its value $f_j(u)$ of the j-th attribute; if you agree that $M = \{1, \ldots, m\}$ (objects are indexed by natural numbers then data are presented by a rectangular matrix $\{z_{ij}\}_{i=1,\ldots,m}^{j=1,\ldots,n}$ where $z_{ij} = f_j(i)$ – the value of j-th attribute for the i-th object. In the simplest case the values are Boolean, i.e. 0 and 1; more generally, each attribute P_j has its domain V_j (e.g. integers, reals, abstract values like colors, etc.) and f_j assigns to each object an element (or a subset, as in Demri and Orlowska(2002)) of V_j. Still alternatively, Boolean data can be represented as the set of pairs $\langle u, I(u) \rangle$ where $u \in M$ and $I(u)$ is the set of all j such that $f_j(i) = 1$ (the itemset of u, cf. Agrawal *et al.*(1996)).

Up to now, everything has been *one-sorted* and *unary* (one sort of objects; all attributes have one argument). One can generalize to attributes of *arbitrary arity* – *relations* (e.g. binary property "– is a parent of –", "– and – have met" etc. etc.) and to *many-sortedness*: objects of various sorts (people, towns, ...) and relations with possibly several arguments, each having a sort ("– has visited

–"). Alternatively, there may be *functions* (arity 1 or more) assigning to each object (tuple of objects) of given sort(s) another object ("birthplace of —").

Admittedly, the mainstream in GUHA as well as in later data mining systems has concerned unary attributes; *relational data mining* in the sense of Džeroski and Lavrač(2001) is one of pioneering works in data mining with data in form of relations of arbitrary arity. (Concerning GUHA with attributes with higher arity see below Sectoion 6 referring on Rauch(1986), Rauch(1996).) We shall mainly discuss the unary case (concentrating to the notion of generalized quantifier) and comment on the general case. The basic formalism and results are from Hájek, Havránek(1978a), complemented by newer results.

One more remark: till now we have discussed *data* as some structures on finite sets of objects. These are *observational* structures in the terminology of Hájek, Havránek(1978a). When data mining is used to find *statistical* hypotheses (which may but need not be the case), one is led to investigation of *theoretical structures* of probability. In Hájek, Havránek(1978a) statistical inference is formalized using a kind of modal (Kripke) structures. See Sect. 8 below.

3 The Language

Our language is a kind of predicate calculus with generalized quantifiers, interpreted in finite structures as described above. First we discuss the case of unary attributes. Thus we have n unary *predicates* $P_1, \ldots, P_n$ (identified with the attributes) an *object variable* x and for each P_j and each non-empty proper subset X of the domain V_j of the j-th attribute a *coefficient* (X); *atomic formulas* have the form $(X)P_j(x)$. Clearly, an object $u \in M$ *satisfies* $(X)P_j(x)$ in the data $\mathbf{M} = \langle M, f_1, \ldots, f_n \rangle$ iff the value $f_j(u)$ of the j-th attribute for the object u is in X ($f_i(u) \in X$). (If $f_j(u)$ is a subset of V_j this can be generalized in several ways, e.g. $f_j(u) \subseteq X$ or $f_j(u) \cap X \neq \emptyset$ etc.) Atomic formulas are also called *literals*. Note that if $V_j = \{0, 1\}$ then we may write $P_j(x)$ for $(1)P_j(x)$ and $\neg P_j(x)$ for $(0)P_j(x)$.

Open formulas are built from literals using logical connectives $\&, \vee, \rightarrow, \neg$ (conjunction, disjunction, implication, negation); *satisfaction* of an open formula $\varphi(x)$ by $u \in M$ (in $\mathbf{M}$) is defined from satisfaction of literals using truth functions of classical logic. Given data $\mathbf{M}$, each open formula $\varphi(x)$ determines two frequencies:

r – the number of objects satisfying $\varphi(x)$ in $\mathbf{M}$,
s – the number of objects not satisfying $\varphi(x)$ in $\mathbf{M}$.
The pair (r, s) may be called the *two-fold table* of φ.

Each pair $\varphi(x), \psi(x)$ of open formulas determines four frequencies:

a – the number of objects satisfying $\varphi(x)\&\psi(x)$ in $\mathbf{M}$,
b – the number of objects satisfying $\varphi(x)\&\neg\psi(x)$ in $\mathbf{M}$,
c – the number of objects satisfying $\neg\varphi(x)\&\psi(x)$ in $\mathbf{M}$,
d – the number of objects satisfying $\neg\varphi(x)\&\neg\psi(x)$ in $\mathbf{M}$.

They are usually presented as a so-called *four-fold table*

$$
\begin{array}{cc|c}
a & b & r \\
c & d & s \\
\hline
k & l & m
\end{array}
$$

where $r = a + b$, $s = c + d$, $k = a + c$, $l = b + d$, $m = a + b + c + d$ (marginal sums). Similarly for triples, quadruples,... of formulas.

A (unary) *quantifier q* of dimension $\delta = 1, 2, \ldots$ is applied to a δ-tuple of formulas and *binds* the variable x; the resulting formula is $(qx)(\varphi_1(x), \ldots, \varphi_\delta(x))$.

The *semantics* of a quantifier q of dimension 1 is given by a truth function t_q associating with each two-fold table (r, s) either 1 (truth) or 0 (falsity). Examples (for (r, s) being the two-fold table of φ):

$(\forall x)\varphi(x)$ $Tr_\forall(r, s) = 1$ iff $s = 0$ (all objects satisfy φ)
$(\exists x)\varphi(x)$ $Tr_\exists(r, s) = 1$ iff $r > 0$ (at least one object satisfies φ)
$(Many_p x)\varphi(x)$ $Tr_{Many,p}(r, s) = 1$ iff $r/(r + s) \geq p$ $(0 < p \leq 1)$

The semantics of a quantifier q of dimension 2 is given by a truth function Tr_q associating with each four-fold table (a, b, c, d) either 1 or 0. Let (a, b, c, d) be the four-fold table of $\varphi(x), \psi(x)$. Observe the use of infix notation (writing $\varphi(x)q^x\psi(x)$ instead of $(qx)(\varphi(x), \psi(x))$. Examples:

$\varphi(x) \Rightarrow^x \psi(x)$ $Tr_\Rightarrow(a, b, c, d) = 1$ iff $b = 0$ (each u satisfying φ satisfies ψ)

$\varphi(x) \Rightarrow_p^x \psi(x)$ $Tr_{\Rightarrow_p}(a, b, c, d) = 1$ iff $a/(a+b) \geq p$ $(0 < p \leq 1$; p-many objects satisfying φ satisfy ψ)

$\varphi(x) \sim^x \psi(x)$ $Tr_{\sim_h}(a, b, c, d) = 1$ iff $ad > bc$ (see below)

$(Eqcx)(\varphi(x), \psi(x))$ $Tr_{Eqc}(a, b, c, d) = 1$ iff $a + b = a + c$ (i.e. $r = k$ and hence $s = l$). (equicardinality: φ, ψ have the same two-fold table).

The definition of semantics of a quantifier of dimension $3, 4, \ldots$ is fully analogous using the notion of a 8-fold, 16-fold,... 2^n-fold table). In particular, if q is a quantifier of dimension 2 (say), we can define its *partialization* as a quantifier of dimension 3 defining the formula $(qx)(\varphi(x), \psi(x))/\chi(x)$ to be true in the data structure $\mathbf{M}$ iff the formula $q(x)(\varphi(x), \psi(x))$ is true in the substructure $M \restriction \chi(x)$ of all objects of $\mathbf{M}$ satisfying $\chi(x)$. (If there is no such object, the truth of the partialiation is defined arbitrarily.)

This can be extended in various ways to incomplete data (some values missing, i.e. the functions f_j may be partial). In Hájek, Havránek(1978a) three semantics of missing information are described (called pessimistic, deleting and optimistic).

The reader may wonder why we work with just one object variable x and do not allow several variables $x, y \ldots$ A *normal form theorem* guarantees that whatever your semantics of quantifiers is, each closed formula (all variables in the scope of a quantifier) in the language with several variables is logically equivalent to a closed formula with just one variable (see Hájek, Havránek(1978a)). This is true for the case of *unary predicates* (and of course fails for the general case).

If we admit predicates of higher arity (binary, ternary,...) all quantifiers discussed till now remain meaningful but become a particular case; a general quantifier has its dimension δ (numbers of formulas to which it is applied) and its *arity* – number variables it binds. A general form of a formula resulting by an application of a quantifier is then

$$(qx_1, \ldots, x_t)(\varphi_1, \ldots, \varphi_\delta)$$

(where q has arity t and dimension δ). We shall not go into details and restrict ourselves also here to unary quantifiers (which enables us to keep the original reading of GUHA, relating now "unary" to the arity of quantifiers). More on the generalized approach is presented below. For *theoretical languages* formalizing statistical influence see Hájek, Havránek(1978a).

4 Associations

In GUHA, one is particularly interested in two-dimensional quantifiers expressing in some sense *association* of two open formulas (positive dependence, influence,...). It should be clear that there may be many quantifiers satisfying this, not just one. This leads to the following definitions:

A two-dimensional quantifier $\sim$ is *associational* if for each pair (a, b, c, d), (a', b', c', d') of four-fold tables
$Tr_\sim(a, b, c, d) = 1, a' \geq a, b' \leq b, c' \leq c, d' \leq d$ implies $Tr_\sim(a', b', c', d') = 1$.
(The inequalities mean that the latter table results from the former by possible increase of coincidences and decrease of differences; recall that a is the number of objects satisfying both formulas in question and d the number of objects dissatisfying both formulas.)

The quantifier $\sim$ is *implicational* (or *multitudinal*) if

$$Tr_\sim(a, b, c, d) = 1, a' \geq a, b' \leq b \text{ implies } Tr_\sim(a', b', c', d') = 1$$

(i.e. the value does not depend on c, d). An implicational quantifier (to be strictly distinguished from the connective of implication) is often denoted by $\Rightarrow$ with some indices etc. The formula $\varphi \Rightarrow^* \varphi$ (for $\Rightarrow^*$ implicational) says in some sense that *many objects satisfying φ satisfy ψ*.

From our examples in the preceding section $\Rightarrow$ and $\Rightarrow_p$ are implicational, $\sim_1$ is associational, Eqc is not. Clearly each implicational quantifier is associational. Seemingly the most popular implicational quantifier used in GUHA from its beginning in the sixties is the quantifier FIMPL of founded implication:

$$Tr_{\Rightarrow_{p,t}}(a, b, c, d) = 1 \text{ iff } a/(a+b) \geq p \text{ and } a \geq t.$$

Thus: p-many objects satisfying φ satisfy ψ and at least t objects satisfy $\varphi \& \psi$ (t is called the *base*).

Remarkably, this quantifier was rediscovered as *the* notion of association by Agrawal *et al.*(1996) when introducing their mining association rules (with the minor difference: instead giving a natural t as base demanding $a \geq t$ they give a number between 0 and 1 as the lower bound for the relative frequence $a/)a + b + c + d))$.

GUHA works since long time ago with two implicational quantifiers of statistical origin LIMPL and UIMPL – see below or Hájek, Havránek(1978a), Hájek *et al.*(1995).

A quantifier $\sim$ is *symmetric* if $Tr_\sim(a, b, c, d) = Tr_\sim(a, c, b, d)$ (hence $\varphi \sim \psi$ is equivalent to $\psi \sim \varphi$). A quantifier is *comparative* if $Tr_\sim(a, b, c, d)$ implies $ad > bc$, equivalently $a/(a + b) > (a + c)/(a + b + c + d)$ – observe that if (a, b, c, d) is the table of φ, ψ then $a/(a + b)$ is the relative frequence $Freq(\psi|\varphi)$ of ψ among objects satisfying φ and $(a + b)/(a + b + c + d)$ is the relative frequence $Freq(\psi)$ of ψ in the whole data. Thus $ad > bc$ may be read: φ *increases the frequence of* ψ. (Caution: this does not mean that $Fr(\psi|\varphi)$ is big; only that $Freq(\psi|\varphi)$ is bigger than $Freq(\psi)$.)

The quantifier SIMPLE $\sim_{u,t}$:

$$Tr_{\sim_{u,t}}(a, b, c, d) = 1 \text{ iff } ad > h.bc \text{ and } a \geq t \text{ (base)}.$$

It is association, symmetric and for $h \geq 1$ it is comparative. Statistical analogs FISHER and CHISQ are described in Hájek, Havránek(1978a), Hájek *et al.* (1995) and below.

5   Some Further Classes of Quantifiers

This section is based on the large unpublished work Rauch(1998b), some main results and definitions are also in Rauch(1998a). The generalized quantifier $\sim$ is *double implicational* if

$$Tr_\sim(a, b, c, d) = 1 \ \wedge \ a' \geq a \ \wedge b' \leq b \ \wedge \ c' \leq c \ \text{ implies } Tr_\sim(a', b', c', d') = 1$$

for each pair of four-fold tables.

The condition $a' \geq a \wedge b' \leq b \wedge c' \leq c$ is *the truth preservation condition for double implicational quantifiers.*

We can see a reason for such a definition in an analogy to propositional logic. If u and v are propositions and both $u \to v$ and $v \to u$ are true, then u is equivalent to v ($\to$ is the propositional connective of implication). Thus we can try to express the relation of equivalence of attributes φ and ψ using "double implicational" quantifier $\Leftrightarrow^*$ such that

$$\varphi \Leftrightarrow^* \psi \ \text{ if and only if } \ \varphi \Rightarrow^* \psi \text{ and } \psi \Rightarrow^* \varphi,$$

where $\Rightarrow^*$ is a suitable implicational quantifier. Thus $\varphi \Leftrightarrow^* \psi$ says *many φ's are ψ's amd many ψ's are φ's*. If we apply the truth preservation condition for implicational quantifier to $\varphi \Rightarrow^* \psi$, we obtain $a' \geq a \ \wedge \ b' \leq b$. If we apply it to

$\psi \Rightarrow^* \varphi$, we obtain $a' \geq a \wedge c' \leq c$, ($c$ is here instead of b). This leads to the truth preservation condition for double implicational quantifiers $a' \geq a \wedge b' \leq b \wedge c' \leq c$. Several quantifiers are defined according to this idea in Hájek *et al.*(1983), an example follows.

Example 1. The quantifier $\Leftrightarrow_{p,s}$ of *founded double implication* for $0 < p \leq 1$ and $s > 0$ is double implicational, see Rauch(1998a). Here $\Leftrightarrow_{p,s} (a, b, c, d) = 1$ if and only if $\frac{a}{a+b+c} \geq p \wedge a \geq s$.

Several further classes of quantifiers are also defined and studied in Rauch(1978), Rauch(1986), Rauch(1998b) and Rauch(1998a), e.g. *pure double implicational, typical double implicational, pure equivalence, Σ- equivalence* and *F-quantifiers*.

It is proved in Rauch(1998a) that quantifier $\Leftrightarrow_{p,s}$ belongs to the class of Σ-double implicational quantifiers:

The generalized quantifier $\sim$ is *Σ-double implicational* if

$$Tr_\sim(a, b, c, d) = 1 \wedge a' \geq a \wedge b' + c' \leq b + c \text{ implies } \sim (a', b', c', d') = 1$$

for each pair of four-fold tables.

It is obvious that each Σ-double implicational quantifier is also double implicational. It follows from the definition that if a quantifier $\Leftrightarrow^*$ belongs to the class of Σ-double implicational quantifiers, then the value $\Rightarrow^* (a, b, c, d)$ does not depend on d.

We have a similar situation for equivalence. If u and v are propositions and both $u \rightarrow v$ and $\neg u \rightarrow \neg v$ are true, then u is equivalent to v. Thus we can try to express the relation of equivalence of attributes φ and ψ using an "equivalence" quantifier $\equiv^*$ such that

$$\varphi \equiv^* \psi \text{ if and only if } \varphi \Rightarrow^* \psi \text{ and } \neg\varphi \Rightarrow^* \neg\psi ,$$

where $\Rightarrow^*$ is a suitable implicational quantifier. Thus "*many φ's are psi's and many $(\neg\varphi)$'s are $\neg\psi$'s*. If we apply the truth preservation condition for implicational quantifier to $\varphi \Rightarrow^* \psi$ we obtain $a' \geq a \wedge b' \leq b$, if we apply it to $\neg\varphi \Rightarrow^* \neg\psi$, we obtain $d' \geq d \wedge c' \leq c$, ($c$ is here instead of b and d instead of a, see Tab.1). This leads to the our *truth preservation condition for associational quantifiers*: $a' \geq a \wedge b' \leq b \wedge c' \leq c \wedge d' \geq d$ (associational quantifiers are called "equivalence quantifiers" in Rauch(1996)).

Let us give the definition of the pure double implicational quantifier as an example:

The generalized quantifier $\Leftrightarrow^*$ is *pure double implicational* if there is the implicational quantifier $\Rightarrow^*$ such that

$$Tr_{\Leftrightarrow^*}(a, b, c, d) = 1 \text{ iff } Tr_{\Rightarrow^*}(a, b) = 1 \text{ and } Tr_{\Rightarrow^*}(a, c) = 1 .$$

Finally, let us mention the quantifier of *conviction* (see Adamo(2001)): $Tr_{\sim_h^{conv}}(a, b, c, d) = 1$ iff $conv(a, b, c, d) \geq h$ where $conv(a, b, c, d) = \frac{(a+b)(b+d)}{b(a+b+c+d)} = \frac{rl}{bm}$. It is shown in Hájek(2003) that the conviction quantifier is associational.

6 A Simple Many-Sorted Language

Here we present an example of a two-sorted data structure and of expressive possibilities of a corresponding language. This approach was studied in Rauch's thesis Rauch(1986) (unpublished); see also Rauch(2002a).

We use two simple data matrices: *Clients* (see Figure 1) and *Transactions* (see Figure 2).

client	Age	Sex	Salary	District	Amount	Repayment	Quality
c_1	45	M	high	Prague	48 000	4 000	good
$\vdots$	$\vdots$	$\vdots$	$\vdots$	$\vdots$	$\vdots$	$\vdots$	$\vdots$
c_{6181}	32	F	average	Plzen	20 000	2 000	bad

Fig. 1. Data matrix *Clients*

In other words, the domain of the attribute *Client* in the matrix of transactions coincides with the set of object of the matrix of client.

Each row of data matrix *Clients* corresponds to a client of a bank. There are 6181 clients. Attributes Age, Sex, Salary and District describe the client. Attributes Amount, Repayment and Quality describe loans of the clients. The first row corresponds to a 45 years old man. This man has a high salary and he lives in the district Prague. He borrowed 48000 Czech crowns. He repays 4000 Czech crowns and the quality of his loan is good.

Transaction	Client	Amount	Bank	Type
t_1	c_1	3 300	A	T1
t_2	c_1	4 500	Z	T4
...	c_1	...	...	...
t_{100}	c_1	6 700	C	T6
$\vdots$	$\vdots$	$\vdots$	$\vdots$	$\vdots$
t_r	c_{6181}	12 900	B	T2

Fig. 2. Data matrix *Transactions*

Rows of data matrix *Transactions* correspond to particular transactions of clients. The client c_1 has 100 transactions $t_1, \ldots, t_{100}$ The data matrices *Clients* and *Transactions* are related by the attribute Client. The attribute Client can be understood as a function from data matrix *Transactions* to data matrix *Clients*. It assigns a row from data matrix *Clients* to each row of data matrix *Transactions*.

Each transaction is described by 3 attributes. The attribute Amount gives the amount of transferred money. The attribute Bank describes the bank from/to which the transaction goes; there are 26 particular banks A,..., Z. The attribute

Type describes the type of transaction; there are 6 types of transactions: T1,...,
T6.

We can use quantifiers to define new attributes of clients. We start with the
formula

$$(A)Bank(y) \Rightarrow^y_{p,s} (T1)Type(y)/(c_1)Client(y)$$

concerning the data matrix *Transactions* and saying: on the submatrix of transactions of the client c_1, most transactions in bank A are of type $T1$. Supress-
ing the bound variable y for transactions we may abbreviate the formula as
$(A)Bank \Rightarrow_{p,s} (T1)Type/(c_1)Client$. It defines a Boolean value for the client
c_1. The same can be made for each other client, thus we get a new attribute for
clients. Formally, allow a *variable* x for clients in the coefficient in the restricting
condition; we get a formula $(A)Bank \Rightarrow_{p,s} (T1)Type/(x)Client$ which defines a
Boolean property $\varphi(x)$ of clients. This formula may now be combined with other
formulas with free variable x to define further composed properties of client
which can then be parts of closed formulas containing a quantifier binding x.

This approach is clearly related to the relational data mining developed by
Džeroski and Lavrač(2001); elaborating this remains a future task. Here let us
only mention (in passing) Chapter 8 of Džeroski and Lavrač(2001) (the authors
of that chapter being Dehaspe and Toivonen) on generating frequent DATALOG
patterns. From purely logical point of view they generate hypotheses like

$$(\exists y)P(x,y) \Rightarrow^* (\exists y)(P(x,y)\&B(y))$$

(their example being "if a customer x has a child then x (usually) has a child
buying Coca-Cola"). The use of binary predicates and predicates of higher arity
is an extremally important feature of relational data mining. Here GUHA can
contribute by offering its generalized quantifiers to generate hypotheses on association (in the general sense of GUHA) between formulas containing predicates
of arbitrary arity.

7 Deduction Rules, Tautologies, Computational Complexity, (Un)definability

Here we present some metamathematical results on the calculi under investigation (the unary case). We assume a language with n unary predicates $P_1,...,P_n$,
a quantifier $\sim$ of dimension 2. Formulas of the form $\varphi \sim^x \psi$ with φ, ψ open with
just one variable x are called *prenex normal*; recall that formulas in normal form
are just boolean combinations of prenex normal formulas. Prenex normal formulas play the role of *hypotheses* in GUHA; alternative name, in the style of
contemporary data mining, is *association rules*.

Simple *deduction rules* for prenex normal formulas are important for GUHA:
they make it possible to identify immediate consequences of hypotheses found
true in given data so that the consequences need not be extra verified in data.

The following are two "classical" examples of rules sound for all implicational
quantifiers $\Rightarrow^*$ ("sound" meaning of course: if the premise is true in given data
then so is the conclusion"):

$$\frac{\varphi \Rightarrow^* \psi}{\varphi \Rightarrow^* \psi \vee \chi}, \qquad \frac{(\varphi \& \psi) \Rightarrow^* \chi}{\varphi \Rightarrow^* \chi \vee \neg\psi}.$$

An example sound for all associational quantifiers:

$$\frac{\varphi \sim \psi}{\varphi \sim (\varphi \& \psi)}$$

This is a relatively simple example of a deduction rule of the form

$$\frac{\varphi \sim \psi}{\varphi' \sim \psi'}$$

where $\varphi, \psi, \varphi', \psi'$ are subjected to some syntactic condition, cf. Rauch(1998a). This condition depends on the class the quantifier $\sim$ belongs to. A more detailed description of deduction rules concerning association rules is out of range of this paper.

In Hájek, Havránek(1978a) the reader may find also deduction rules containing besides formulas of the form $\varphi \sim \psi$ ($\sim$ associational) also formulas with an auxiliary "improving" quantifier.

To say that a rule $\frac{\varphi \sim \psi}{\varphi' \sim \psi'}$ is Q-sound (sound for each quantifier from the class Q of quantifiers is the same as to say that the formula $(\varphi \sim \psi) \to (\varphi' \sim \psi')$ (where $\to$ is the connective of implication) is a Q-tautology.

A formula (in the normal form) is a Q-*tautology* if it is true in each data matrix for each quantifier $\sim$ from the class Q.

How complex is to decide whether a formula is a Q-tautology? It is proved in Hájek, Havránek(1978a) that the set of all associational tautologies is recursive (3.2.29); but now we know more: both the set of associational tautologies and the set of implicational tautologies is co-NP-complete. See Hájek(2001b), Hájek(2003) ; for general information on computational complexity see e.g. Garrey and Johnson(1979).

Let us also mention the question of definability of the partialized quantifier from the non-partialized one: we present a positive and negative result from Hájek(2003)).

For each implicational quantifier $\Rightarrow^*$, the formula $\varphi(x) \Rightarrow^* \psi(x)/\chi(x)$ is logically equivalent to $(\varphi(x) \& \chi(x)) \Rightarrow^* \psi x)$ (easy). But there are associational quantifiers $\sim$ such that the formula $\varphi(x) \sim \psi(x)/\chi(x)$ is not logically equivalent to any formula containing only the non-partialized quantifier $\sim$ (SIMPLE can be taken as an example.)

Note once more that the above concerns the case with unary predicates; for calculi with predicates of higher arity the situation is different. To mention a classical example, recall Trakhtenbrot's result saying that the set of formulas of classical predicate logic with at least one at least unary predicate true in all finite structures (thus being tautologies of finite model theory) is not recursively enumerable.

8   GUHA and Statistics

One of the key areas to which observational calculus has been applied is *statistical hypotheses testing* Hájek, Havránek(1978a). In such applications, the rows of a data matrix are realizations of a random vector, independently generated by some probability distribution $\mathbf{P}$, and the matrix itself is a realization of a random sample from $\mathbf{P}$. Through appropriate constraints on $\mathbf{P}$, it is possible to capture probabilistic relationships between the properties corresponding to the columns of that data matrix. Assuming that $\mathbf{P}$ is a priori known to belong to some set $\mathcal{D}$ of probability distributions, a *statistical test* assesses the validity of the constraint $H_0 : \mathbf{P} \in \mathcal{D}_0$ for a given $\mathcal{D}_0 \subset \mathcal{D}$, called *null hypothesis*, against the alternative $H_1 : \mathbf{P} \in \mathcal{D}_1 = \mathcal{D} \setminus \mathcal{D}_0$. The test is performed using some random variable t, the test statistic, and some borel set C_α, the critical region of the test on a significance level $\alpha \in (0, 1)$, which are connected to H_0 through the condition $(\forall \mathbf{P} \in \mathcal{D}_0)\mathbf{P}(t \in C_\alpha) \leq \alpha$. In addition, the test often makes use of the fact that if the considered model is a realization of a random sample, then also the matrix M of interpretations of open formulae $\varphi_1, \ldots, \varphi_m$ is a realization of a random sample, more precisely a realization of a multinomial random sample (random sample from a multinomial distribution). That fact allows to choose some test statistic t that is a function of such a multinomial random sample, and to define a generalized quantifier $\sim$ corresponding to the constraint H_1 by means of the truth function

$$\mathrm{Tf}_\sim(M) = 1 \text{ iff } t_M \in C_\alpha, \tag{1}$$

where t_M is the realization of t provided the realization of the random sample from P yields the matrix of interpretations M.

The existing implementations of GUHA include about a dozen quantifiers defined according to (1). All of them are two-dimensional, and their truth functions have the special property that the dependence of the truth function on the two-column matrix M of interpretations reduces to its dependence on the four-fold table corresponding to M. Formally, there exists a mapping τ from four-fold tables to reals such that for each two-column matrix of interpretations M, the following holds

$$t_M = \tau(a_M, b_M, c_M, d_M), \tag{2}$$

where a_M, b_M, c_M and d_M are the counts of the pairs $(1, 1)$, $(1, 0)$, $(0, 1)$ and $(0, 0)$, respectively, among the rows of M (four-fold table).

Recall the definitions of an associational, implicational and symmetric associational quantifier above.

A typical implicational quantifier $\Rightarrow^*$ is suitable to capture a high probability that the random variable corresponding to ψ assumes the value 1 provided the random variable corresponding to φ assumes the value 1, whereas a symmetric associational quantifier $\sim$ is suitable to capture a positive correlation between those random variables.

If the binary generalized quantifier $\sim$ in a sentence $\varphi \sim \psi$ represents a statistical test of a null hypothesis H_0 against an alternative H_1, given some initial assumptions A, then the statistical inference based on that test can be represented by means of the following inference rule:

$$\frac{A, \varphi \sim \psi}{H_1}, \tag{3}$$

where A is some "frame assumption", which is justified, assuming A and the incompatibility of H_0, H_1, by the fact that if H_0 held, then the observed validity of $\varphi \sim \psi$ would be improbable. More precisely, its probability would not exceed the prescribed significance level α: $\Pr(\varphi \sim \psi | A \ \& \ H_0) \leq \alpha$. (See Hájek, Havránek(1978a) for some details.)

The most fundamental examples of symmetric associational quantifiers are $\sim_\alpha^F$, the *Fisher quantifier*, corresponding to the one-sided Fisher exact test of independence in four-fold tables with the significance level $\alpha \in (0, 1)$ [1], and the *chi-square quantifier* $\sim_\alpha^{\chi^2}$, corresponding to the χ^2 asymptotic test of independence in four-fold tables with the significance level α. The truth functions of their most simple versions are defined as follows:

$$\mathrm{Tf}_{\sim_\alpha^F}(M) = 1 \text{ iff } a_M d_M > b_M c_M \wedge \sum_{i=a_M}^{\min(r_M, k_M)} \frac{\binom{k_M}{i}\binom{\ell_M}{r_M - i}}{\binom{m_M}{r_M}} \leq \alpha, \tag{4}$$

$$\mathrm{Tf}_{\sim_\alpha^{\chi^2}}(M) = 1 \text{ iff } a_M d_M > b_M c_M \wedge \frac{m_M(a_M d_M - b_M c_M)^2}{r_M s_M k_M \ell_M} \geq \chi_1^2(1 - 2\alpha), \tag{5}$$

where $\chi_f^2(t)$ for $t > 0$ denotes the t-quantile of the χ^2 distribution with $f \in \mathcal{N}$ degrees of freedom, $r_M = a_M + b_M$, $s_M = c_M + d_M$, $k_M = a_M + c_M$, $\ell_M = b_M + d_M$, and $m_M = a_M + b_M + c_M + d_M$, i.e., m_M is the number of rows of the matrix M.

The basic example of an implicational quantifier is the *lower critical implication* $\rightarrow_\alpha^\theta$ with the threshold $\theta \in (0, 1)$, which corresponds to the binomial test with a significance level α. The truth function of the most simple version of this quantifier is defined

$$\mathrm{Tf}_{\rightarrow_\alpha^\theta}(M) = 1 \text{ iff } \sum_{i=a_M}^{r_M} \binom{r_M}{i} p^i (1 - p)^{r_M - i} \leq \alpha. \tag{6}$$

In statistical hypotheses testing, the test is usually performed in such a way that the condition $\Pr(\varphi \sim \psi | A \ \& \ H_0) \leq \alpha$ is fulfilled for a particular pair (H_0, H_1) of tested hypotheses. Therefore, (3) is justified only for each such pair separately, saying nothing about the simultaneous validity of other pairs of tested hypotheses. All implementations of GUHA have been restricted only to this most simple way of statistical hypotheses testing, in statistical terms called *testing*

---

[1] In more details, if the sentence $\varphi \sim_\alpha^F \psi$ is true in our data then we may assume that the probabilities of φ, ψ, $\varphi \& \psi$ (in the universe from which our data present a random sample) satisfy $P(\varphi \& \psi) > P(\varphi) \cdot P(\psi)$ (given some probabilistic frame assumption).

on a local significance level. In practical applications of GUHA, however, this restriction does not prevent the possibility to make use of methods for *hypotheses testing on a multiple significance level* and to obtain sets of simultaneously valid hypotheses.

To get an idea about how hypotheses testing on a multiple significance level can be treated within the GUHA framework, consider pairs of null and alternative hypotheses $(H_0^{(1)}, H_1^{(1)}), \ldots, (H_0^{(n)}, H_1^{(n)})$, such that for each $i = 1, \ldots, n$, there is a significance level $\alpha_i \in (0,1)$ and a binary generalized quantifier $\sim_i$ defined for $(H_0^{(i)}, H_1^{(i)})$ on the significance level α_i and used to construct a sentence $\varphi_i \sim_i \psi_i$. It is immaterial whether the open formulae φ_i, $i = 1, \ldots, n$, respectively ψ_i, $i = 1, \ldots, n$, in the sentences $\varphi_1 \sim_1 \psi_1, \ldots, \varphi_n \sim_1 \psi_n$ differ for different $i = 1, \ldots, n$. Suppose, for simplicity, that the pairs of hypotheses are ordered in such a way that $\alpha_1 \leq \ldots \leq \alpha_n$, and that the initial assumptions A, the null hypotheses $H_0^{(1)}, \ldots, H_0^{(n)}$ and the quantifiers $\sim_1, \ldots, \sim_n$ fulfil the following condition: $(\forall \mathcal{I} \subset \{1, \ldots, n\})$ $\mathcal{I} \neq \emptyset \Rightarrow \Pr(\bigwedge_{i \in \mathcal{I}}(\varphi_i \sim_i \psi_i) | A \ \& \ \bigvee_{i \in \mathcal{I}} H_0^{(i)}) \leq \alpha$. Taking as axioms the system $A \ \& \ \neg(\bigwedge_{i \in \mathcal{I}} H_1^{(i)}) \rightarrow \bigvee_{i \in \mathcal{I}} H_0^{(i)}$ for $\emptyset \neq \mathcal{I} \subset \{1, \ldots, n\}$, the condition $\Pr(\varphi \sim \psi | A \ \& \ H_0) \leq \alpha$ yields an inference rule representing the statistical inference in multiple hypotheses testing: $\{\frac{A, \bigwedge_{i \in \mathcal{I}}(\varphi_i \sim_i \psi_i)}{\bigwedge_{i \in \mathcal{I}} H_1^{(i)}}, \emptyset \neq \mathcal{I} \subset \{1, \ldots, n\}\}$. Existing methods for multiple hypotheses testing differ with respect to how the validity of the condition (8) is ensured (Hochberg and Tamhane(1987), Samuel-Cahn(1996), Westfall(1997)). In the applications of GUHA Coufal *et al.*(1999), Holeňa *et al.*(1999), the following methods have been employed:

a) Bonferroni method, requiring $(\forall i \in \{1, \ldots, n\})$ $\alpha_i = \frac{\alpha}{n}$.
b) Holm method, in which $(\forall i \in \{1, \ldots, n\})$ $\alpha_i = \frac{\alpha}{n+1-i}$.
c) Simes method, the initial assumptions of which include the independence of the achieved significance levels of the tests represented by $\sim_1, \ldots, \sim_n$, whereas $\alpha_1, \ldots, \alpha_n$ are required to fulfil $(\forall i \in \{1, \ldots, n\})$ $\alpha_i = \frac{i\alpha}{n}$.

9 Generalization of GUHA to Fuzzy Hypotheses

As was shown in the preceding section, important examples of generalized quantifiers encountered in the GUHA method correspond to common tests of statistical hypotheses. However, there is an *inner contradiction* inherent to using common statistical tests in GUHA. Those tests always require a *precise formulation* of the tested hypothesis, i.e. precisely stated restrictions on the distribution of the underlying random variables. On the other hand, due to its exploratory character, GUHA is used predominantly in situations in which the user has a priori only a rather *vague knowledge* about the underlying random variables, thus being unable to make a competent a priori choice of assumptions concerning their distributions.

Therefore, it would be advantageous to reflect the vagueness of the user's a priori knowledge in the formulation of the generated hypotheses. To generalize the GUHA approach in such a way is the objective of its recently proposed extension, which successfully tackles the case of GUHA implications Holeňa(1996a), Holeňa(1998). That extension provides the possibility to replace a precisely formulated traditional statistical hypothesis with a *fuzzy hypothesis*, viewed as a normalized fuzzy set $\tilde{\Pi}$ on a set Π of admissible parameters, $\tilde{\Pi} = \{(p, \mu(p)) : p \in \Pi\}$ with $\mu : \Pi \to \langle 0, 1 \rangle$.

Consider, for example, a sentence $\varphi \to_p^! \psi$, in which the quantifier of a lower critical implication $\to_p^!$ with $p \in (0, 1)$ is applied to open formulae φ and ψ. Due to (6), generating $\varphi \to_p^! \psi$ corresponds to testing the statistical null hypothesis

$$\Pr(\psi|\varphi) \leq p, \tag{7}$$

where $\Pr(\psi|\varphi)$ denotes the conditional probability of ψ being valid conditioned on φ being valid Holeňa(1998). The constant p in (6) and (7) is a precise number and must be chosen in advance. On the other hand, the extension of GUHA proposed in Holeňa(1996a), Holeňa(1998) allows testing, instead of (7), the fuzzy hypothesis

$$\Pr(\psi|\varphi) \text{ is low.} \tag{8}$$

Needless to say, (8) is much more feasible than (7), at least for a user who has only poor previous knowledge of the data.

Similarly, generating a sentence $\varphi \to_p^? \psi$, where $\to_p^?$ is the quantifier of a upper critical implication, corresponds to testing the statistical null hypothesis

$$\Pr(\psi|\varphi) \geq p. \tag{9}$$

Instead, the proposed extension allows testing the fuzzy hypothesis

$$\Pr(\psi|\varphi) \text{ is high.} \tag{10}$$

Both (8) and (10) can be viewed as fuzzy sets on the open interval $(0, 1)$, which is the set of admissible values for the parameter p in (7) and (9). A number of examples of membership functions that correspond to (8) or to (10) can be found in Holeňa(1996a), Holeňa(1998).

For testing fuzzy hypotheses such as (8) or (10), a suitable generalization of the concept of implicational quantifiers is needed. Such a generalization is now briefly sketched.

Let a fuzzy hypothesis be viewed as a normalized fuzzy set $\tilde{\Pi}$ on a set Π of admissible parameters. Then a binary generalized quantifier $\Rightarrow$ is called *fuzzy-implicational* w.r.t. $\tilde{\Pi}$ if for each two-column models M_1, M_2, the following implication is valid:

$$a_{M_2} \geq a_{M_1} \ \& \ b_{M_2} \leq b_{M_1} \Rightarrow$$
$$\Rightarrow \mathrm{Ftf}_\to^a(M_2) \geq \mathrm{Ftf}_\to^a(M_1) \ \& \ \mathrm{Ftf}_\to^r(M_2) \geq \mathrm{Ftf}_\to^r(M_1). \tag{11}$$

Here, $\mathrm{Ftf}_\to^a$ and $\mathrm{Ftf}_\to^r$ are two $\langle 0, 1 \rangle$-valued functions defined in Holeňa(1996a) and called *accepting fuzzy truth function* of $\to$ w.r.t. $\tilde{\Pi}$, and *rejecting fuzzy associated function* of $\to$ w.r.t. $\tilde{\Pi}$, respectively. Several important properties of these associated functions have been proven and their relationship to the truth function Tf$_\rightarrow$ from (1) has been studied in Holeňa(1998).

At present, another approach to fuzzy hypotheses testing is being elaborated, in which fuzzy hypotheses are viewed not as fuzzy sets, but as predicates of a many-sorted predicate calculus Holeňa(2001a), Holeňa(2001b). Differently to the approach outlined here, that alternative approach allows to tackle fuzzy hypotheses testing with purely logical means.

10   GUHA Procedures and Their Implementations

Under a GUHA procedure we understand a concrete algorithm of generation GUHA hypotheses given the input data and values of input parameters defining the syntax and semantics of hypotheses to be generated and evaluated; at the same time we call an implementation of such an algorithm also a GUHA procedure.

This is entirely general; but historically, (almost) all existing GUHA procedures generated GUHA-*associations*, i.e. hypotheses of the form $\varphi \sim \psi$ where φ, ψ are open formulas (satisfying some syntactical restrictions given by input parameter) and $\sim$ is a fixed associational quantifier, taken from an offer of a list of possible choices, parameters of the quantifier chosen being a part of input parameters of the procedure.

We shall not discuss the "pre-historical" implementations and mention only three last ones:

GUHA-ASSOC for personal computers, described in Hájek *et al.*(1995); for long time this was the most used implementation.

GUHA+-, implemented by a group of students, working under WINDOWS, described and freely available on web – see 2;

4ft-Miner is a part of the system LISP-Miner[2]. We shall describe this last implementation.

The GUHA procedure 4ft-Miner mines for association rules $\varphi \sim \psi$ and for conditional association rules $\varphi \sim \psi/\chi$. Quantifiers of fourteen types can be used. Open formulas φ, ψ and χ are conjunctions of literals and are called *antecedent succedent* and *condition* respectively. Input and output of 4ft-Miner are described below.

A *conditional association rule* is an expression of the form $\varphi \sim \psi/\chi$ where φ, ψ and χ are open formulas. The intuitive meaning is that φ and ψ are in relation given by the quantifier $\sim$ when the condition χ is satisfied (see above).

10.1   Quantifiers of 4ft-Miner

Fourteen types of quantifiers are implemented in procedure 4ft-Miner. There are three implicational quantifiers: $\Rightarrow_{p,s}$, $\Rightarrow^{!}_{p,\alpha,s}$ and $\Rightarrow^{?}_{p,\alpha,s}$.

---

[2] LISP-Miner is an academic software system for Knowledge Discovery in Databases research and teaching. It is suitable namely for mid-size, pilot and students projects. The system LISp-Miner is developed at University of Economics, Prague and it is freely available, see 1.

Further there are three analogous double implication quantifiers: $\Leftrightarrow_{p,s}$, $\Leftrightarrow^!_{p,\alpha,s}$, $\Leftrightarrow^?_{p,\alpha,s}$ and three analogous equivalence quantifiers: $\equiv_{p,s}$ (see above). $\equiv^!_{p,\alpha,s}$, $\equiv^?_{p,\alpha,s}$, and also Rauch(1978).

There are also symmetric associational quantifiers corresponding to χ^2 test and to Fisher's test and the simple quantifier defined by the condition $ad > h.bc$ where $\delta \geq 0$. The quantifier $\rightarrow_{C,S}$ corresponding to Agrawal's association rule and the quantifier corresponding to condition $\max(\frac{b}{a+b}, \frac{c}{d+c}) < \gamma$ where $0 \leq \gamma < 1$, see Zembowicz and Zytkow(1996) is also implemented.

10.2   Antecedent, Succedent and Condition

Antecedent, succedent and condition are conjunctions of literals of the form $A[\sigma]$ where A is a predicate and σ is a subset of a set of possible values of A (above we wrote $(\sigma)A$ for $A[\sigma]$.) An example of the conditional association rule:

$$A_1[1,3,4] \wedge A_3[5,6] \Leftrightarrow_{0.9} A_5[8,12] \wedge A_7[11,12,14] \wedge A_8[2] \,/\, A_9[4,5,6,7,8].$$

10.3   Input of 4ft-Miner

An input of 4ft-Miner is given by: **(i):** *The analyzed data matrix.* **(ii):** *The quantifier.* **(iii):** *Simple definition of all antecedents to be automatically generated.* It consists of: **(iii-a):** A list of all predicates from which literals of antecedent will be automatically generated. **(iii-b):** Simple definition of the set of all literals to be automatically generated from each particular predicate. **(iii-c):** Minimal and maximal number of literals in each generated antecedent. **(iv):** Analogous definitions of all succedents and of all conditions to be automatically generated.

The set of all literals to be generated from a particular predicate is given by a *type of subsets* and by the minimal and the maximal number of particular values in the subset. There are five types of subsets to be generated: *all subsets, intervals, left cuts, right cuts* and *cuts.*

Examples of a definition of the set of basic Boolean attributes for column A with possible values $\{1,2,3,4,5\}$: (1) *all subsets with values* defines basic Boolean attributes $A[1,2]$, $A[1,3]$, ..., $A[4,5]$, $A[1,2,3]$, $A[1,2,4]$, ..., $A[3,4,5]$; (2) *intervals with 2-3 values* defines basic Boolean attributes $A[1,2]$, $A[2,3]$, $A[3,4]$, $A[4,5]$, $A[1,2,3]$, $A[2,3,4]$ and $A[3,4,5]$; (3) *left cuts with most 3 values* defines basic Boolean attributes $A[1]$, $A[1,2]$, $A[1,2,3]$.

10.4   Output of 4ft-Miner

4ft-Miner automatically generates all (GUHA-)association rules or all conditional (GUHA-)association rules given by the conditions (ii)–(iv) (usually 10^5–10^7) *and verifies them* in data matrix given by (i). Output of 4ft-Miner is the set of all association rules (all conditional association rules) true in data matrix given by (i).

Usual output of 4ft-Miner consists of tens or hundreds of true association rules (true conditional association rules). There are strong tools for dealing with output of 4ft-Miner. It is possible to sort output 4ft association rules by various criterions. Flexible conditions can be used to define subsets of output 4ft association rules. It is also possible to export defined subsets in several formats.

10.5   Some Further Features of 4ft-Miner

4ft-Miner works under WINDOWS, analyzed data matrix can be stored in a database (ODBC is applied). New values can be defined for particular columns (e.g. intervals or groups of original values). New columns of data matrix can be also defined (SQL - like) and used in conditions (iii) - (v).

4ft-Miner works very fast. Usual task (data matrix with 10^4 rows, several millions of 4ft association rules to be generated and verified) requires only several minutes at PC with Pentium II and with 128 MB of operational memory. Several optimisation techniques and deep theoretical results are used, e.g. bit strings for representation of analyzed data matrix Rauch(1978) and deduction rules. Note that the method of first finding frequent patterns (frequent conjunctions of literals) powerful in Agrawal style data mining cannot be immediately used in our general case; but the paper Hájek(2001a) contains a comparison of that method with the approach of GUHA and suggests a variant of the former approach appliable in GUHA.

11   GUHA and Information Relations

Information relations on objects of a data structures are defined and studied in Demri and Orlowska(2002) Chapt. 3 (where the reader also finds references to relevant previous works). There are particular (finitely many) binary relations on the set of objects definable by quantifying over *subsets of predicates* (not objects), eg. strong indiscernibility:

for any set A of predicates and $x, y \in M$, $\langle x, y \rangle \in \inf(A)$ iff $(\forall P \in A)(f_P(x) = f_P(y))$ (the underlying data being $\mathbf{M} = \langle M, (f_P)_{P \, predicate} \rangle$.

Another example, for the case that f_P maps objects onto *subsets* of V_P: forward inclusion –

$$\langle x, y \rangle \in fin(A) \text{ iff } (\forall P \in A)(f_P(x) \subseteq f_P(y)).$$

There are examples of six indistinguishability relations; there are also six distinguishability relations. If you represent the data by a matrix, then the fact of $\langle x, y \rangle$ is or is not in the relation is fully determined by the pair of rows/vectors corresponding to x, y, more precisely to their subvectors with coordinates in A.

But this can be understood as dual – in some sense – to the situation in GUHA, just interchanging the role of objects and predicates (transposing the data matrix). For simplicity, assume Boolean data $f_P(x) \in \{0, 1\}$ (where $0 = \emptyset$ – empty set, $1 = \{\emptyset\}$) and let $\Leftrightarrow$ be the quantifier of classical equivalence; let P_1, P_2 be predicates and a_1, a_2 objects. Then $P_1(x) \Leftrightarrow^x P_2(x)$ is true in our data $\mathbf{M}$ iff for all objects u, $f_{P_1}(u) = f_{P_2}(u)$. And $P(a_1) \Leftrightarrow^P P(a_2)$ (where P is a variable ranging over predicates iff for all predicates P, $f_P(a_1) = f_P(a_2)$. (This is the case of $A = M$; similarly for arbitrary A.) This leads to the following *research topic. Investigate information relations* defined using generalized quantifiers. We give some examples, still assuming $A = M$ (the case of $A \subseteq M$ is handled by relativized quantifiers as presented above).

y has many properties that x has: $P(x) \Rightarrow_{p,s}^P P(y)$. Here $\Rightarrow_{p,s}$ is FIMPL and the quantified variable is P. Thus we work with the four-fold table (a, b, c, d) where a is the number of properties presented both by x and by y, b is the number of properties presented by x but not by y, etc. The formula is true if $a/(a + b) \geq p$ and $a \geq s$ (think of $p = 0.9, s = 10$).

x is (simply) associated with y: $P(x) \sim_{h,s}^P P(y)$.

Here $\sim_{h,s}$ is SIMPLE; think of $h = 2$ and $s = 10$. The formula is true if $a \geq 10$ (x, y posses at least 10 properties in common) and $ad > 2bc$ (in this sense coincidence of x, y in properties dominates their difference). Also statistically motivated quantifiers may be considered (saying e.g. that x, y are *significantly positively dependent* with respect to their properties); of course, some care is necessary in interpreting such relations.

12   Conclusion

The GUHA method and its underlying logical and statistical theory has been presented and its relations to mining association rules, theory of information relations and to relational data mining have been discussed. Further research topics include:

- processing of extremely large data sets by GUHA,
- more on GUHA and fuzzy logic,
- GUHA with more general data (many sorted, non-unary predicates but unary quantifiers) and connection with relational data mining,
- connection of GUHA with non-invasive data mining and information relations.

It is hoped that this program is inspiring for various partners of the COST Action 274–TARSKI.

References

Adamo, J. M. [2001], *Data mining for associational rules and sequential patterns, Sequential and parallel algorithms*. Springer.

Agrawal, R., H. Manilla, R. Sukent, A. Toivonen and A. Verkamo [1996], "Fast discovery of association rules." In: *Advance in Knowledge Discovery and Data Mining*, AAA Press, pp. 307–328.

Coufal, D. [2001], "GUHA analysis of air pollution data. in: Artificial neural nets and genetic algorithms." In: *Proceedings of the International conference ICAN-NGA'2001*, edited by V. Kůrková, N. C. Steele, R. Neruda and M. Kárný, Springer, Wien, pp. 465–468.

Coufal, D., M. Holeňa and A. Sochorová [1999], "Coping with discovery challenge by GUHA." In: *Workshop Notes on Discovery Challenge. PKDD'1999*, Prague, pp. 7–12.

Demri, S. P. and E. Orlowska [2002], *Incomplete information: Structure, inference, complexity.* Springer-Verlag.

Düntsch, I. and G. Gediga [2000], *Rough set data analysis - a road to non-invasive data analysis.* Methodos.

Džeroski, S. and N. Lavrač [2001], *Relational data mining.* Springer.

Feglar, T. [2001], "The GUHA architecture." In: *Proc. Relmics 6*, edited by H. de Swart, Katholieke Universiteit Brabant, Tilburg, The Netherlands, pp. 358–364.

Garrey, M. R. and D. S. Johnson [1979], *Computers and intractability.* W. J. Freeman and Co., New York.

Hájek, P. (guest editor) [1978], *International Journal for Man-Machine Studies*, vol. 10, No 1 (special issue on GUHA).

Hájek, P. (guest editor) [1981], *International Journal for Man-Machine Studies*, vol. 15, No 3 (second special issue on GUHA).

Hájek, P. [1984], "The new version of the GUHA procedure ASSOC." In: *COMPSTAT 1984*, Physica-Verlag Wien, pp. 360–365.

Hájek, P. [2001a], "The GUHA method and mining association rules." In: *Proc. CIMA'2001*, Bangor, Wales, pp. 533–539.

Hájek, P. [2001b], "Relations in GUHA style data mining." In: *Proc. Relmics 6*, edited by H. de Swart, Katholieke Universiteit Brabant, Tilburg, The Netherlands, pp. 91–96.

Hájek, P. [2002], "Metoda GUHA – současný stav." In: *Proc. ROBUST 2002*, Hejnice.

Hájek, P. [2003], "Generalized quantifiers, finite sets and data mining." In: *Intelligent Information Processing and Web Mining*, edited by Klopotek et al., Physica Verlag, pp. 489–496.

Hájek, P., K. Bendová and Z. Renc [1971], "The GUHA method and three-valued logic." *Kybernetika*, **7**, pp. 421–431.

Hájek, P., I. Havel and M. Chytil [1966a], "The GUHA method of automatic hypotheses determination." *Computing*, **1**, pp. 293–308.

Hájek, P., I. Havel and M. Chytil [1966b], "Metoda GUHA automatického zjišťování hypotéz I." *Kybernetika*, **2**, pp. 31–47.

Hájek, P., I. Havel and M. Chytil [1967], "Metoda GUHA automatického zjišťování hypotéz II." *Kybernetika*, **3**, pp. 430–437.

Hájek, P., I. Havel and M. Chytil [1983], *Metoda GUHA – automatická tvorba hypotéz.* Academia, Prague, in Czech.

Hájek, P. and T. Havránek [1978a], *Mechanizing hypothesis formation (mathematical foundations for a general theory).* Springer-Verlag, Berlin-Heidelberg-New York.

Hájek, P. and T. Havránek [1978b], "Mechanizing hypothesis formation (mathematical foundations for a general theory)." . Internet edition. http://www.cs.cas.cz/~hajek/guhabook.

Hájek, P. and M. Holeňa [2003], "Formal logics of discovery and hypothesis formation by machine." *Theoretical Computer Science*, **393**, pp. 345–358.

Hájek, P., J. Rauch, T. Feglar and D. Coufal [2002], "The GUHA method, data preprocessing and mining." In: *Proc. DTDM02 (Database technologies for data mining)*, Prague, pp. 29–36.

Hájek, P., A. Sochorová and J. Zvárová [1995], "GUHA for personal computers." *Comp. Stat., Data Arch.*, **19**, pp. 149–153.

Hálová, J. and P. Žák [2000], "Coping discovery challenge of mutagenes discovery with GUHA+/- for windows." In: *The Fourth Pacific-Asia Conference on Knowledge Discovery and Data Mining. Workshop KDD Challenge 2000*, Kyoto, pp. 55–60.

Havránek, T. [1971], "The statistical modification and interpretation of GUHA method." *Kybernetika*, **7**, pp. 13–21.

Hochberg, Y. and A. C. Tamhane [1987], *Multiple Comparison Procedures*. John Wiley and Sons, New York.

Holeňa, M. [1996a], "Exploratory data processing using a fuzzy generalization of the GUHA approach." In: *Fuzzy Logic*, edited by J. Baldwin, John Wiley and Sons, New York, pp. 213–229.

Holeňa, M. [1996b], "Exploratory data processing using a fuzzy generalization of the GUHA approach." In: *Fuzzy Logic*, edited by J. Baldwin, John Wiley and Sons, New York, pp. 213–229.

Holeňa, M. [1998], "Fuzzy hypotheses for GUHA implications." *Fuzzy Sets and Systems*, **98**, pp. 101–125.

Holeňa, M. [2001a], "A fuzzy logic framework for testing vague hypotheses with empirical data." In: *Proceedings of the Fourth International ICSC Symposium on Soft Computing and Intelligent Systems for Industry*, ICSC Academic Press, Sliedrecht, pp. 401–407.

Holeňa, M. [2001b], "A fuzzy logic generalization of a data mining approach." *Neural Network World*, **11**, pp. 595–610.

Holeňa, M., A. Sochorová and J. Zvárová [1999], "Increasing the diversity of medical data mining through distributed object technology." In: *Medical Informatics Europe '99*, edited by P. Kokol, B. Zupan, J. Stare, M. Premik and R. Engelbrecht, IOS Press, Amsterdam, pp. 442–447.

Pecen, L., E. Pelikán, H. Beran and D. Pivka [1996], "Short-term fx market analysis and prediction." *Neural Networks in Financial Engeneering*, pp. 189–196.

Pokorný, J. and J. Rauch [1981], "The GUHA-DBS database system." *International Journal of Man-Machine Studies*, **15**, pp. 289–298.

Rauch, J. [1978], "Some remarks on computer realisations of GUHA procedures." *International Journal of Man-Machine Studies*, **10**, pp. 23–28.

Rauch, J. [1981], "Main problems and further possibilities of the computer realizations of GUHA procedures." *International Journal of Man-Machine Studies*, **15**, pp. 283–287.

Rauch, J. [1986], *Logical foundations of mechanizing hypotheses formation from databases*. Ph.D. thesis, Mathematical Institute of Czechoslovak Academy of Sciences, in Czech.

Rauch, J. [1988], "Logical problems of statistical data analysis in databases." In: *Proc. Eleventh Int. Seminar on Database Management Systems*, pp. 53–63.

Rauch, J. [1996], "GUHA as a data mining tool." In: *Practical Aspects of Knowledge management*, Schweizer Informatiker Gesellshaft, Basel, p. 10 pp.

Rauch, J. [1997], "Logical calculi for knowledge discovery." Springer Verlag, Berlin, pp. 47–57.

Rauch, J. [1998a], "Classes of four-fold table quantifiers." In: *Principles of Data Mining and Knowledge Discovery*, edited by J. Zytkow and M. Quafafou, Springer Verlag, pp. 203–211.

Rauch, J. [1998b], *Contribution to logical foundations of KDD*. Ph.D. thesis, University of Economics, Prague, in Czech.

Rauch, J. [1998c], "Four-fold table calculi and missing information." In: *JCIS'98 Proceedings, Association for Intelligent Machinery*, edited by P. P. Wang, pp. 375–378.

Rauch, J. [2001], "Association rules and mechanizing hypothesis formation." In: *Working notes of ECML'2001 Workshop: Machine Learning as Experimental Philosophy of Science*, See also http://www.informatik.uni-freiburg.de/ ml/ecmlpkdd/.

Rauch, J. [2002a], "Interesting association rules and multi-relational association rules." *Communications of Institute of Information and Computing Machinery*, **5**(2), pp. 77 –82.

Rauch, J. [2002b], "Mining for scientific hypotheses." In: *Dealing with the data flood. Mining Data, Text and Multimedia*, edited by J. Meij, STT/Beweton, The Hague, pp. 73–84.

Rauch, J. and M. Šimůnek [2000], "Mining for 4ft association rules." In: *Proc. Discovery Science 2000*, Springer Verlag, Kyoto, pp. 268–272.

Rauch, J. and M. Šimůnek [2001a], "Mining for 4ft association rules by 4ft-miner." In: *INAP 2001, The Proceeding of the International Rule-Based Data Mining*, Tokyo.

Rauch, J. and M. Šimůnek [2001b], "Mining for statistical association rules." In: *Proc. PAKDD 2001*, Hong Kong, pp. 149–158.

Samuel-Cahn, E. [1996], "Is the Simes improved Bonferroni procedure conservative?" *Biometrika*, **83**, pp. 928–933.

Šebesta, V. and L. Straka [1998], "Determination of suitable markers by the GUHA method for the prediction of bleeding at patients with chronic lymphoblastic leukemia." In: *Medicon 98, Mediterranean Conference on Medical and Biological Engineering and Computing*, Lemesos, Cyprus.

Westfall, P. H. [1997], "Multiple testing of general contrasts using logical constraints and correlations." *Journal of the American Statistical Association*, **92**, pp. 299–306.

Zembowicz, R. and J. Zytkow [1996], "From contingency tables to various forms of knowledge in databases." In: *Advances in Knowledge Discovery and Data Mining*, edited by U. M. Fayyad, AAAI Press/ The MIT Press, pp. 329–349.

Zvárová, J., J. Preiss and A. Sochorová [1995], "Analysis of data about epileptic patients using GUHA method." In: *EuroMISE 95: Information, Health and Education, TEMPUS International Conference*, edited by J. Zvárová and I. Malá, Prague, EuroMISE Center, Prague, Czech Republic, p. 87.

http://lispminer.vse.cz.

GUHA+- – project web site http://www.cs.cas.cz/ics/software.html.

Mechanised Reasoning and Model Generation for Extended Modal Logics

Renate A. Schmidt[1] and Ullrich Hustadt[2]

[1] Department of Computer Science, University of Manchester
Manchester M13 9PL, United Kingdom
schmidt@cs.man.ac.uk
[2] Department of Computer Science, University of Liverpool
Liverpool L69 7ZF, United Kingdom
U.Hustadt@csc.liv.ac.uk

Abstract. The approach presented in this overview paper exploits that modal logics can be seen to be fragments of first-order logic and deductive methods can be developed and studied within the framework of first-order resolution. We focus on a class of extended modal logics very similar in spirit to propositional dynamic logic and closely related to description logics. We review and discuss the development of decision procedures for decidable extended modal logics and look at methods for automatically generating models.

1 Introduction

Over the last nearly ten years a variety of methods for reasoning with modal and description logics have been developed, implemented and applied in several case studies, cf. for example [36, 41, 46, 47, 42]. Though the logics involved are very similar, the reasoning methods used and proof search strategies employed can differ considerably. Various empirical studies have been undertaken mainly for basic multi-modal logic or its corresponding description logic $\mathcal{ALC}$. Many of these studies are competitive in nature or study the effects of various optimisation methodologies on the performance of provers. While such work is vital, it is only a beginning. From current studies it is difficult to extrapolate general conclusions for different reasoning methods and different logics not considered in such studies. The literature on formal logic and proof theory does not give much guidance either. Particular forms of reasoning methods tend to get favoured over others, mostly due to ease of presentation, without there being theoretical or empirical evidence for the practical usefulness of the considered proof methods. Thus currently it is still difficult to choose between the different methods and provers; what is lacking is a general body of knowledge which would support well-judged choices. In the area of automated reasoning for propositional logics there have been extensive analytical and empirical evaluations of different proof methods. Similar research developments for modal and description logic reasoning systems are only starting to get off the ground.

H. de Swart et al. (Eds.): TARSKI, LNCS 2929, pp. 38–67, 2003.
© Springer-Verlag Berlin Heidelberg 2003

The aim of this paper is to give an overview of some recent advances in the area. We concentrate on decision procedures developed in the framework of first-order resolution and focus on translation-based resolution methods for modal logics. This means that we take a modal formula, translate it into first-order logic through the Kripke-semantics, and then apply some variant of resolution to it. Using the combination of a translation method and resolution has some obvious advantages. Any modal logic which can be embedded into first-order logic can be treated. The translations are straightforward, and can be performed in time $O(n \log n)$, so no engineering effort is needed here. For the resolution part, standard resolution provers can be used, or otherwise they can be used with small adaptations. Modern resolution provers are among the most sophisticated and fastest theorem provers available. The translation approach is generic, it can handle first-order modal logics, undecidable (first-order definable) modal logics, and combinations of modal and non-modal logics. In all cases soundness and completeness of the approach is immediate from the soundness and completeness of the translation mapping and the resolution calculus. Resolution provers provide decision procedures for a large class of (extended) modal logics and description logics. Often the same refinements that decide modal and description logics decide also more expressive first-order generalisations such as the guarded fragment or Maslov's class K [25, 48].

This survey focusses on the extended modal logic $K_{(m)}(\cap, \cup, \bar{\ }, \smile)$, first considered in De Nivelle, Hustadt and Schmidt [16], and subsystems thereof as well as extensions with relational theories. $K_{(m)}(\cap, \cup, \bar{\ }, \smile)$ is a PDL-like logic which permits complex formulae as parameters of the modal operators. This is useful for application domains in artificial intelligence and computational linguistics. For example, if e denotes the eats relation and p is the set of plants, then $\langle e \rangle p$ can be interpreted as denoting the set of plant eaters, while $[e]p$ denotes the set of vegetarians, who eat nothing but plant matter[1]. An expression which requires complex relational parameters is the set of cheese lovers, given by $\langle e \wedge l \rangle c \wedge \neg \langle \neg \langle e \wedge c \rangle \rangle c$, where l denotes the 'likes' relation and c is interpreted as the set of cheeses[2].

Formally, $K_{(m)}(\cap, \cup, \bar{\ }, \smile)$ is the multi-modal logic defined over families of relations closed under intersection, union, complementation and converse. It extends Boolean modal logic (due to Gargov and Passy [29]) with converse on relations. $K_{(m)}(\cap, \cup, \bar{\ }, \smile)$ is very expressive. It subsumes standard modal logics such as K, KT, KD, KB, KTB, and KDB, their independent joins, as well as the basic tense logic K_t. Global satisfiability of these logics can be embedded in $K_{(m)}(\cap, \cup, \bar{\ }, \smile)$ and it subsumes modal logics extended with the universal modality. Logics of philosophical interest such as logics expressing inaccessibility, sufficiency, or both necessity and sufficiency [30, 43, 44] can be embedded

---

[1] $x \models [e]p$ iff for any y, such that $R_e(x, y)$ we have that $y \models p$. Thus, the meaning of $x \models [e]p$ is 'everything that x eats is plant matter'.

[2] Observe $x \models \langle e \wedge l \rangle c$ iff x eats and likes (some) cheese. Further, $x \models \neg \langle \neg \langle e \wedge c \rangle \rangle c$ iff for any $y \models c$, both $R_e(x, y)$ and $R_l(x, y)$ are true. Therefore, cheese lovers are people who eat and like every cheese.

in $K_{(m)}(\cap, \cup, \bar{\ }, \smile)$. Certain forms of interactions and correspondence properties, for example, inclusions among relations and symmetry, are covered as well. $K_{(m)}(\cap, \cup, \bar{\ }, \smile)$ subsumes a large class of well-known description logics [16, 34]. It is most closely related to the description logic $\mathcal{ALB}$ introduced in [50].

In this paper we focus on first-order logic fragments induced by the standard relational translation of modal logics. Other translation methods exist (see Section 12) but, as yet, it is not known how to treat modal logics with complex modal parameters within the context of these translation methods.

Regardless as to which translation method is adopted, a crucial decision is the choice of a suitable refinement of the basic resolution calculus for first-order logic. Depending on our aims we have various options. Ordering refinements provide decision procedures for very expressive logics, while if we are interested in generating models for satisfiable formulae selection-based refinements (or hyperresolution) are more natural (Fermüller et al. [23, 22], Leitsch [55], Hustadt et al [35, 50, 49, 52]). We discuss an ordered resolution decision procedure for a class of clauses induced by $K_{(m)}(\cap, \cup, \bar{\ }, \smile)$ in Section 6. In Section 7 we describe a refinement which relies solely on the selection of negative literals for certain extensions of $K_{(m)}(\cap, \cup, \smile)$. This refinement has the property that for many modal logics its derivations resemble those of tableau calculi. We consider the polynomial simulation of single-step prefix tableau by selection-based resolution in Section 8. Such simulation results do not only say something about the relative complexity of resolution and tableaux, they can also be exploited to transfer proof procedures, extra inferences rules, search strategies, simplification criteria and optimisation techniques between the different approaches. Moreover, the relationship can be exploited for extracting new tableau calculi from resolution in a more or less automatic way. As a case analysis, in Section 9, we define a semantic tableau calculus for the logic $K_{(m)}(\cap, \cup, \smile)$ which is derived from the selection-based resolution procedure. Soundness, completeness and termination results are then mere corollaries of corresponding results for the resolution refinement. The selection-based refinement also has the property that, like tableau-based procedures, it can be used for the automatic construction of models for satisfiable formulae and the models are finite if the procedure is a decision procedure. This is the topic of Section 10. In Section 11 we mention automated reasoning tools that implement the procedures described in this paper.

Before we can proceed to the main part of this paper, namely Sections 6–11, we need to define the class of logics under consideration, how they translate to first-order logic and describe the resolution framework. This is done in Sections 3– 5. Section 2 summarises the notational conventions used in this paper. The final section mentions some important topics not covered in this paper because of space limitations.

2   Notational Convention

Throughout the notational convention is the following. The letters x, y, z are reserved for first-order variables, s, t, u, v for terms, a, b for constants, f, g, h for function symbols, and p, q, r for propositional symbols, and P, Q, R for predicate symbols. A is the letter reserved for atoms, L for literals, and C, D for clauses. For sets of clauses the letter N is used. The Greek letters φ, ψ, ϕ are reserved for modal or first-order formulae, and α, β, γ are reserved for relational formulae.

3   Extended Modal Logics

The language of $K_{(m)}(\cap, \cup, {}^-, {}^\smile)$ is defined over countably many propositional variables $p, p_1, p_2, \ldots$, and countably many relational variables $r, r_1, r_2, \ldots$. A *propositional atom* is a propositional variable, $\top$ or $\bot$. A *modal formula* is either a propositional atom or a formula of the form $\neg\varphi$, $\varphi \wedge \psi$, $\varphi \vee \psi$, $\langle\alpha\rangle\varphi$ and $[\alpha]\varphi$, where φ is a modal formula and α is a relational formula. A *relational formula* is a relational variable or has one of the following forms: $\alpha \wedge \beta$, $\alpha \vee \beta$, $\neg\alpha$, and $\alpha^\smile$ (converse), where α and β are relational formulae. Other connectives are defined to be abbreviations, for example, $\varphi \to \psi = \neg\varphi \vee \psi$ or the universal modality is $[u] = [r_j \vee \neg r_j]$, for some relational variable r_j.

The semantics of $K_{(m)}(\cap, \cup, {}^-, {}^\smile)$ is defined in terms of relational structures or frames. A frame is a tuple (W, R) of a non-empty set W (of worlds) and a mapping R from relational formulae to binary relations over W satisfying:

$$R_{\alpha\wedge\beta} = R_\alpha \cap R_\beta \qquad R_{\alpha\vee\beta} = R_\alpha \cup R_\beta \qquad R_{\neg\alpha} = \overline{R_\alpha} \qquad R_{\alpha^\smile} = R_\alpha^\smile.$$

Here and in the rest of the paper we prefer to use the notation R_α instead of $R(\alpha)$. The defining class of frames of a modal logic determines, and is determined by, a corresponding class of models. A model (an interpretation) is given by a triple $\mathcal{M} = (W, R, \iota)$, where (W, R) is a frame and ι is a mapping from modal formulae to subsets of W satisfying:

$$\iota(\bot) = \emptyset \qquad\qquad \iota(\top) = W \qquad\qquad \iota(\neg\varphi) = \overline{\iota(\varphi)}$$
$$\iota(\varphi \wedge \psi) = \iota(\varphi) \cap \iota(\psi) \quad \iota(\langle\alpha\rangle\varphi) = \{x \mid \exists y \in W\, ((x, y) \in R_\alpha \wedge y \in \iota(\varphi))\}$$
$$\iota(\varphi \vee \psi) = \iota(\varphi) \cup \iota(\psi) \quad \iota([\alpha]\varphi) = \{x \mid \forall y \in W\, ((x, y) \in R_\alpha \to y \in \iota(\varphi))\}.$$

A modal formula is satisfiable iff an $\mathcal{M}$ exists such that for some x in W, $x \in \iota(\varphi)$.

We also consider logics with fewer relational operations, as well as logics restricted by relational theories consisting of additional frame properties. A logic $K_{(m)}(\star_1, \ldots, \star_k)$ *in-between* $K_{(m)}$ and $K_{(m)}(\cap, \cup, {}^-, {}^\smile)$, where the $\star_i$ are distinct operations from $\{\cap, \cup, {}^-, {}^\smile\}$ ($m \geq 1$, $0 \leq i \leq k \leq 4$), is defined to be the multi-modal logic defined over relations closed under $\star_1, \ldots, \star_k$. If L is a logic in-between $K_{(m)}$ and $K_{(m)}(\cap, \cup, {}^-, {}^\smile)$ and Δ is a set of relational frame properties then $L\Delta$ denotes the logic characterised by the class of L-frames which satisfy the conjunction of properties in Δ [3]. Examples of relational frame properties and the corresponding modal axiom schemas are given in Figure 1.

It is well-known that an implication between relational formulae can be defined by $(\alpha \to \beta) = [\alpha \wedge \neg\beta]\bot$ in Boolean modal logic [70] and therefore also

---

[3] Used in a formula Δ is assumed to represent the conjunction of relational properties.

Seriality, for $D = \langle r \rangle \top$: $\forall x \exists y\, R(x, y)$

Reflexivity, for $T = [r]p \rightarrow p$: $\forall x\, R(x, x)$

Symmetry, for $B = \langle r \rangle [r]p \rightarrow p$: $\forall x \forall y\, (R(x, y) \rightarrow R(y, x))$

Inclusion, for $[r_1]p \rightarrow [r_2]p$: $\forall x \forall y\, (R_2(x, y) \rightarrow R_1(x, y))$

For $[r_3]p \rightarrow ([r_1]p \vee [r_2^{\smile}]p)$: $\forall x \forall y\, (R_1(x, y) \wedge R_2(y, x) \rightarrow R_3(x, y))$

Fig. 1. Some correspondence properties.

in $K_{(m)}(\cap, \cup, {}^{-}, {}^{\smile})$. For example in $K_{(m)}(\cap, \cup, {}^{-}, {}^{\smile})$ the symmetry of the accessibility relation R_1 associated with r_1 can be specified by $r_1 \rightarrow r_1^{\smile}$. We also observe that a modal logic L including relational complementation and one relational conjunction and relational disjunction allows for the definition of the universal modality. If r is a relational name in the language of the logic L then the universal modality $[u]$ can be defined by $[u]\varphi = [r \vee \neg r]\varphi = [\neg(r \wedge \neg r)]\varphi$.

4   Translation to First-Order Logic

Modal formulae will be mapped to first-order logic formulae by two transformations: a translation of the modal formula into first-order logic, in this case, a semantics-based translation, followed by a structural transformation.

The *standard semantics-based translation* of $K_{(m)}(\cap, \cup, {}^{-}, {}^{\smile})$ into first-order logic is determined by the definition of the semantics of the logical operators. For modal formulae the translation is specified by the following.

$$\pi(\top, x) = \top \qquad\qquad\qquad \pi(\bot, x) = \bot$$
$$\pi(p_i, x) = P_i(x) \qquad\qquad\quad \pi(\neg \varphi, x) = \neg \pi(\varphi, x)$$
$$\pi(\varphi \star \psi, x) = \pi(\varphi, x) \star \pi(\psi, x) \quad \text{for } \star \in \{\wedge, \vee, \rightarrow, \leftrightarrow\}$$
$$\pi(\langle \alpha \rangle \varphi, x) = \exists y\, (\tau(\alpha, x, y) \wedge \pi(\varphi, y)) \quad \pi([\alpha]\varphi, x) = \forall y\, (\tau(\alpha, x, y) \rightarrow \pi(\varphi, y))$$

Relational formulae are translated according to the following.

$$\tau(r_j, x, y) = R_j(x, y) \qquad \tau(\neg\alpha, x, y) = \neg\tau(\alpha, x, y) \qquad \tau(\alpha^{\smile}, x, y) = \tau(\alpha, y, x)$$
$$\tau(\alpha \star \beta, x, y) = \tau(\alpha, x, y) \star \tau(\beta, x, y) \quad \text{for } \star \in \{\wedge, \vee\}$$

In the translation each propositional or relational variable (p_i or r_j) is uniquely associated with a unary or binary predicate variable, denoted by the corresponding capital letter (P_i or R_j).

By definition, Π maps any modal formula φ to $\exists x\, \pi(\varphi, x)$.

Theorem 1. *Let L be a logic in-between $K_{(m)}$ and $K_{(m)}(\cap, \cup, {}^{-}, {}^{\smile})$ and Δ a (possibly empty) set of relational frame properties. For any modal formula φ, φ is satisfiable in $L\Delta$ iff $\Delta \wedge \Pi(\varphi)$ is first-order satisfiable.*

The purpose of the structural transformation is to convert the first-order translation into a more manageable form. Before we describe it formally, we need to state some definitions of basic notions.

The polarity of (occurrences of) modal or first-order subformulae is defined as usual. Any occurrence of a proper subformula of an equivalence has *zero polarity*. For occurrences of subformulae not below a '↔' symbol, an occurrence of a subformula has *positive polarity* if it is one inside the scope of an even number of (explicit or implicit) negations, and it has *negative polarity* if it is one inside the scope of an odd number of negations.

For any first-order formula φ, if λ is the position of a subformula in φ, then $\varphi|_\lambda$ denotes the subformula of φ at position λ and $\varphi[\psi \mapsto \lambda]$ is the result of replacing $\varphi|_\lambda$ at position λ by ψ. The set of all the positions of subformulae of φ is denoted by $\mathrm{Pos}(\varphi)$.

Structural transformation, also referred to as *renaming*, associates with each element λ of $\Lambda \subseteq \mathrm{Pos}(\varphi)$ a predicate symbol Q_λ and a literal $Q_\lambda(\overline{x})$, where $\overline{x} = x_1, \ldots, x_n$ are the free variables of $\varphi|_\lambda$, the symbol Q_λ does not occur in φ and two symbols Q_λ and $Q_{\lambda'}$ are equal only if $\varphi|_\lambda$ and $\varphi|_{\lambda'}$ are equivalent formulae[4]. Let $\mathrm{Def}_\lambda^+(\varphi) = \forall\overline{x}\,(Q_\lambda(\overline{x}) \to \varphi|_\lambda)$ and $\mathrm{Def}_\lambda^-(\varphi) = \forall\overline{x}\,(\varphi|_\lambda \to Q_\lambda(\overline{x}))$. The *definition* of Q_λ is the formula

$$
\mathrm{Def}_\lambda(\varphi) = \begin{cases} \mathrm{Def}_\lambda^+(\varphi) & \text{if } \varphi|_\lambda \text{ has positive polarity,} \\ \mathrm{Def}_\lambda^-(\varphi) & \text{if } \varphi|_\lambda \text{ has negative polarity,} \\ \mathrm{Def}_\lambda^+(\varphi) \wedge \mathrm{Def}_\lambda^-(\varphi) & \text{otherwise.} \end{cases}
$$

The corresponding clauses are called *definitional clauses*. Now, define $\mathrm{Def}_\Lambda(\varphi)$ inductively by:

$$
\mathrm{Def}_\emptyset(\varphi) = \varphi \quad \text{and}
$$
$$
\mathrm{Def}_{\Lambda \cup \{\lambda\}}(\varphi) = \mathrm{Def}_\Lambda(\varphi[Q_\lambda(\overline{x}) \mapsto \lambda]) \wedge \mathrm{Def}_\lambda(\varphi),
$$

where λ is maximal in $\Lambda \cup \{\lambda\}$ with respect to the prefix ordering on positions. A *definitional form* of φ is $\mathrm{Def}_\Lambda(\varphi)$, where Λ is a subset of all positions of subformulae (usually, non-atomic or non-literal subformulae).

Theorem 2 (e.g. [9, 71]). *Let φ be a first-order formula. (i) φ is satisfiable iff $\mathrm{Def}_\Lambda(\varphi)$ is satisfiable, for any $\Lambda \subseteq \mathrm{Pos}(\varphi)$. (ii) $\mathrm{Def}_\Lambda(\varphi)$ can be computed in linear time.*

5   First-Order Resolution

Basics. The usual definition of clausal logic is assumed. A *literal* is an atom or the negation of an atom. The former is said to be a *positive literal* and the latter a *negative literal*. If the predicate symbol of a literal has arity one (two) then we call this literal a *unary literal* (*binary literal*). A clause with one literal is a *unit clause* (or unit). If this literal is a unary (binary) literal then the clause will be called a *unary* (*binary*) *unit clause*. In this paper *clauses* are assumed

---

[4] In practice, one may want to use the same symbols for variant subformulae, or subformulae which are obviously equivalent, for example, $\varphi \vee \psi$ and $\psi \vee \varphi$.

to be multisets of literals, and will be denoted by $P(x) \vee P(x) \vee \neg R(x,y)$, for example. The empty clause will be denoted by $\emptyset$. The components in the variable partition of a clause are called *variable-disjoint* or *split components*, that is, split components do not share variables. A clause which cannot be split further will be called a *maximally split clause*. A *positive* (resp. *negative*) clause contains only *positive* (resp. *negative*) literals.

Two formulae or clauses are said to be *variants* of each other if they are equal modulo variable renaming. Variant clauses are assumed to be equal.

We say an expression is *functional* if it contains a constant or a non-nullary function symbol. Otherwise it is called *non-functional*.

Resolution. Now, we briefly recall the definition of ordered resolution extended with a selection function from Bachmair et al [4–6]. Derivations are controlled by an admissible ordering $\succ$ and a selection function. Basically the idea is that inferences are restricted to literals maximal under the ordering $\succ$ while the selection function is used to override the ordering, and give preference to inferences with negative literals. A third parameter in our presentation is a normalisation function NORM.

By definition, an ordering $\succ$ is *admissible*, if (i) it is a total, well-founded ordering on the set of ground literals, (ii) for any atoms A and B, it satisfies: $\neg A \succ A$, and $B \succ A$ implies $B \succ \neg A$, and (iii) it is stable under the application of substitutions. An ordering is said to be *liftable* if it satisfies (iii). The multiset extension of $\succ$ provides an admissible ordering on clauses. A literal L is said to be *(strictly) maximal* with respect to a clause C if for any literal L' in C, $L' \not\succ L$ ($L' \not\succeq L$). Let M be a set and $\succ_c$ an arbitrary ordering on M. Assume that with every literal L we associate a complexity measure $c_L \in M$. An ordering is *compatible with a given complexity measure* c_L on ground literals, if $c_L \succ_c c_{L'}$ implies $L \succ L'$ for any two ground literals L and L'.

A *selection* function S assigns to each clause a possibly empty set of occurrences of negative literals. If C is a clause, then the literal occurrences in $S(C)$ are *selected*. No restrictions are imposed on the selection function. The minimal requirement for the normalisation function is that NORM(C) is a clause which is logically equivalent to C and NORM$(C) \preceq C$. Many resolution decision procedures rely on condensing (defined below) as the minimal normalisation function.

Let R be the resolution calculus defined by the rules of Figure 2. As is usual we implicitly assume that the premises of the resolution rule have no common variables. The premise $C \vee A_1$ of the resolution rule and premise of the factoring rule will be referred to as a *positive premise*, while the premise $\neg A_2 \vee D$ of the resolution rule will be referred to as a *negative premise*. The literals resolved upon and factored upon are called *eligible literals*.

The *splitting rule* is a rule familiar from DPLL algorithms and tableau calculi. Instead of trying to refute $N \cup \{C \vee D\}$ one tries to refute $N \cup \{C\}$ and $N \cup \{D\}$ (or $N \cup \{C\}$ and $N \cup \{D, \neg C\}$, if C is a ground clause). The splitting rule is don't know non-deterministic and requires backtracking. However, in the resolution context splitting can be simulated by introducing a new propositional symbol. If $C \vee D$ is a clause that can be split into two split components C and D, then

Deduce:
$$\frac{N}{N \cup \{\text{NORM}(C)\}}$$
if C is a factor or resolvent of premises in N.

Delete:
$$\frac{N \cup \{C\}}{N}$$
if C is redundant.

Split:
$$\frac{N \cup \{C \vee D\}}{N \cup \{C\} \mid N \cup \{D\}}$$
if C and D are variable-disjoint.

Resolvents and factors are computed with:

Ordered resolution:
$$\frac{C \vee A_1 \quad \neg A_2 \vee D}{(C \vee D)\sigma}$$

provided (i) σ is the most general unifier of A_1 and A_2, (ii) no literal is selected in C, and $A_1\sigma$ is strictly $\succ$-maximal with respect to $C\sigma$, and (iii) $\neg A_2$ is either selected, or $\neg A_2\sigma$ is maximal with respect to $D\sigma$ and no literal is selected in D.

Ordered factoring:
$$\frac{C \vee A_1 \vee A_2}{(C \vee A_1)\sigma}$$

provided (i) σ is the most general unifier of A_1 and A_2, and (ii) no literal is selected in C and $A_1\sigma$ is $\succ$-maximal with respect to $C\sigma$.

Fig. 2. The calculus R.

it is possible to replace $C \vee D$ by two clauses $C \vee q$, and $\neg q \vee D$. q is made minimal in the ordering $\succ$, and $\neg q$ is selected [14, 75]. In most cases this is easier to implement than the full splitting rule.

R forms a complete refutation system for clause sets. In general, the calculus R can be enhanced with standard simplification rules such as tautology deletion and subsumption deletion, in fact, it can be enhanced by any simplification rule which is compatible with a general notion of redundancy [5, 6]. Essentially, a ground clause is redundant in a set N with respect to the ordering $\succ$ if it follows from smaller instances of clauses in N, and a non-ground clause is redundant in N if all its ground instances are redundant in N. A set N of clauses is *saturated up to redundancy* with respect to a particular refinement of resolution if the conclusion of every inference from non-redundant premises in N is either contained in N, or else is redundant in N. Subsumption and condensing are instances of redundancy elimination. A clause D *subsumes* a clause C iff there exists a substitution σ such that $D\sigma \subseteq C$ (strictly speaking, in our framework $D\sigma \subset C$ has to hold). The *condensation* COND(C) of a clause C is a minimal[5] multiple factor of C which subsumes C. A clause C is *condensed* if there is no proper subclause of C which is a factor of C.

A *derivation* in R from a set of clauses N is a finitely branching, ordered tree T with root N and nodes which are sets of clauses. The tree is constructed by applications of the expansion rules to the leaves. We assume that no resolution or factoring inference is computed twice on the same branch of the derivation.

---

[5] Minimality is with respect to the number of literals in the clause.

Any path $N(= N_0), N_1, \ldots$ in a derivation T is called a *closed branch* in T iff the clause set $\bigcup_{j \geq 0} N_j$ contains the empty clause, otherwise it is called an *open branch*. We call a branch B in a derivation tree *complete* (with respect to R) iff no new successor nodes can be added to the endpoint of B by R, otherwise it is called an *incomplete branch*. A derivation T is a *refutation* iff every path $N(= N_0), N_1, \ldots$ in it is a closed branch, otherwise it is called an *open derivation*.

A derivation T from N is called *fair* iff for any path $N(= N_0), N_1, \ldots$ in T, with *limit* $N_\infty = \bigcup_{j \geq 0} \bigcap_{k \geq j} N_k$, it is the case that each clause C which can be deduced from non-redundant premises in N_∞ is contained in some N_j. Intuitively, fairness means that no non-redundant inferences are delayed indefinitely. For a finite complete branch $N(= N_0), N_1, \ldots N_n$, the limit N_∞ is equal to N_n.

Theorem 3 ([6]). *Let T be a fair R derivation from a set N of clauses. Then: (i) If $N(= N_0), N_1, \ldots$ is a path with limit N_∞, then N_∞ is saturated (up to redundancy). (ii) N is satisfiable if and only if there exists a path in T with limit N_∞ such that N_∞ is satisfiable. (iii) N is unsatisfiable if and only if for every path $N(= N_0), N_1, \ldots$ the clause set $\bigcup_{j \geq 0} N_j$ contains the empty clause.*

It should be noted that inferences with ineligible literals are not unsound, but are provably redundant. In other words, only inferences with eligible literals need to be performed for soundness and completeness.

6 Decision Procedures Using Ordered Resolution

Many modal logics naturally translate into decidable fragments of first-order logic. For example the basic modal logic K translates into the two-variable fragment, into the guarded fragment [1], into Maslov's class K [58], and into fluted logic [72, 73] (cf. [34]). By constructing decision procedures for these decidable fragments, one obtains generic decision procedures for modal logics and the corresponding description logics. Resolution decision procedures have been developed for the guarded fragment [15, 25], for Maslov's class K [48], for fluted logic [78] and various other classes related to modal logics, see e.g. [22, 34, 45]. In this paper we consider only the relationship to a fragment of clausal logic based on the two-variable fragment. The fragment is called DL^* [16]. It is a variation of the class of DL-clauses, that was introduced in [50] with the purpose of handling expressive description logics.

In order to simplify the definition of the fragment DL^* of clausal logic all clauses are assumed to be maximally split. The notions can be easily adopted for clauses with more than one split component. A maximally split clause C is a DL^*-*clause* iff the following conditions are satisfied.

1. All literals are unary, or binary.
2. There is no nesting of function symbols.
3. Every functional term in C contains all the variables of C. (This condition implies that if C contains a functional ground term, then C is ground.)
4. Every binary literal (even if it has no functional terms) contains all the variables of C.

The first-order translation of the modal formula $[\neg r_1 \wedge r_2]\langle \neg r_1 \wedge r_2\rangle p$ is

$$\exists x \forall y\, ((\neg R_1(x,y) \wedge R_2(x,y)) \to \exists z\, (\neg R_1(y,z) \wedge R_2(y,z) \wedge P(z))).$$

The structural transformation results in the set of formulae on the left, while the clausal form is given on the right. (Here α is used as an abbreviation for $\neg r_1 \wedge r_2$.)

$$\begin{array}{ll}
\exists x\, Q_{[\alpha]\langle \alpha \rangle p}(x) & Q_{[\alpha]\langle \alpha \rangle p}(a)^* \\[4pt]
\forall x\, (Q_{[\alpha]\langle \alpha \rangle p}(x) \to & \neg Q_{[\alpha]\langle \alpha \rangle p}(x) \vee \neg Q_\alpha(x,y)^* \vee Q_{\langle \alpha \rangle p}(y) \\[4pt]
\qquad \forall y\, (Q_\alpha(x,y) \to Q_{\langle \alpha \rangle p}(y))) & \neg Q_{\langle \alpha \rangle p}(x) \vee Q_\alpha(x,f(x))^* \\[4pt]
\forall x\, (Q_{\langle \alpha \rangle p}(x) \to \exists y\, (Q_\alpha(x,y) \wedge P(y))) & \neg Q_{\langle \alpha \rangle p}(x) \vee P(f(x))^* \\[4pt]
\forall xy\, (Q_\alpha(x,y) \to (\neg R_1(x,y) \wedge R_2(x,y))) & \neg Q_\alpha(x,y)^* \vee \neg R_1(x,y)^* \\[4pt]
\forall xy\, ((\neg R_1(x,y) \wedge R_2(x,y)) \to Q_\alpha(x,y)) & \neg Q_\alpha(x,y)^* \vee R_2(x,y)^* \\[4pt]
& R_1(x,y)^* \vee \neg R_2(x,y)^* \vee Q_\alpha(x,y)^*
\end{array}$$

Fig. 3. A sample transformation of a modal logic formula to DL^*.

Examples of DL^*-clauses include ground clauses, and the following.

$$\begin{array}{ll}
\neg Q_0(x) \vee Q_1(x) \vee \neg Q_2(x) & Q_0(x) \vee \neg R_0(x,y) \vee Q_1(y) \\[4pt]
\neg Q_0(x) \vee Q_1(f(x)) & R_0(x,y) \vee \neg R_1(y,x) \vee R_2(x,y) \\[4pt]
\neg Q_0(x) \vee \neg R_0(f(x),x) & R_0(x,y) \vee \neg R_1(x,f(x,y)) \vee R_2(f(x,y),y)
\end{array}$$

The clauses $R_0(x,y) \vee R_0(x,f(x))$, $Q_0(x,x,x) \vee Q_1(f(f(x)))$ and $R_0(x,x) \vee R_1(x,y)$ do not belong to the class of DL^*-clauses. The clause $Q_0(x) \vee Q_1(a)$ does not belong to DL^*, since it is not maximally split.

Theorem 4 ([16, 50]). *Over a finite signature[6] there are only finitely many maximally split DL^*-clauses (modulo variable renaming).*

The proof can be obtained by first observing that there is a fixed upper bound for the maximal number of variables in a clause. Then there are only a finite number of possible literals. Because every clause is a subset of the set of possible literals, there is a finite set of possible clauses.

Theorem 5 ([16]). *The number of possible DL^*-clauses is bounded by $2^{2^{f(s)}}$, where f is of order $s\log(s)$ and s is the size of the signature.*

The reduction of modal formulae to sets of DL^*-clauses makes use of a structural transformation introducing new names for subformulae corresponding to non-atomic subexpressions of the original modal formula [16, 50]. The reduction is illustrated in Figure 3. It is not difficult to verify that the generated clauses are all DL^* clauses. In general, it can be proved that:

---

[6] The supply of function symbols and predicate symbols is finite, while there are possibly infinite but countably many variables.

Theorem 6 ([16]). *Let φ' be a first-order formula that results from translation of a modal formula φ in $K_{(m)}(\cap, \cup, \bar{\ }, \smile)$. Every clause in the clausal normal form of $\mathrm{Def}_\Lambda(\varphi')$ is a DL*-clause, where $\Lambda = \{\lambda \mid$ there is a non-atomic subexpression $\varphi|_\lambda$ of φ and $\varphi'|_\lambda = \Pi(\varphi|_\lambda)\}$.*

In order to decide the class DL^*, we use the following ordering which is similar to the recursive path ordering. First we define an order $>_d$ on terms: $s >_d t$ if s is deeper than t, and every variable that occurs in t, occurs deeper in s. Then we define $P(s_1, \ldots, s_n) \succ Q(t_1, \ldots, t_m)$ as $\{s_1, \ldots, s_n\} >_d^{\mathrm{mul}} \{t_1, \ldots, t_m\}$. Here $>_d^{\mathrm{mul}}$ is the multiset extension of $>_d$. So we have $P(f(x)) \succ P(a), P(x)$ and $P(x,y) \succ Q(x)$, but not $P(f(x)) \succ P(f(a))$. The ordering $>_d$ originates from Fermüller et al. [23]. The selection function S does not select any negative literal in any clause. We denote this particular instance of the resolution calculus R by $\mathsf{R}^{\mathrm{ord}}$.

In the example in Figure 3 the maximal literals are marked with *. These are the literals that can potentially be resolved or factored upon.

In order to prove that the procedure $\mathsf{R}^{\mathrm{ord}}$ is indeed a decision procedure we have to show that it is complete, and terminating. The completeness follows from Theorem 3. Termination is a consequence of Theorem 4, and the fact that the restriction derives only clauses that are within DL^*, or splittable clauses with split components in DL^* (cf. [45, 50]).

Theorem 7. *Let L be a logic in-between $K_{(m)}$ and $K_{(m)}(\cap, \cup, \bar{\ }, \smile)$. Let Δ be a finite set of relational properties expressible in DL^*. Let N be the clausal form of $\Delta \wedge \mathrm{Def}_\Lambda \Pi(\varphi)$, where φ is any modal formula in L and Λ is defined as in Theorem 6. Then: (i) Any derivation from N in $\mathsf{R}^{\mathrm{ord}}$ (up to redundancy) terminates in double exponential time. (ii) φ is unsatisfiable in L iff there is a refutation of N in $\mathsf{R}^{\mathrm{ord}}$.*

Relational properties expressible in DL^* include the Boolean combination of relational inclusions or equivalences expressed over intersection, union, complementation and converse. Moreover, reflexivity and irreflexivity can be expressed in DL^*. It is usually the case that when studying modal decidability problems by analysing the decidability of related clausal classes one comes to realise that stronger results are possible than initially anticipated. For instance, it is not difficult to see that modal and relational formulae with positive occurrences of relational composition can also be embedded into the class DL^*. This means that Theorem 7 can be strengthened. Let $K_{(m)}(\cap, \cup, \bar{\ }, \smile, \uparrow, ;^{\mathsf{pos}})$ denote the multimodal logic in which relational formulae may also have the forms: $\alpha \uparrow \varphi$ (domain restriction), and $\alpha \,; \beta$ (composition), but the latter may occur positively only (that is, occur in the scope of an even number of explicit and implicit negation symbols). The semantics of the new operators are defined (as expected) by:

$$R_{\alpha \uparrow \varphi} = \{(x,y) \mid (x,y) \in R_\alpha \wedge x \in \iota(\varphi)\}, \quad \text{and}$$
$$R_{\alpha \,; \beta} = R_\alpha \,; R_\beta = \{(x,y) \mid \exists z \, ((x,z) \in R_\alpha \wedge (z,y) \in R_\beta)\}.$$

Observe that the range restriction of a relation can be represented in terms of domain restriction and converse, by $(\alpha^\smile \uparrow \varphi)^\smile$.

Det	$[\beta]p \to \langle\gamma\rangle p$	$\forall x \exists y\, (R_\beta(x,y) \wedge R_\gamma(x,y))$
Sym	$\langle\alpha\rangle[\beta]p \to p$	$\forall x \forall y\, (R_\alpha(x,y) \to R_\beta(y,x))$
Gr	$[\beta]p \to [\alpha]p$	$\forall x \forall y\, (R_\alpha(x,y) \to R_\beta(x,y))$
Conf	$\langle\alpha\rangle[\beta]p \to \langle\gamma\rangle p$	$\forall x \forall y\, (R_\alpha(x,y) \to \exists z\, (R_\alpha(x,z) \wedge R_\beta(z,y)))$

Fig. 4. Modal axioms and their correspondence properties.

Theorem 8. *Let L be a logic in-between $K_{(m)}$ and $K_{(m)}(\cap, \cup, ^-, \smile, 1, ;^{pos})$. Let Δ be a finite set of relational properties expressible in DL^*. Let N be the clausal form of $\Delta \wedge \mathrm{Def}_\Lambda \Pi(\varphi)$, where φ is any modal formula in L and Λ is defined as in Theorem 6. Then: (i) Any derivation from N in R^{ord} (up to redundancy) terminates in double exponential time. (ii) φ is unsatisfiable in L iff there is a refutation of N in R^{ord}.*

This theorem cannot be strengthened further by removing the restriction on compositions. From the undecidability result of the equational theory of Boolean algebras with composition in [54] it follows that allowing arbitrary occurrences of composition leads to undecidability.

Theorem 9. *The satisfiability problem in every logic in-between $K_{(m)}(\cap, \cup, ^-, ;)$ and $K_{(m)}(\cap, \cup, ^-, \smile, 1, ;)$ is undecidable.*

Theorem 10. *Every logic in-between $K_{(m)}(\cap, \cup, ^-)$ and $K_{(m)}(\cap, \cup, ^-, \smile, 1)$ is NEXPTIME-complete.*

Proof. A consequence of the NEXPTIME-completeness of the satisfiability of Boolean modal logic and FO^2 formulae [37, 56].

From Theorem 8 we can obtain some decidability results for propositional modal logics. In the following let α, β and γ denote either a relational variable or a relational formula built from relational variables using disjunction and composition. Let Σ be a set modal formulae in the language of multi-modal $K_{(m)}$ and let $K_{(m)}\Sigma$ be the extension of $K_{(m)}$ closed under the formulae in Σ. For example, the axiom schema listed in Figure 4 determine classes of logics considered in Catach [12] and Baldoni [7].

Theorem 11. *Let Σ be any finite set of instances of formulae in Figure 4, and let Δ^Σ be the set of associated first-order properties as specified in Figure 4. Then: For any modal formula φ, φ is satisfiable in $K_{(m)}\Sigma$ iff $\Delta^\Sigma \wedge \Pi(\varphi)$ is first-order satisfiable.*

Proof. By noting that disjunction and composition in the relational parameters of the modal operators can be normalised away, it is not difficult to see that all formulae in Figure 4 are Sahlqvist formulae. Using the SCAN algorithm [24] one can prove the properties associated with the modal formulae in Figure 4 are in fact their correspondence properties. Thus, the theorem follows from the well-known Sahlqvist Theorem [76].

This theorem is also an easy consequence of a more general theorem by Catach [12].

Theorem 12. *Let Σ be any finite set of instances of formulae in Figure 4, with the restriction that in each case α is a relational formula built from relational variables and disjunction only, while β and γ denote either a relational variable or a relational formula built from relational variables using disjunction and composition. Then, the satisfiability problem in $K_{(m)}\Sigma$ is decidable, and it can be decided by a resolution procedure based on the translation into DL^* and any ordering refinement compatible with $>_d$.*

Proof. The restriction that α is a relational formula built from relational disjunction only ensures that relational composition occurs only positively in the first-order correspondence properties for the axioms in Σ. This implies all correspondence properties can be formulated in DL^*. The result then follows by Theorem 7.

Finally, we observe that decidability is preserved if a relational property ψ in the theory Δ for a logic in-between $K_{(m)}$ and $K_{(m)}(\cap, \cup, ^-, \smile, 1, ;^{\text{pos}})$ may also include binary literals of the form $(\neg)R(x,x)$ or $(\neg)R(y,y)$, if x and y are the two universally quantified variables in the property ψ. An example is $\forall xy\,(R_1(x,y) \to R_2(x,x) \vee R_3(y,y))$. Although the clausal form is not an DL^* clause it can easily be transformed into a set of DL^* clauses using the renaming techniques used in Hustadt and Schmidt [48] for deciding Maslov's dual class K by ordered resolution.

7   Decision Procedures Using Selection-Based Resolution

$K_{(m)}(\cap, \cup, \smile)$ and logics below it have the property that they can be decided by a refinement of resolution which is defined solely by a selection function of negative literals [16, 50]. The transformation to clausal form is based on the standard translation and we use the same structural transformation as described in Section 4, except that Def_Λ introduces the same symbol for variant subformulae with the same polarity. For simplicity it is assumed that φ is in negation normal form, that is, in every subformula of the form $\neg\psi$, ψ is a propositional variable. As a consequence all occurrences of non-atomic subformulae of φ' with one free variable have positive polarity. This means that $\text{Def}_\lambda(\varphi') = \text{Def}_\lambda^+(\varphi')$ for the positions λ associated with these occurrences. But subformulae corresponding to relational formulae (subformulae with two free variables) can occur both positively and negatively. For these Def_Λ introduces one symbol for all variant occurrences of subformulae corresponding to non-atomic relational subformulae with positive polarity and a different symbol for all variant occurrences with negative polarity. For example, Def_Λ maps $[\alpha]\langle\alpha\rangle p$ with $\alpha = r_1 \wedge r_2$ to the conjunction of the following formulae.

$$\mathcal{P}(a)$$

$$\neg Q_\psi(x)^+ \vee \neg P_i(x)^+ \qquad\qquad\text{if } \psi = \neg p_i$$

$$\neg Q_\psi(x)^+ \vee \mathcal{P}(x) \,[\vee\, \mathcal{P}(x)] \qquad\qquad\text{if } \psi = \phi_1 \wedge[\vee]\, \phi_2$$

$$\neg Q_\psi(x)^+ \vee \neg \mathcal{R}(x,y)^+ \,[\vee\, \mathcal{P}(y)] \qquad\text{if } \psi = [\alpha]\phi \;[\psi = [\alpha]\bot]$$

$$\neg Q_\psi(x)^+ \vee \mathcal{P}(f(x))$$
$$\qquad\qquad\qquad\qquad\qquad\qquad\qquad\quad\text{if } \psi = \langle\alpha\rangle\phi$$
$$\neg Q_\psi(x)^+ \vee \mathcal{R}(x,f(x))$$

$$\neg Q_\alpha^p(x,y)^+ \vee \mathcal{R}(x,y) \,[\vee\, \mathcal{R}(x,y)] \qquad\text{if } \alpha = \beta_1 \wedge[\vee]\, \beta_2 \text{ has pos. polarity}$$

$$Q_\alpha^n(x,y) \vee \neg\mathcal{R}(x,y)^+ \,[\vee\, \neg\mathcal{R}(x,y)^+] \qquad\text{if } \alpha = \beta_1 \wedge[\vee]\, \beta_2 \text{ has neg. polarity}$$

Fig. 5. Schematic clausal forms for $K_{(m)}(\cap, \cup, \smile)$.

$$\exists x\, Q_{[\alpha]\langle\alpha\rangle p}(x)$$
$$\forall x\, (Q_{[\alpha]\langle\alpha\rangle p}(x) \to \forall y\, (Q_\alpha^n(x,y) \to Q_{\langle\alpha\rangle p}(y)))$$
$$\forall x\, (Q_{\langle\alpha\rangle p}(x) \to \exists y\, (Q_\alpha^p(x,y) \wedge P(y)))$$
$$\forall xy\, (Q_\alpha^p(x,y) \to (R_1(x,y) \wedge R_2(x,y)))$$
$$\forall xy\, ((R_1(x,y) \wedge R_2(x,y)) \to Q_\alpha^n(x,y)).$$

The symbol Q_α^n (resp. Q_α^p) is associated with the negative (resp. positive) occurrence of α.

Subsequently, introduced predicate symbols are denoted by Q_ψ and Q_α^p or Q_α^n, where Q_ψ represents an occurrence of a modal subformula ψ and Q_α^p (Q_α^n) represents a positive (negative) occurrence of a relational subformula α. Let

$$\mathcal{P}(s) \quad \text{denote some literal in } \{P_i(s), Q_\psi(s)\}_{i,\psi}, \text{ and let}$$

$$\mathcal{R}(s,t) \quad \text{denote some literal in } \{R_j(s,t), R_j(t,s), Q_\alpha^{p/n}(s,t), Q_\alpha^{p/n}(t,s)\}_{j,\alpha}.$$

Note two occurrences of $\mathcal{P}(s)$ or $\mathcal{R}(s,t)$ need not be identical. For example, $\neg Q_\psi(x) \vee P_i(x) \vee Q_\chi(x)$ is an instance of $\neg Q_\psi(x) \vee \mathcal{P}(x) \vee \mathcal{P}(x)$, while $\neg Q_\psi(x) \vee \neg R_j(y,x) \vee Q_\chi(y)$ and $\neg Q_\psi(x) \vee \neg Q_\alpha^n(x,y) \vee Q_\chi(y)$ are instances of $\neg Q_\psi(x) \vee \neg \mathcal{R}(x,y) \vee \mathcal{P}(y)$.

All input clauses have one of the forms described in Figure 5 [16,50]. The literals marked with $^+$ are selected in the clauses by the specific selection function we use.

The calculus is based on maximal selection of negative literals. This means the selection function selects exactly the set of all negative literals in any non-positive clause. An ordering refinement is optional. In this case, the ordered resolution rule of R can be replaced by the following rule.

Resolution with maximal selection:

$$\frac{C_1 \vee A_1 \quad \cdots \quad C_n \vee A_n \quad \neg A_{n+1} \vee \ldots \vee \neg A_{2n} \vee D}{(C_1 \vee \ldots \vee C_n \vee D)\sigma}$$

provided for any $1 \le i \le n$, (i) σ is the most general unifier of A_i and A_{n+i}, (ii) $C_i \vee A_i$ and D are positive clauses, (iii) no A_i occurs in C_i, and (iv) A_i,

$\neg A_{n+i}$ are selected. The *negative premise* is $\neg A_{n+1} \vee \ldots \vee \neg A_{2n} \vee D$ and the other premises are the *positive premises*. The literals A_i and A_{n+i} are the *eligible literals*.

Let R^{hyp} be the calculus based on maximal selection and no ordering. This means the rules are the above resolution rule, positive unordered factoring and splitting. The normalisation function is not needed (but could of course be added without losing completeness), that is, we assume NORM is the identity mapping. For simplification tautology deletion is used. All derivations in R^{hyp} are generated by strategies in which no application of the resolution or factoring with identical premises and identical consequence may occur twice on the same path in any derivation. In addition, deletion rules, splitting, and the deduction rules are applied in this order, except that splitting is not applied to clauses which contain a selected literal.

As all non-unit clauses of a typical input set contain a selected literal all definitional clauses can only be used as negative premises of resolution steps. To begin with there is only one candidate for a positive premise, namely, the ground unit clause $Q_{\varphi}(a)$ representing the input formula φ. Inferences with such ground unary unit clauses produce ground clauses consisting of positive literals only, which will be split into ground unit clauses.

Lemma 1 ([50]). *Maximally split (non-empty) inferred clauses have one of two forms: $\mathcal{P}(s)$, or $\mathcal{R}(s, f(s))$, where s is a ground term.*

In general, s will be a nested non-constant functional ground term, which is typically avoided in resolution decision procedures based on an ordering refinement, because in most situations nesting causes unbounded computations. However, it can be shown that for the class of clauses under consideration any derived clause is smaller than its positive parent clauses with respect to a well-founded ordering which reflects the structure of the formula.

Theorem 13 ([50]). *Let L be a logic in-between $K_{(m)}$ and $K_{(m)}(\cap, \cup, \smile)$. Let φ be any L-formula and let N be the clausal form of $\mathrm{Def}_A \Pi(\varphi)$. Then: (i) Any R^{hyp}-derivation from N terminates. (ii) φ is unsatisfiable in L iff there is a refutation of N by R^{hyp}.*

Theorem 14 ([16]). *For any logic in-between $K_{(m)}$ and $K_{(m)}(\cap, \cup, \smile)$, the space complexity for testing the satisfiability of a modal formulae φ with R^{hyp} is bounded by $O(nd^m)$, where n is the number of symbols in φ, d is the number of different diamond subformulae in φ, and m is the modal depth of φ[7].*

Formulae in $K_{(m)}(\cap, \cup, \smile)$ translate by Π into the guarded fragment, while there are formulae in $K_{(m)}(\cap, \cup, ^-, \smile)$ which do not [16]. It is not difficult to see that formulae in $K_{(m)}(\cap, \cup, \smile)$ are in fact translated into the subfragment $GF1^-$, introduced by Lutz, Sattler and Tobies [57]. In contrast to the guarded fragment, $GF1^-$ permits the development of PSPACE decision procedures [31, 57]. Under

---

[7] By definition the *modal depth* of a formula φ is the maximal nesting of modal operators $\langle \alpha \rangle$ or $[\alpha]$ in φ.

the assumption that either (i) there is a bound on the arity of predicate symbols in $GF1^-$ formulae, or (ii) that each subformula of a $GF1^-$ formula has a bounded number of free variables, the satisfiability problem of $GF1^-$ is the same as for $K_{(m)}$ [57]. Thus, we can conclude:

Theorem 15. *The computational complexity of the satisfiability problem of any modal logic in-between $K_{(m)}$ and $K_{(m)}(\cap, \cup, \smile)$ is PSPACE-complete.*

In [31] it is shown that R^{hyp} can be implemented as a modification of the main procedure of a standard (saturation based) first-order theorem prover with splitting (e.g. SPASS) to provide a space optimal decision procedure for $GF1^-$. A direct consequence is the following.

Theorem 16. *R^{hyp} can be turned into a polynomial space resolution decision procedure for logics in-between $K_{(m)}$ and $K_{(m)}(\cap, \cup, \smile)$.*

So far in this section we have considered only logics with empty relational theory. It is natural to try and strengthen the results obtained. We might ask whether the results can be generalised, and if it is indeed possible, to try and determine for which theories the above results can be generalised. Generalisations of Theorem 13 have been considered in [16, 50]. We quote here a generalised theorem established in [16].

Theorem 17 ([16]). *Let L be a logic in-between $K_{(m)}$ and $K_{(m)}(\cap, \cup, \smile)$. Let Δ be a finite R^{hyp}-saturated set of clauses consisting of two kinds of split components.*

1. *Clauses with at most two free variables, which are built from finitely many binary predicate symbols R_j, no function symbols, and containing at least one guard literal (that is, this literal is negative and includes all the variables of the clause).*
2. *Clauses built from one variable, finitely many function symbols (including constants), and finitely many binary predicate symbols R_j, with the restriction that (a) the argument multisets of all non-ground literals coincide, and (b) each literal which contains a constant is ground.*

Suppose φ is an L-formula and N is the clausal form of $\mathrm{Def}_A \Pi(\varphi)$. Then: (i) Any R^{hyp}-derivation from $N \cup \Delta$ terminates. (ii) φ is unsatisfiable in $L\Delta$ iff there is a refutation of $N \cup \Delta$ by R^{hyp}.

Relational frame properties covered by this result include reflexivity, irreflexivity, seriality, symmetry, inclusions among relations, for example, $R_1 \subseteq R_2$ or $R_1 \subseteq (R_2^{\smile} \cap R_3)$, as well as, for example, $\forall x \exists y \, \neg R(x, y)$, $\forall x \exists y \, (R(x, y) \vee R(y, x))$, or $\forall xy \, (R(x, y) \rightarrow R(x, x))$. Of the properties in Figure 4 the properties Det, Sym and Gr are covered, provided that the relational parameters α, β and γ are formed from relational variables and disjunction. Thus, familiar logics covered by the above results include KT, KD, KB, KTB, and KDB, but also the basic tense logic K_t.

The results of this section also cover the corresponding description logics, for example, the basic description logic $\mathcal{ALC}$ possibly extended with role conjunction, role disjunction and inverse roles. Acyclic TBox statements, and both concept and role ABox statements are also in the scope of the last theorem.

8   Simulating Tableaux

Selection refinements of resolution (and hyperresolution) are closely related to standard modal tableau calculi and description logic systems [16, 22, 49, 50, 52]. In this section we investigate simulation relationships between the selection-based resolution procedure R^{hyp} and Massacci's single-step prefixed tableau calculi [59].

There are three notions of simulation [16]: polynomial simulation of derivations, polynomial simulation of search, and step-wise simulation. By definition, a proof system $\mathcal{A}$ *p-simulates (polynomially simulates) derivations* of a proof system $\mathcal{B}$ iff there is a function g, computable in polynomial time, which maps derivations in $\mathcal{B}$ for any given formula φ, to derivations in $\mathcal{A}$ for φ. A system $\mathcal{A}$ *p-simulates search* of a system $\mathcal{B}$ iff there is a polynomial function g such that for any formula φ, g maps derivations from φ in $\mathcal{A}$ to derivations from φ in $\mathcal{B}$. The first notion generalises the notion of p-simulation found in [13], who are only concerned with the p-simulation of proofs (that is, successful derivations leading to a proof). Simulation of search is a relationship in the opposite direction. It implies that $\mathcal{A}$ does not perform any inference steps for which no corresponding inference steps exist in $\mathcal{B}$. To show that $\mathcal{A}$ p-simulates proofs or derivations of $\mathcal{B}$ it is sufficient to prove that for every formula φ and every derivation $D_{\mathcal{B}}$ of φ in $\mathcal{B}$, there exists a derivation $D_{\mathcal{A}}$ of φ in $\mathcal{A}$ such that the number of applications of inference rules in $D_{\mathcal{A}}$ is polynomially bounded by the number of applications of inference rules in $D_{\mathcal{B}}$. This can be achieved by showing that there exists a number n such that each application of an inference rule in $D_{\mathcal{A}}$ corresponds to at most n applications of inference rules in $D_{\mathcal{B}}$. It follows that the length of derivation $D_{\mathcal{B}}$ is polynomially bounded by the length of $D_{\mathcal{A}}$. This is known as a *step-wise simulation* of $\mathcal{B}$ by $\mathcal{A}$ [20]. Note that a step-wise simulation is independent of whether the considered derivations are proofs or not.

The single-step prefixed tableau calculi of Massacci [59] for subsystems of $S5$ are defined by Figures 6 and 7. The basic entities are formulae labelled with prefixes. A labelled (prefixed) formula has the form $\sigma : \varphi$, where σ is a sequence of positive integers and φ is a modal formula. σ represents a world in which φ is true. Tableau derivations in the single-step prefixed tableau calculi have a tree structure and begin with the formula, $1 : \varphi$ in the root node. Successor nodes are then constructed by the application of the expansion rules in Figure 6. The prefixes in the expansion rules, except for $\sigma.n$ of the $(\diamond)$-rule, are assumed to be present on the current branch.

Theorem 18 ([59]). *Let $\Sigma \subseteq \{D, T, B, 4, 5\}$. A formula φ is satisfiable in a logic $K\Sigma$ iff a tableau containing a branch $\mathcal{B}$ can be constructed by the tableau calculus for $K\Sigma$ such that $\mathcal{B}$ does not contain the falsum and further rule applications are redundant.*

$$(\bot) \frac{\sigma : \psi, \sigma : \neg\psi}{\sigma : \bot} \qquad (\wedge) \frac{\sigma : \psi \wedge \phi}{\sigma : \psi, \sigma : \phi} \qquad (\vee) \frac{\sigma : \psi \vee \phi}{\sigma : \psi \mid \sigma : \phi}$$

$$(\Diamond) \frac{\sigma : \Diamond\psi}{\sigma.n : \psi} \text{ with } \sigma.n \text{ new to the current branch}$$

$$(\Box) \frac{\sigma : \Box\psi}{\sigma.n : \psi} \qquad (D) \frac{\sigma : \Box\psi}{\sigma : \Diamond\psi} \qquad (T) \frac{\sigma : \Box\psi}{\sigma : \psi}$$

$$(B) \frac{\sigma.n : \Box\psi}{\sigma : \psi} \qquad (4) \frac{\sigma : \Box\psi}{\sigma.n : \Box\psi} \qquad (4^r) \frac{\sigma.n : \Box\psi}{\sigma : \Box\psi}$$

$$(4^d) \frac{\sigma.n : \Box\psi}{\sigma.n.m : \Box\psi} \qquad (5) \frac{1.n : \Box\psi}{1 : \Box\Box\psi}$$

Fig. 6. Single step prefixed tableau expansion rules for subsystems of $S5$.

$K:$	(K)	$K5:$	$(K),(4^r),(4^d),(5)$
$KD:$	$(K),(D)$	$KDB:$	$(K),(D),(B)$
$KT:$	$(K),(T)$	$KD4:$	$(K),(D),(4)$
$KB:$	$(K),(B)$	$KTB:$	$(K),(T),(B)$
$K4:$	$(K),(4)$	$S4:$	$(K),(T),(4)$

$KD5:$	$(K),(D),(4^r),(4^d),(5)$
$KB4:$	$(K),(B),(4),(4^r)$
$K45:$	$(K),(4),(4^r),(4^d)$
$KD45:$	$(K),(D),(4),(4^r),(4^d)$
$S5:$	$(K),(T),(4),(4^r)$

Fig. 7. Tableau calculi for subsystems of $S5$. (K) denotes the sequence of rules $(\bot),(\wedge),(\vee),(\Diamond),(\Box)$.

Theorem 19 ([52]). *Let $\Sigma \subseteq \{D,T,B,4,5\}$. There is a p-simulation of single step prefix tableau derivations for $K\Sigma$ using R^{hyp}.*

The proof exploits the step-wise simulation of tableau inference steps by resolution inference steps where the theories are given by the clausal form of the conjunction of the first-order correspondence properties of the axioms. For the modal logics $K\Sigma$ with $\Sigma \subseteq \{D,T,B\}$ simulation in the other direction can also be proved.

Theorem 20 ([52]). *R^{hyp} p-simulates search in single step prefix tableaux for $K\Sigma$ with $\Sigma \subseteq \{D,T,B\}$.*

This is a consequence of a near bisimulation between the tableau derivations and R^{hyp} derivations for the logics under consideration. If factoring rules are added to the single step prefix tableau calculus then this calculus can also p-simulate R^{hyp} derivations.

Because R^{hyp} does not terminate in the presence of the transitivity clause or Euclideanness, Theorem 20 does not extend to transitive and Euclidean modal logics. For 4 and 5 termination in single step prefixed tableaux is ensured by a loop checking mechanism [59]. Once a loop is detected in a branch no further rules are applied. In R^{hyp} further inference steps will be performed. To prevent this we would have to provide a means by which the resolution procedure can recognise the redundancy of further inference steps. This can be realised with a blocking inference rule, used in [49], which has an effect similar to loop checking. Using soft typing described in [27] might provide an alternative solution.

Similar simulation results can be obtained for other forms of modal tableau calculi, including caluli with implicit or explicit accessibility relation and analytic modal KE tableaux, e.g. [46, 59], or even sequent proof systems. Simulation results of tableau calculi for description logics by resolution can be found in Hustadt and Schmidt [49, 50].

9   Developing Tableaux via Resolution

In general, resolution (refutation) proofs for the first-order translation of a modal formula have little resemblance to proofs in the modal source logic. This is because the modal form is usually lost during the transformation to clausal form and subsequent deduction. It is therefore difficult to translate first-order resolution proofs back into modal proofs. By using a different translation method, for example, translation methods based on the functional translation where accessibility is encoded in terms of paths [3, 40, 69], this problem can be reduced and eliminated for certain logics, cf. [11, 17]. A solution to the problem of backward translation of resolution proofs is provided by the structural translation used in Section 7 and 8, and the tableau simulating resolution refinement R^{hyp} [52]. It makes it easy to convert resolution proofs into tableau style (or natural deduction style) modal proofs.

Taking this idea a step further, the approach using R^{hyp} can be exploited for systematically developing sound and complete tableau proof systems. For instance, De Nivelle et al [16] show how a tableau system for $K_{(m)}(\cap, \cup, \smile)$ can be extracted from the resolution method described in Section 7. The idea is to express a R^{hyp} resolution inference step by a tableau rule, or if this is not possible, as is the case for conjunctive subformulae, to express a group of R^{hyp} resolution inference steps as a tableau rule.

A *tableau* is a finitely branching tree whose nodes are sets of labelled formulae. Given that φ is a formula to be tested for satisfiability the root node is the set $\{a : \varphi\}$. Successor nodes are constructed in accordance with a set of φ expansion rules. A rule $\frac{X}{X_1 \mid ... \mid X_n}$ fires for a selected formula F in a node if F is an instance of the numerator X, or more generally, F together with other formulae in the node are instances of the formulae in X. n successor nodes are created which contain the formulae of the current node and the appropriate instances of X_i. It is assumed that no rule is applied twice to the same instance of the numerator.

Assume that φ is a formula in negation normal form. Recall, the inference in R^{hyp} starts with an inference with the only positive clause $Q_\varphi(a)$. Accordingly, the root of the tableau is given by $\{a : \varphi\}$. Subsequent R^{hyp} inference steps can be translated more or less directly into the tableau expansion rules listed in Figure 8 (for details see [16]). The rules for $K_{(m)}(\cap, \cup, \smile)$ include the clash rule $(\bot)$, seven 'elimination' rules $(\wedge)$, $(\vee)$, $(\Diamond)$, $(\Box)$, $(\smile)$, $(\wedge^r)$, and $(\vee^r)$ for positive occurrences of subformulae, and three 'introduction' rules $(\smile_I)$, $(\wedge_I^r)$ and $(\vee_I^r)$ for negative occurrences of subformulae. The side conditions for the introduction rules ensure

$$(\bot) \; \frac{s : \psi, \, s : \neg\psi}{s : \bot} \qquad\qquad (\wedge) \; \frac{s : \psi \wedge \phi}{s : \psi, \, s : \phi} \qquad\qquad (\vee) \; \frac{s : \psi \vee \phi}{s : \psi \mid s : \phi}$$

$$(\Diamond) \; \frac{s : \langle\alpha\rangle\psi}{(s,t) : \alpha, \, t : \psi} \;\; \text{with } t \text{ new to the branch} \qquad (\Box) \; \frac{(s,t) : \alpha, \, s : [\alpha]\psi}{t : \psi}$$

$$(\smile) \; \frac{(s,t) : \alpha^{\smile}}{(t,s) : \alpha} \qquad\qquad (\wedge^r) \; \frac{(s,t) : \alpha \wedge \beta}{(s,t) : \alpha, \, (s,t) : \beta} \qquad (\vee^r) \; \frac{(s,t) : \alpha \vee \beta}{(s,t) : \alpha \mid (s,t) : \beta}$$

$$(\smile_I) \; \frac{(t,s) : \alpha}{(s,t) : \alpha^{\smile}} \qquad\qquad (\wedge_I^r) \; \frac{(s,t) : \alpha, \, (s,t) : \beta}{(s,t) : \alpha \wedge \beta} \qquad (\vee_I^r) \; \frac{(s,t) : \alpha}{(s,t) : \alpha \vee \beta}$$

Fig. 8. Tableau expansion rules for $K_{(m)}(\cap, \cup, \smile)$. For the rules $(\smile_I)$, $(\wedge_I^r)$ and $(\vee_I^r)$ the side conditions are that the formulae in the denumerator, i.e. $\alpha^{\smile}$, $\alpha \wedge \beta$ or $\alpha \vee \beta$, occur as subformulae of the parameter γ of a box formula $s : [\gamma]\psi$ on the current branch.

that formulae are not introduced unnecessarily. Conjunction and disjunction are assumed to be associative and commutative operations. Only the disjunction rules are don't know nondeterministic and require the use of backtracking. For any logic L in-between $K_{(m)}$ and $K_{(m)}(\cap, \cup, \smile)$ the expansion rules are given by appropriate subsets.

Unnecessary duplication and superfluous inferences can be kept to a minimum by adopting a notion of redundancy which is in the spirit of Bachmair and Ganzinger [4]. A labelled formula F is redundant in a node if the node contains labelled formulae $F_1, \ldots, F_n$ (for $n \geq 0$) which are smaller than F and $\models_L (F_1 \wedge \ldots \wedge F_n) \to F$. In this context a formula ψ is smaller than a formula ϕ if ψ is a subformula of ϕ, but a more general definition based on an admissible ordering in the sense of [4,5] may be chosen. The application of a rule is redundant if its premise(s) or its conclusion(s) is (are) redundant in the current node. For example, for any s, $s : \top$ is redundant, and if a node includes $s : \psi$ and $s : \psi \vee \phi$, then the $(\vee)$ rule need not be applied, and no new branches are introduced.

Theorem 21 ([16]). *A formula φ is satisfiable in $K_{(m)}(\cap, \cup, \smile)$ iff a tableau containing a branch $\mathcal{B}$ can be constructed with the rules of Figure 8 such that $\mathcal{B}$ does not contain falsum ($s : \bot$ for some s) and each rule application is redundant.*

Corollary 1 ([16]). *The appropriate subsets of the rules from Figure 8 provide sound, complete and terminating tableau calculi for logics in-between $K_{(m)}$ and $K_{(m)}(\cap, \cup, \smile)$.*

10 Generating Herbrand Models

A problem closely related to the satisfiability problem is the problem of generating (counter-)models. It is well-known that hyperresolution can be employed with dual purpose, namely, as a reasoning method and a Herbrand model builder [22]. Thus, in this section we briefly discuss the use of R^{hyp} as a procedure for automatically constructing Herbrand models for extended modal logics. The results are actually consequences of properties of classes of range restricted clause sets. A clause C is said to be *range restricted* iff the set of variables of the positive part of C is a subset of the set of variables of the negative part of C. A clause set is range restricted iff it contains only range restricted clauses. This means that a positive clause is range restricted only if it is a ground clause.

A *Herbrand interpretation* is a set of ground atoms. By definition a ground atom A is *true* in an interpretation H iff $A \in H$ and it is *false* in H iff $A \notin H$, $\top$ is true in all interpretations and $\bot$ is false in all interpretations. A literal $\neg A$ is true in H iff A is false in H. A conjunction of two ground atoms A and B is true in an interpretation H iff both A and B are true in H and respectively, a disjunction of ground atoms is true in H iff at least one of A or B is true in the interpretation. A clause C is true in H iff for all ground substitutions σ there is a literal L in $C\sigma$ which is true in H. A set N of clauses is true in H iff all clauses in N are true in H. If a set N of clauses is true in an interpretation H then H is referred to as a *Herbrand model* of N.

For range restricted clause sets the procedure R^{hyp} implicitly generates Herbrand models [10, 31, 35]. For a class of solvable range restricted clauses, if R^{hyp} terminates on a clause set N without having produced a refutation then a model can be extracted from any complete, open branch in the derivation. The model is given by the set of ground unit clauses in the limit of the branch, i.e. the clause set at the leaf of the branch.

Theorem 22 ([52]). *Let N be the clausal form of a $K_{(m)}(\cap, \cup, \smile)$ formula φ (as defined in Section 7), and let N_∞ be the limit of an arbitrary branch $\mathcal{B}$ in a R^{hyp} derivation tree with root N. Let $[\![\mathcal{B}]\!]$ be the set of positive ground unit clauses in N_∞. If N_∞ does not contain the empty clause, then $[\![\mathcal{B}]\!]$ is a finite (Herbrand) model of N.*

Now a modal model $\mathcal{M} = (W, R, \iota)$ can be easily constructed from $[\![\mathcal{B}]\!]$ for φ. Essentially, the set of worlds is defined by the set of ground terms occurring in $[\![\mathcal{B}]\!]$. The interpretation of relational formulae is determined by the set of R_i literals in $[\![\mathcal{B}]\!]$. For any R_i, if $R_i(s, t)$ is in $[\![\mathcal{B}]\!]$ then $(s, t) \in R_{r_i}$, which can be extended to a homomorphism for complex relational formulae. The interpretation of modal formulae can be defined similarly. For any unary literal $P_i(s)$ (resp. $Q_\psi(s)$) in $[\![\mathcal{B}]\!]$, $s \in \iota(p_i)$ (resp. $s \in \iota(\psi)$), that is, p_i (resp. ψ) is true in the world s. This is homomorphically extended as expected. Consequently:

Theorem 23 ([16, 52]). *Let L and Δ be as in Theorem 17. For any modal formula satisfiable in $L\Delta$ a finite modal model can be effectively constructed on the basis of R^{hyp}.*

Corollary 2 ([16, 52]). *Let L and Δ be as in Theorem 17. Then, $L\Delta$ has the finite model property.*

These results extend also to methods closely related to R^{hyp} such as the tableau calculus introduced in the previous section. For example:

Corollary 3 ([16]). *If L is a logic in-between $K_{(m)}$ and $K_{(m)}(\cap, \cup, \smile)$, and φ is satisfiable in L then a finite modal model can be effectively constructed on the* *basis of the tableau calculus for L given by the appropriate subset of rules from Figure 8.*

Besides hyperresolution there exist a number of other methods for creating Herbrand models, or *representations* of Herbrand models. For references see Fermüller et al [22, §4.2]. It would be worth studying the application of these methods to translations of modal logics.

11   Mechanisation

There are a number of first-order theorem provers that implement the Bachmair-Ganzinger framework of resolution and equality reasoning. These include the state-of-the-art theorem provers: E [80,81], SATURATE [28], SPASS [82,83] and VAMPIRE [74]. The theorem prover OTTER [60] also implements ordered resolution and hyperresolution. These provers are sophisticated programs which have been developed over many years. Of these provers, SPASS forms the basis of the theorem prover MSPASS which has been used to study the practical properties of R^{ord} and R^{hyp} for automating modal logic reasoning and simulating tableau procedures [47, 51–53, 77]. These studies have been mainly for basic multi-modal logic. Furthermore, with one exception, a variation of the optimised functional translation method was used, since MSPASS has shown better performance for the optimised functional translation method compared to the relational translation method on the benchmark problems used in the studies. However, according to current knowledge the practical scope of the optimised functional translation method is limited to modal logics with K or KD modalities.

The main difference between SPASS and MSPASS is that MSPASS accepts also modal logic, description logic and relational formulae as input. Modal formulae and description logic formulae are built from a vocabulary of propositional symbols of two disjoint types, namely, propositional (Boolean or concept) and relational (role). The repertoire of logical constructs includes: (i) the standard Boolean operators on both propositional and relational formulae, (ii) multiple modal operators, permitting complex relational parameters, i.e. $\langle \_ \rangle$ and $[\_]$, as well as the domain and range operators, (iii) the relational operators, composition, relative sum, converse, identity, diversity, and (iv) the test operator of PDL, domain restriction and range restriction. MSPASS supports non-logical axioms which are true in every possible worlds (that is, global satisfiability, or generalised concept or role terminological axioms for description logics). In addition, it is possible to specify additional frame properties, or any other first-order restrictions on the translated formulae.

Among the mentioned theorem provers the following features make (M)SPASS the most flexible and well-suited theorem prover for reasoning within extended modal logics and description logics. (M)SPASS includes an advanced converter of first-order logic formulae into clausal form. Special features of the converter include optimised Skolemisation, strong Skolemisation, and an improved implementation of renaming [64]. (M)SPASS supports splitting and branch condensing (branch condensing resembles branch pruning or backjumping). Ordered inference, splitting, and condensing are of particular importance concerning the performance for satisfiable formulae, and for randomly generated formulae, unit propagation and branch condensing are important as well.

Mechanisation of the approaches described in this paper is not difficult. All that is needed is to select a correct set of flag settings to turn MSPASS into implementations of a particular combination of a translation, and R^{ord} or R^{hyp}.

Although MSPASS does not provide a decision procedure for all the modal logics one might be interested in, for example, PDL or graded modal logic are not supported, an attractive feature of MSPASS is the possibility to specify arbitrary first-order theories. Anything which can be encoded into first-order logic with equality can be expressed with MSPASS. This allows for its use as a flexible tool for the investigation of combinations of interacting non-classical logics or description logics, which have not been been studied in depth before, and which may not have been anticipated by the implementors. In this context it is useful that, on termination, MSPASS does not only produce a 'yes'/'no' answer, but it also outputs a proof or a saturated set of clauses (depending on whether input problem is unsatisfiable or satisfiable). A finite saturated set of clauses provides a characterisation of a class of models for the input problem. In the case R^{hyp} is used the generated ground clauses define a Herbrand model (whenever all clauses are range-restricted).

12   Topics Not Covered

The combination of translation and first-order inference methods provides a powerful and versatile approach for studying and mechanising reasoning, model generation and other aspects of modal logic. Due to space restrictions we had to be selective in our choice of topics covered in this overview. Some important topics omitted in this overview include the following.

Non-standard Translation Approaches. Non-standard translation methods include reductions derived from the functional semantics of normal modal logics with unparameterised modalities, namely the functional translation [3, 40, 65], the optimised functional translation [40, 69, 84] and the semi-functional translation [62]. In addition, a tree layered translation was introduced by Areces et al [2] for basic modal logic. It is not difficult to see that methods based on this translation can be simulated with the optimised functional translation method. Surveys of the different translation methods of modal logics and other non-classical logics are [66–68]. The above mentioned non-standard translation approaches are all implemented in MSPASS [51]. Experience shows that the performance of first-order theorem provers is best when a variation of the optimised functional translation can be used [47].

A recent development is the introduction in [79] of a translation principle, called the *axiomatic translation* principle, which promises to make it easier to develop inference calculi and automated decision procedures for extensions of the modal logic $K_{(m)}$ with modal axioms. The axiomatic translation reduces propositional modal logics with relational background theories, including triangular properties such as transitivity, Euclideanness and functionality, not covered in this paper, into the two-variable guarded fragment. In [79] it is shown that any resolution procedure based on R^{ord} decides the satisfiability problem of modal logics for which the axiomatic translation can be shown to be complete. These include the logics $K4$, KT, KD, KB, $K\mathrm{alt}_1$, $K5$, $K4B$, $KT4B$, and their fusions, as well as extensions of K with certain generalised axioms. Another reduction to first-order logic of interest is due to Demri and De Nivelle [18]. They show that a certain class of modal logics, the class of regular grammar logics with converse, are decidable by reduction to the two-variable guarded fragment.

Deciding Modal Logics with Transitive Modalities. To decide extensions of $K4$ another possibility besides using the axiomatic translation is to use the semi-functional translation method in combination with ordered resolution [45]. Rather than modifying the translation function, the approach of Ganzinger et al [26] modifies the calculus and use ordered chaining rules for transitive relations. Both these approaches provide decision procedures for the logics $K4$, $S4$ and $KD4$.

Mechanising Correspondence Theory. Computing the first-order correspondence property, if it exists, for a modal formula amounts to the elimination of the universal monadic second-order quantifiers expressing the validity in a frame of that formula or, equivalently, the elimination of the existential monadic second-order quantifiers expressing the satisfiability of the formula. One of the best known algorithms for elimination of existential second-order quantifiers is SCAN, developed by Ohlbach and Gabbay [24]. The SCAN algorithm is based on constraint resolution and was implemented by Engel [21] as an extension of the OTTER theorem prover. The SCAN algorithm is known to be sound, meaning that whenever the algorithm terminates successfully the resulting formula is equivalent to the original formula. Unfortunately, it is provably impossible for such a reduction (by SCAN or any other method) to be always successful, even if there is a simpler equivalent formula for a second-order logic formula. However, even though SCAN cannot be complete in this general sense, SCAN has been shown to be useful for computing first-order frame correspondents for modal axiom schemata. It is known that SCAN can compute the frame correspondence properties for very many well-known axioms such as T, 4, 5, etcetera.

SCAN is not the only quantifier elimination algorithm, though to date it is the only algorithm based on resolution. Another algorithm is the DLS algorithm, due to Doherty, Lukaszewicz and Szalas [19] which is also suitable for computing modal correspondence properties. The DLS algorithm was implemented by Gustafsson [38]. Both SCAN and the DLS algorithm can be used remotely. An overview of these and other quantifier elimination algorithms is [63].

Generating Minimal Herbrand Models. In general Herbrand models are not unique and can be large. Therefore there is a need for generating minimal models. There are various approaches to generating minimal Herbrand models with hyperresolution [8, 10, 39, 61]. With a moderate extension of R^{hyp} it is possible to guarantee the generation of all and only minimal Herbrand models for any modal and description logics reducible to a decidable class of range restricted clauses. This follows from [10] and recent investigations of $GF1^-$ and the class BU [31–33]. An alternative approach proposed in [31–33] uses a variant of a local minimality test developed for propositional logic.

Acknowledgements

We thank Hans de Nivelle, coauthor of the survey [16] on which a part of this text is based. We also thank the reviewer for helpful comments. The work is supported by EU COST Action 274, and research grants GR/M36700 and GR/M88761 from the UK Engineering and Physical Sciences Research Council.

References

1. H. Andréka, I. Németi, and J. van Benthem. Modal languages and bounded fragments of predicate logic. *Journal of Philosophical Logic*, 27(3):217–274, 1998.
2. C. Areces, R. Gennari, J. Heguiabehere, and M. de Rijke. Tree-based heuristics in modal theorem proving. In W. Horn, editor, *Proceedings of the Fourteenth European Conference on Artificial Intelligence (ECAI 2000)*, pages 199–203. IOS Press, 2000.
3. Y. Auffray and P. Enjalbert. Modal theorem proving: An equational viewpoint. *Journal of Logic and Computation*, 2(3):247–297, 1992.
4. L. Bachmair and H. Ganzinger. Rewrite-based equational theorem proving with selection and simplification. *Journal of Logic and Computation*, 4(3):217–247, 1994.
5. L. Bachmair and H. Ganzinger. Resolution theorem proving. In A. Robinson and A. Voronkov, editors, *Handbook of Automated Reasoning*, volume I, chapter 2, pages 19–99. Elsevier, 2001.
6. L. Bachmair, H. Ganzinger, and U. Waldmann. Superposition with simplification as a decision procedure for the monadic class with equality. In G. Gottlob, A. Leitsch, and D. Mundici, editors, *Proceedings of the Third Kurt Gödel Colloquium (KGC'93)*, volume 713 of *Lecture Notes in Computer Science*, pages 83–96. Springer, 1993.
7. M. Baldoni. Normal multimodal logics with interaction axioms. In D. Basin, M. D'Agostino, D. M. Gabbay, and L. Vigano, editors, *Labelled Deduction*, pages 33–57. Kluwer, 2000.
8. P. Baumgartner, J. D. Horton, and B. Spencer. Merge path improvements for minimal model hyper tableaux. In N. V. Murray, editor, *Proceedings of the Eighth International Conference on Automated Reasoning with Analytic Tableaux and Related Methods (TABLEAUX'99)*, volume 1617 of *Lecture Notes in Artificial Intelligence*, pages 51–65. Springer, 1999.
9. T. Boy de la Tour. An optimality result for clause form translation. *Journal of Symbolic Computation*, 14:283–301, 1992.
10. F. Bry and A. Yahya. Positive unit hyperresolution tableaux for minimal model generation. *Journal of Automated Reasoning*, 25(1):35–82, 2000.

11. R. Caferra and S. Demri. Cooperation between direct method and translation method in non classical logics: Some results in propositional S5. In R. Bajcsy, editor, *Proceedings of the Thirteenth International Joint Conference on Artificial Intelligence (IJCAI'93)*, pages 74–79. Morgan Kaufmann, 1993.

12. L. Catach. Normal multimodal logics. In *Proceedings of the Seventh National Conference on Artificial Intelligence (AAAI'88)*, pages 491–495. AAAI Press/MIT Press, 1988.

13. S. A. Cook and R. A. Reckhow. The relative efficiency of propositional proof systems. *Journal of Symbolic Logic*, 44(1):36–50, 1979.

14. H. de Nivelle. Splitting through new proposition symbols. In R. Nieuwenhuis and A. Voronkov, editors, *Proceedings of the Eighth International Conference on Logic for Programming, Artificial Intelligence, and Reasoning (LPAR 2001)*, volume 2250 of *Lecture Notes in Artificial Intelligence*, pages 172–185. Springer, 2001.

15. H. de Nivelle and M. de Rijke. Deciding the guarded fragment by resolution. *Journal of Symbolic Computation*, 35(1):21–58, 2003.

16. H. de Nivelle, R. A. Schmidt, and U. Hustadt. Resolution-based methods for modal logics. *Logic Journal of the IGPL*, 8(3):265–292, 2000.

17. S. Demri. A hierarchy of backward translations: Applications to modal logics. In M. de Glas and Z. Pawlak, editors, *Proceedings of the Second World Conference on the Fundamentals of Artificial Intelligence (WOCFAI'95)*, pages 121–132. Angkor, Paris, 1995.

18. S. Demri and H. de Nivelle. Deciding regular grammar logics with converse through first-order logic. Research Report LSV-03-4, Spécification et Vérification, CNRS & ENS de Cachan, France, 2003.

19. P. Doherty, W. Lukaszewicz, and A. Szalas. Computing circumscription revisited: A reduction algorithm. *Journal of Automated Reasoning*, 18(3):297–336, 1997.

20. E. Eder. *Relative Complexities of First Order Calculi*. Artificial Intelligence. Vieweg, Wiesbaden, 1992.

21. T. Engel. Quantifier elimination in second-order predicate logic. Diplomarbeit, Fachbereich Informatik, Univ. des Saarlandes, Saarbrücken, 1996.

22. C. Fermüller, A. Leitsch, U. Hustadt, and T. Tammet. Resolution decision procedures. In A. Robinson and A. Voronkov, editors, *Handbook of Automated Reasoning*, volume II, chapter 25, pages 1791–1849. Elsevier, 2001.

23. C. Fermüller, A. Leitsch, T. Tammet, and N. Zamov. *Resolution Method for the Decision Problem*, volume 679 of *Lecture Notes in Computer Science*. Springer, 1993.

24. D. M. Gabbay and H. J. Ohlbach. Quantifier elimination in second-order predicate logic. *South African Computer Journal*, 7:35–43, 1992. Also published in B. Nebel, C. Rich, W. R. Swartout, editors, *Proceedings of the Third International Conference on Principles of Knowledge Representation and Reasoning (KR'92)*, pages 425–436. Morgan Kaufmann, 1992.

25. H. Ganzinger and H. de Nivelle. A superposition decision procedure for the guarded fragment with equality. In *Proceedings of the Fourteenth Annual IEEE Symposium on Logic in Computer Science (LICS'99)*, pages 295–303. IEEE Computer Society Press, 1999.

26. H. Ganzinger, U. Hustadt, C. Meyer, and R. A. Schmidt. A resolution-based decision procedure for extensions of K4. In M. Zakharyaschev, K. Segerberg, M. de Rijke, and H. Wansing, editors, *Advances in Modal Logic, Volume 2*, volume 119 of *Lecture Notes*, chapter 9, pages 225–246. CSLI Publications, Stanford, 2001.

27. H. Ganzinger, C. Meyer, and C. Weidenbach. Soft typing for ordered resolution. In W. McCune, editor, *Proceedings of the Fourteenth International Conference on Automated Deduction (CADE-14)*, volume 1249 of *Lecture Notes in Artificial Intelligence*, pages 321–335. Springer, 1997.

28. H. Ganzinger and R. Nieuwenhuis. The Saturate system. `http://www.mpi-sb.mpg.de/SATURATE/Saturate.html`, 1994.

29. G. Gargov and S. Passy. A note on Boolean modal logic. In P. P. Petkov, editor, *Mathematical Logic: Proceedings of the 1988 Heyting Summerschool*, pages 299–309. Plenum Press, New York, 1990.

30. G. Gargov, S. Passy, and T. Tinchev. Modal environment for Boolean speculations. In D. Skordev, editor, *Mathematical Logic and its Applications: Proceedings of the 1986 Gödel Conference*, pages 253–263. Plenum Press, New York, 1987.

31. L. Georgieva, U. Hustadt, and R. A. Schmidt. Computational space efficiency and minimal model generation for guarded formulae. In R. Nieuwenhuis and A. Voronkov, editors, *Proceedings of the Eighth International Conference on Logic for Programming, Artificial Intelligence, and Reasoning (LPAR 2001)*, volume 2250 of *Lecture Notes in Artificial Intelligence*, pages 85–99. Springer, 2001.

32. L. Georgieva, U. Hustadt, and R. A. Schmidt. A new clausal class decidable by hyperresolution. Preprint series, University of Manchester, UK, 2002. Long version of [33].

33. L. Georgieva, U. Hustadt, and R. A. Schmidt. A new clausal class decidable by hyperresolution. In A. Voronkov, editor, *Proceedings of the Eighteenth International Conference on Automated Deduction (CADE-18)*, volume 2392 of *Lecture Notes in Artificial Intelligence*, pages 260–274. Springer, 2002.

34. L. Georgieva, U. Hustadt, and R. A. Schmidt. On the relationship between decidable fragments, non-classical logics, and description logics. In I. Horrocks and S. Tessaris, editors, *Proceedings of the 2002 International Workshop on Description Logics (DL'2002)*, pages 25–36. CEUR Workshop Proceedings, Vol. CEUR-WS/Vol-53, 2002.

35. L. Georgieva, U. Hustadt, and R. A. Schmidt. Hyperresolution for guarded formulae. *Journal of Symbolic Computation*, 36(1–2):163–192, 2003.

36. E. Giunchiglia, F. Giunchiglia, R. Sebastiani, and A. Tacchella. Sat vs. translation based decision procedures for modal logics: A comparative evaluation. *Journal of Applied Non-Classical Logics*, 10(2):145–172, 2000.

37. E. Grädel, P. Kolaitis, and M. Vardi. On the decision problem for two-variable first-order logic. *Bulletin of Symbolic Logic*, 3:53–69, 1997.

38. J. Gustafsson. An implementation and optimization of an algorithm for reducing formulas in second-order logic. Technical Report LiTH-MAT-R-96-04, Department of Mathematics, Linköping University, Sweden, 1996.

39. Hasegawa, H. R., Fujita, and M. Koshimura. Efficient minimal model generation using branching lemmas. In D. McAllester, editor, *Proceedings of the Seventeenth International Conference on Automated Deduction (CADE-17)*, volume 1831 of *Lecture Notes in Artificial Intelligence*, pages 184–199. Springer, 2000.

40. A. Herzig. *Raisonnement automatique en logique modale et algorithmes d'unification*. PhD thesis, Univ. Paul-Sabatier, Toulouse, 1989.

41. A. Heuerding, G. Jäger, S. Schwendimann, and M. Seyfried. The Logics Workbench LWB: A snapshot. *Euromath Bulletin*, 2(1):177–186, 1996.

42. I. Horrocks and P. F. Patel-Schneider. FaCT and DLP. In H. de Swart, editor, *Proceedings of the Seventh International Conference on Automated Reasoning with Analytic Tableaux and Related Methods (TABLEAUX'98)*, volume 1397 of *Lecture Notes in Computer Science*, pages 27–30. Springer, 1998.

43. I. L. Humberstone. Inaccessible worlds. *Notre Dame Journal of Formal Logic*, 24(3):346–352, 1983.

44. I. L. Humberstone. The modal logic of 'all and only'. *Notre Dame Journal of Formal Logic*, 28(2):177–188, 1987.

45. U. Hustadt. *Resolution-Based Decision Procedures for Subclasses of First-Order Logic*. PhD thesis, Univ. d. Saarlandes, Saarbrücken, Germany, 1999.

46. U. Hustadt and R. A. Schmidt. Simplification and backjumping in modal tableau. In H. de Swart, editor, *Proceedings of the 7h International Conference on Automated Reasoning with Analytic Tableaux and Related Methods (TABLEAUX'98)*, volume 1397 of *Lecture Notes in Artificial Intelligence*, pages 187–201. Springer, 1998.

47. U. Hustadt and R. A. Schmidt. An empirical analysis of modal theorem provers. *Journal of Applied Non-Classical Logics*, 9(4):479–522, 1999.

48. U. Hustadt and R. A. Schmidt. Maslov's class K revisited. In H. Ganzinger, editor, *Proceedings of the Sixteenth International Conference on Automated Deduction (CADE-16)*, volume 1632 of *Lecture Notes in Artificial Intelligence*, pages 172–186. Springer, 1999.

49. U. Hustadt and R. A. Schmidt. On the relation of resolution and tableaux proof systems for description logics. In T. Dean, editor, *Proceedings of the Sixteenth International Joint Conference on Artificial Intelligence (IJCAI'99)*, volume 1, pages 110–115. Morgan Kaufmann, 1999.

50. U. Hustadt and R. A. Schmidt. Issues of decidability for description logics in the framework of resolution. In R. Caferra and G. Salzer, editors, *Automated Deduction in Classical and Non-Classical Logics*, volume 1761 of *Lecture Notes in Artificial Intelligence*, pages 191–205. Springer, 2000.

51. U. Hustadt and R. A. Schmidt. MSPASS: Modal reasoning by translation and first-order resolution. In R. Dyckhoff, editor, *Proceedings of the Nineth International Conference on Automated Reasoning with Analytic Tableaux and Related Methods (TABLEAUX 2000)*, volume 1847 of *Lecture Notes in Artificial Intelligence*, pages 67–71. Springer, 2000.

52. U. Hustadt and R. A. Schmidt. Using resolution for testing modal satisfiability and building models. *Journal of Automated Reasoning*, 28(2):205–232, 2002.

53. U. Hustadt, R. A. Schmidt, and C. Weidenbach. Optimised functional translation and resolution. In H. de Swart, editor, *Proceedings of the Seventh International Conference on Automated Reasoning with Analytic Tableaux and Related Methods (TABLEAUX'98)*, volume 1397 of *Lecture Notes in Artificial Intelligence*, pages 36–37. Springer, 1998.

54. Á. Kurucz, I. Németi, I. Sain, and A. Simon. Undecidable varieties of semilattice-ordered semigroups, of Boolean algebras with operators and logics extending Lambek calculus. *Bulletin of the IGPL*, 1(1):91–98, 1993.

55. A. Leitsch. *The Resolution Calculus*. EATCS Texts in Theoretical Computer Science. Springer, 1997.

56. C. Lutz and U. Sattler. The complexity of reasoning with boolean modal logics. In F. Wolter, H. Wansing, M. de Rijke, and M. Zakharyaschev, editors, *Advances in Modal Logics Volume 3*. CSLI Publications, Stanford, 2002.

57. C. Lutz, U. Sattler, and S. Tobies. A suggestion of an *n*-ary description logic. In P. Lambrix, A. Borgida, M. Lenzerini, R. Möller, and P. Patel-Schneider, editors, *Proceedings of the 1999 International Workshop on Description Logics (DL'99)*, pages 81–85. Linköping University, 1999.

58. S. Ju. Maslov. The inverse method for establishing deducibility for logical calculi. In V. P. Orevkov, editor, *The Calculi of Symbolic Logic I: Proc. of the Steklov Institute of Mathematics edited by I. G. Petrovskiĭ and S. M. Nikol'skiĭ, Nr. 98 (1968)*, pages 25–96. Amer. Math. Soc., Providence, Rhode Island, 1971.

59. F. Massacci. Single step tableaux for modal logics: Computational properties, complexity and methodology. *Journal of Automated Reasoning*, 24(3):319–364, 2000.

60. W. McCune. The Otter theorem prover, 1995. `http://www.mcs.anl.gov/AR/otter/`.

61. I. Niemelä. A tableau calculus for minimal model reasoning. In *Proceedings of the Fifth International Conference on Automated Reasoning with Analytic Tableaux and Related Methods (TABLEAUX'96)*, volume 1071 of *Lecture Notes in Artificial Intelligence*, pages 278–294. Springer, 1996.

62. A. Nonnengart. First-order modal logic theorem proving and functional simulation. In *Proceedings of the Thirteenth International Joint Conference on Artificial Intelligence (IJCAI'93)*, volume 1, pages 80–85. Morgan Kaufmann, 1993.

63. A. Nonnengart, H. J. Ohlbach, and A. Szalas. Quantifier elimination for second-order predicate logic. To appear in *Logic, Language and Reasoning: Essays in honour of Dov Gabbay*, Part I, Kluwer.

64. A. Nonnengart, G. Rock, and C. Weidenbach. On generating small clause normal forms. In C. Kirchner and H. Kirchner, editors, *Proceedings of the Fifteenth International Conference on Automated Deduction (CADE-15)*, volume 1421 of *Lecture Notes in Artificial Intelligence*, pages 397–411. Springer, 1998.

65. H. J. Ohlbach. Semantics based translation methods for modal logics. *Journal of Logic and Computation*, 1(5):691–746, 1991.

66. H. J. Ohlbach. Translation methods for non-classical logics: An overview. *Bulletin of the IGPL*, 1(1):69–89, 1993.

67. H. J. Ohlbach. Combining Hilbert style and semantic reasoning in a resolution framework. In C. Kirchner and H. Kirchner, editors, *Proceedings of the Fifteenth International Conference on Automated Deduction (CADE-15)*, volume 1421 of *Lecture Notes in Artificial Intelligence*, pages 205–219. Springer, 1998.

68. H. J. Ohlbach, A. Nonnengart, M. de Rijke, and D. Gabbay. Encoding two-valued nonclassical logics in classical logic. In A. Robinson and A. Voronkov, editors, *Handbook of Automated Reasoning*, volume II, chapter 21, pages 1403–1486. Elsevier Science, 2001.

69. H. J. Ohlbach and R. A. Schmidt. Functional translation and second-order frame properties of modal logics. *Journal of Logic and Computation*, 7(5):581–603, 1997.

70. S. Passy and T. Tinchev. PDL with data constants. *Information Processing Letters*, 20:35–41, 1985.

71. D. A. Plaisted and S. Greenbaum. A structure-preserving clause form translation. *Journal of Symbolic Computation*, 2:293–304, 1986.

72. W. V. Quine. Variables explained away. In *Proceedings of the American Philosophy Society*, volume 104, pages 343–347, 1960.

73. W. V. Quine. Algebraic logic and predicate functors. In R. Rudner and I. Scheffler, editors, *Logic and Art: Esssays in Honor of Nelson Goodman*. Bobbs-Merrill, Indianapolis, 1971.

74. A. Riazanov and A. Voronkov. Vampire. In H. Ganzinger, editor, *Proceedings of the Sixteenth International Conference on Automated Deduction (CADE-16)*, volume 1632 of *Lecture Notes in Artificial Intelligence*, pages 292–296. Springer, 1999.

75. A. Riazanov and A. Voronkov. Splitting without backtracking. In B. Nebel, editor, *Proceedings of the Seventeenth International Joint Conference on Artificial Intelligence (IJCAI 2001)*, pages 611–617. Morgan Kaufmann, 2001.

76. H. Sahlqvist. Completeness and correspondence in the first and second order semantics for modal logics. In S. Kanger, editor, *Proceedings of the Third Scandinavian Logic Symposium, 1973*, pages 110–143. North-Holland, 1975.

77. R. A. Schmidt. MSPASS. http://www.cs.man.ac.uk/~schmidt/mspass/.

78. R. A. Schmidt and U. Hustadt. A resolution decision procedure for fluted logic. In D. McAllester, editor, *Proceedings of the Seventeenth International Conference on Automated Deduction (CADE-17)*, volume 1831 of *Lecture Notes in Artificial Intelligence*, pages 433–448. Springer, 2000.

79. R. A. Schmidt and U. Hustadt. A principle for incorporating axioms into the first-order translation of modal formulae. In F. Baader, editor, *Automated Deduction—CADE-19*, volume 2741 of *Lecture Notes in Artificial Intelligence*, pages 412–426. Springer, 2003.

80. S. Schulz. System Abstract: E 0.3. In H. Ganzinger, editor, *Proceedings of the Sixteenth International Conference on Automated Deduction (CADE-16)*, volume 1632 of *Lecture Notes in Artificial Intelligence*, pages 297–301. Springer, 1999.

81. S. Schulz. E: A Brainiac theorem prover. *Journal of AI Communications*, 2002. To appear.

82. C. Weidenbach. SPASS, 1999. http://spass.mpi-sb.mpg.de.

83. C. Weidenbach. Combining superposition, sorts and splitting. In A. Robinson and A. Voronkov, editors, *Handbook of Automated Reasoning*, volume II, chapter 27, pages 1965–2013. Elsevier, 2001.

84. N. K. Zamov. Modal resolutions. *Soviet Mathematics*, 33(9):22–29, 1989. Translated from *Izv. Vyssh. Uchebn. Zaved. Mat.* **9** (328) (1989) 22–29.

Theory Extraction in Relational Data Analysis

Gunther Schmidt*

Institute for Software Technology, Department of Computing Science
Federal Armed Forces University Munich
Schmidt@Informatik.UniBw-Muenchen.DE

1 Introduction

From numerical mathematics we know that a linear equation $Ax = b$ may be solved more efficiently if a reduction of A as $A = \begin{pmatrix} B & O \\ C & D \end{pmatrix}$ is known beforehand. For the task $\begin{pmatrix} B & O \\ C & D \end{pmatrix} \cdot \begin{pmatrix} y \\ z \end{pmatrix} = \begin{pmatrix} c \\ d \end{pmatrix}$, one will solve $By = c$ first and then $Dz = d - Cy$. Having an a priori knowledge of this kind is also an advantage in many other application fields. We here deal with a diversity of techniques to decompose relations according to some criteria and embed these techniques in a common framework. The results of decompositions obtained may be used in decision making, but also as a support for teaching, as they often give visual help.

Our starting point will always be a concretely given relation, i.e., a Boolean matrix. In most cases, we will look for a partition of the set of rows and the set of columns, respectively, that arises from some algebraic condition. From these partitions, a rearranged matrix making these partitions easily visible shall be computed as well as the permutation matrix necessary to achieve this.

The current article presents results of the report [Sch02] obtainable via
    http://ist.unibw-muenchen.de/People/schmidt/DecompoHomePage.html
which gives a detailed account of the topic. The report is not just a research report but also a Haskell program in literate style. In contrast, the present article only gives hints as to these programs. Therefore, some details are omitted.

This article is organized as follows. Chapter 2 presents the idea of extracting theories as proposed in this paper. Then Ch. 3 will mention some prerequisites. The hints concerning the relational language used are given in Ch. 4, followed by Ch. 5 with models and interpretations in Haskell. With Ch. 6 the first decomposition based on the strongly connected component ontology is elaborated in some detail to further clarify the idea. Theoretical basics of the more sophisticated Galois decompositions are explained in Ch. 7 before these are made ready for programming in Ch. 8.

---

* Cooperation and communication around this research was partly sponsored by the European COST Action 274: TARSKI (Theory and Application of Relational Structures as Knowledge Instruments), which is gratefully acknowledged.

H. de Swart et al. (Eds.): TARSKI, LNCS 2929, pp. 68–86, 2003.

2   The Idea of Theory Extraction

If some concrete relations are given as boolean matrices, one may talk about these in terms of a logical language and theory. We provide names for the relations and names for row and column entries which will be interpreted so that the name of the relation gets assigned the matrix as its interpretation, etc. This already gives us a sparse language and a sparse theory without any specific theorems to hold.

When a relational decomposition is reached, the language will have to contain also the necessary predicates and theorems expressing the algebraic idea behind the decomposition. Which predicates and theorems we need will depend on the decomposition we are aiming at and which we will call the given ontology. One ontology we have in mind is the game ontology. It considers the given homogeneous relation as the graph of a two-player game as described in [SS89,SS93], e.g. One wishes to solve the game, i.e., to qualify the positions as to being a position of win, of draw, or of loss for the player about to move. Other ontologies that have been handled include irreducibility, difunctionality, matching, etc.

The following diagram shows the idea. We start with the given model of the sparse theory (which is the relation originally given) and some ontology-enhanced theory. What we are constructing is in a sense the pushout.

$$\text{Sparse theory} \qquad \models \qquad \text{Given model}$$
$$\downarrow \qquad\qquad\qquad\qquad\qquad\qquad \downarrow$$
$$\text{Ontology-enhanced theory} \qquad \models \qquad \text{Result model}$$

After decomposition, the given model may be viewed with the mechanism our ontology has provided. Then the theory will also contain certain formulae describing what holds between the new ontology-dependent items.

Consider the following trivial example. We start with the left relation B and aim at the game ontology decomposition. Then we will obtain two uniquely determined sets a, b satisfying the relational theorems $a = \overline{B;b}$ and $\overline{B;a} = b$. In the game ontology they may be interpreted as $loss := a$, $win := \overline{b}$, and $draw := b \cap \overline{a}$, below arranged as partition into win, $draw$, $loss$. They will furthermore allow us to obtain a permutation so as to rearrange the matrix B to the right-hand side form, which makes the algebraic laws easily visible.

$$
\begin{array}{c}
\begin{array}{ccccccc}
1 & 2 & 3 & 4 & 5 & 6 & 7
\end{array}\\
\begin{array}{c}1\\2\\3\\4\\5\\6\\7\end{array}
\left(\begin{array}{ccccccc}
0 & 1 & 0 & 0 & 0 & 1 & 0\\
0 & 0 & 0 & 0 & 0 & 0 & 0\\
0 & 0 & 0 & 0 & 0 & 1 & 0\\
1 & 0 & 0 & 0 & 0 & 0 & 0\\
1 & 0 & 0 & 1 & 0 & 0 & 0\\
0 & 0 & 0 & 0 & 1 & 0 & 1\\
0 & 0 & 1 & 0 & 1 & 0 & 0
\end{array}\right)
\end{array}
\quad
\begin{array}{c}
\begin{array}{c}1\\2\\3\\4\\5\\6\\7\end{array}
\left(\begin{array}{c}1\\0\\0\\0\\1\\0\\0\end{array}\right)
\left(\begin{array}{c}0\\0\\1\\0\\0\\1\\1\end{array}\right)
\left(\begin{array}{c}0\\1\\0\\1\\0\\0\\0\end{array}\right)
\end{array}
\quad
\begin{array}{c}
\begin{array}{ccccccc}
2 & 4 & 3 & 6 & 7 & 1 & 5
\end{array}\\
\begin{array}{c}2\\4\\3\\6\\7\\1\\5\end{array}
\left(\begin{array}{cc|ccc|cc}
0 & 0 & 0 & 0 & 0 & 0 & 0\\
0 & 0 & 0 & 0 & 0 & 1 & 0\\
0 & 0 & 0 & 1 & 0 & 0 & 0\\
0 & 0 & 0 & 0 & 1 & 0 & 1\\
0 & 0 & 1 & 0 & 0 & 0 & 1\\
1 & 0 & 0 & 1 & 0 & 0 & 0\\
0 & 1 & 0 & 0 & 0 & 1 & 0
\end{array}\right)
\end{array}
$$

Original game matrix, partition and rearranged matrix

3   Preliminaries

A major aspect in this work is, thus, to be able to handle permutations, to derive them for special purposes, to treat them as relations when appropriate, and to make them fully available in the Haskell program. Permutations may be considered as a function, decomposed into cycles, or as a permutation matrix. Everybody who has worked in functional programming will admit that transition from one form to the other is a simple programming task. As everything is worked out in the report [Sch02], we need not explain it here in detail. The additional general reference is [SS89,SS93], which will not be mentioned every time.

The permutations shall mostly be determined from partitions of a set so as to convert it to a list of elements with elements of an equivalence class side aside. Again, one will concede that this is a solvable task for a functional programmer; so we avoid mentioning this in detail here. One should, however, observe that sometimes rows and columns are permuted simultaneously and sometimes independently. These two situations need rather different treatment.

As space is limited, we cannot collect all the well-known and often recalled material in order to make this article self-contained. We restrict to mentioning that relations here are conceived as subsets $R \subseteq X \times Y$ of the Cartesian product of some sets X, Y between which they are defined to hold. Operations are composition $;$, transposition $^\mathsf{T}$, identities $\mathbb{I}$, union $\cup$, intersection $\cap$, complementation $^-$, null relations $\mathbb{L}$, universal relations $\mathbb{T}$, and containment $\subseteq$ of relations.

When expressing the basic relational operators in Haskell, the programming language chosen, we will handle relations as rectangular boolean matrices. Often we represent their entries `True` by `1` and `False` by `0` when showing matrices in the text. The following basic relational operators `|||`, `&&&`, `***`, `<==` for union, intersection, composition and containment of relations are all formulated in Haskell.

4   Relational Language

To formulate the theorems of an ontology-enhanced theory, we will need the language generated by the denotations of these operations, and we will not always immediately execute them. So we distinguish, e.g., between the denotation `:***:` as a Haskell infix constructor, and its interpretation, the operation `***`. Therefore, a relational language is presented allowing us to talk about elements, vectors (or sets), and (binary) relations. We are going to work in a typed or heterogeneous setting, which means that we start from a category.

```
data CatObject = Obj String | ...
```

So we will be able to give names to the category objects using the constructor `Obj` and a chosen string. The category will later stay the same when an extended theory is extracted guided by some ontology.

Then we need denotations for individual variables, constants, functions, and predicates. In our setting, we always bind these together with their typing, and we restrict to unary predicate constants which we call vectors and binary predicate constants, which we call relations. A relational constant is nothing more than a name, the string. Its type are the `CatObjects` between which the relation is supposed to hold. They are, however, not concretely given as we stay – so far – on the syntactical side.

```
data ElemConst = Elem String CatObject
data VectConst = Vect String CatObject
data RelaConst = Rela String CatObject CatObject
```

On all this, we now build first-order predicate logic, introducing individual variables, terms, and formulae.

```
data ElemVari = VarE String CatObject
data VectVari = VarV String CatObject
data RelaVari = VarR String CatObject CatObject
```

Vectors are here supposed to be column vectors. From the beginning, we distinguish element terms, vector terms, and relation terms. Null, universal, and identity relation constants may uniformly be denoted throughout as indicated.

```
data ElemTerm = EC ElemConst | EV ElemVari | ...
data VectTerm = VC VectConst | VV VectVari | RelaTerm :****: VectTerm |
                VectTerm :||||: VectTerm | VectTerm :&&&&: VectTerm |
                NegVect VectTerm | NullV CatObject | UnivV CatObject |
                VFctAppl RelaFct RelaTerm | ...
data RelaTerm = RC RelaConst | RV RelaVari | RelaTerm :***: RelaTerm |
                NegaRela RelaTerm | Transp RelaTerm | Ident CatObject |
                NullR CatObject CatObject | UnivR CatObject CatObject |
                RelaTerm :|||: RelaTerm | RelaTerm :&&&: RelaTerm | ...
data ElemFct  = EFCT ElemVari ElemTerm
data VectFct  = VFCT VectVari VectTerm
data RelaFct  = RFCT RelaVari RelaTerm
```

The 0-ary operations for $\bot, \mathbb{I}, \top$ require giving domains or codomains, respectively. Typical checks such as `relaTermIsWellFormed`, `typeOfVectTerm` are provided for well-formedness, type control, etc.

In order to further facilitate maintenance of ubiquitous typing three forms of formulae are distinguished. In case $x \in v$, this is represented by the vector formula `VE v x`. In a similar way, `REE r x y` means $(x, y) \in r$.

```
data ElemForm = Equation ElemTerm  ElemTerm | ...
data VectForm = VectTerm :<===: VectTerm | VE VectTerm ElemTerm | ...
data RelaForm = RelaTerm :<==: RelaTerm |
                REE RelaTerm ElemTerm ElemTerm | ...
data Formula  = EF ElemForm | VF VectForm | RF RelaForm |
                Negated  Formula | Implies  Formula Formula |
                Disjunct Formula Formula | Conjunct Formula Formula | ...
```

Again, typing and well-formedness are defined as usual. The types of all kinds of formulae are intended to be `Bool`. Free variables are defined as usual, and then a theory (fragment) may be formulated as a data structure in Haskell.

```
data Theory = TH String        -- name of the theory
               [CatObject]      -- carrier set denotations
               [ElemConst]      -- element denotations
               [VectConst]      -- subset denotations
               [RelaConst]      -- relation denotations
               [VectFct]        -- vector functions
               [RelaFct]        -- relation functions
               [Formula]        -- formulae demanded to hold
```

Of course, some testing is provided with `checkTheoryWellDefined`. The following is an example with just a denotation for a base set and a denotation for a relation intended to be defined on it, but only empty denotation lists [] provided for elements, subsets, functions, and formulae.

```
verySparseTheory = TH "Example" [o] [] [] [Rela "B" o o] [] [] []
                                  where o = Obj "BaseSet"
```

5 Models and Interpretations

While we have so far only been concerned with syntax, we will now offer the opportunity to interpret the language, and the theories we have defined, in a model. Therefore, we show how an interpretation may be given. In our approach, theory and model are both represented in Haskell, so the distinction between the two will sometimes be difficult.

Via an interpretation, the objects get assigned sets in this model, however, we just mention the cardinalities of the sets. So – in a rather trivial sense – numbers can be viewed as names of the rows and columns. Also, vector and relation denotations are assigned concrete versions by the model.

```
data InterpreteObjs = Carrier  CatObject  Int
data InterpreteCons = InterCon ElemConst  Int
data InterpreteVect = InterVec VectConst  [Bool]
data InterpreteRela = InterRel RelaConst  [[Bool]]
data InterpreteVFct = InterVFc VectFct    ([Bool]   ->  [Bool])
data InterpreteRFct = InterRFc RelaFct    ([[Bool]] -> [[Bool]])
```

Only in rare cases as, e.g., when studying rooted graphs with the root distinguished, will we have element constants. We provide an automatic interpretation for null relations, universal relations, and identity relations. Putting this together, a model is defined as follows:

```
data Model = MO String              -- name of the model
              [InterpreteObjs]      -- cardinalities of carrier sets
              [InterpreteCons]      -- numbers of corresponding elements
              [InterpreteVect]      -- subset-interpreting boolean vectors
              [InterpreteRela]      -- relation-interpreting matrices
              [InterpreteVFct]      -- interpreted vector functions
              [InterpreteRFct]      -- interpreted relation functions
```

We provide some generic mechanisms on the model side in order to check whether the sets in question are assigned to objects consistently by the interpretations. Lots of technicalities are necessary to ensure that this works as it is supposed to.

Before an interpretation is possible, we need valuations of the individual variables, i.e., an environment as a list of variable/value pairs.

```
type InterpreteElemVari = (ElemVari,   Int)
type InterpreteVectVari = (VectVari,  [Bool])
type InterpreteRelaVari = (RelaVari, [[Bool]])
type ElemValuations     = [InterpreteElemVari]
type VectValuations     = [InterpreteVectVari]
type RelaValuations     = [InterpreteRelaVari]
type Environment        = (ElemValuations,VectValuations,RelaValuations)
```

Using a rather primitive lookup function, we may then write

```
valuation env v = ...
```

to get the value of v in the environment env. Now terms and formulae may be interpreted according to the following examples.

```
interpreteVectTerm :: Model -> Environment -> VectTerm -> [Bool]
interpreteVectTerm m env vt =
   let MO _ os _ vs _ _ _ = m
       (evs,vvs,rvs) = env
   in  case vt of
           VFctAppl vf vt2 -> let ivf = interpreteVectFct  m env vf
                                  ivt = interpreteVectTerm m env vt2
                              in  ivf ivt
       ...


interpreteVectFct ::  Model -> Environment -> VectFct -> [Bool] -> [Bool]
interpreteVectFct m env vf bv =
   let VFCT vv vt = vf
       (evs,vvs,rvs) = env
       viWITHbm = (evs,(vv,bv) : vvs,rvs)
   in  interpreteVectTerm m viWITHbm vt
```

```
interpreteRelaTerm :: Model -> Environment -> RelaTerm -> [[Bool]]
interpreteRelaTerm m env rt =
   let MO _ _ _ _ rs = m
   in  case rt of
          RC rc              -> (\(InterRel _ b) -> b) $ head $
                                dropWhile (\(InterRel e _) -> rc /= e) rs
          rt1 :***: rt2    -> let int1 = interpreteRelaTerm m env rt1
                                    int2 = interpreteRelaTerm m env rt2
                                in  int1 *** int2
          . . .


interpreteRelaForm :: Model -> Environment -> RelaForm -> Bool
interpreteRelaForm m env rf =
   case rf of
       rt1 :<==: rt2 -> let int1 = interpreteRelaTerm m env rt1
                              int2 = interpreteRelaTerm m env rt2
                          in  int1 <== int2
       . . .
```

Once a model for a theory is given – which is at the same time finite as well as sufficiently small –, it will be possible to check the model property against the theory with checkIsModelForTheory mo th.

6   Strongly Connected Component Ontology

These concepts of language, theory, model, and theory extraction shall now be exemplified in a field for which the theoretical background is well-known. Starting from a homogeneous relation, we plan to permute rows and columns simultaneously.

Let some relation R be given and look for its reflexive transitive closure R^* and for the equivalence $R^* \cap R^{*\top}$ generated by this closure, the equivalence classes of which give the strongly connected components. Permuting rows and columns of R simultaneously so as to have them grouped according to these equivalence classes gives much insight into the structure of R.

First, a sparse theory is formulated, and the schema for an ontology-enhanced theory. In the sparse theory upon start, we simply know that a node set is given together with a relation R on it and no formulae are supposed to hold. So we provide denotations for a single category object s0 and a relation constant r on it.

```
strongConnCompSparseTheory :: CatObject -> RelaConst -> Theory
strongConnCompSparseTheory s0 r =
     TH "StrongConnCompSparseTheory" [s0] [] [] [r] [] [] []
```

Later, we will start with a model depending on a boolean matrix gR, the given relation, as a parameter.

```
strongConnCompGivenModel :: CatObject -> RelaConst -> [[Bool]] -> Model
strongConnCompGivenModel s0 r gR =
    MO "StrongConnCompGivenModel" [Carrier s0 (rows gR)]
                              [] [] [InterRel r gR] [] []
```

No provisions have been made to denote single elements of the node set. As the number of connected components is not known beforehand, we provide the name for the list of partitioning connected components as non-empty subsets of entries as pl. The list of vector denotations of the ontology is then used in the formulae that describe what has been achieved when decomposition is executed once the model is known.

```
strongConnCompOntolEnhancedTheorySchema ::
                    CatObject -> RelaConst -> [VectConst] -> Theory
strongConnCompOntolEnhancedTheorySchema s0 r pl =
    let rt  = RC r
        d   = domRC r
        vts = map VC pl
        nullVect = NullV d
        subsetNonEmpty s = Negated (VF (s :<===: nullVect))
        partitionSetsNonEmpty = map subsetNonEmpty vts
        subsDisjoint (s1,s2) = VF (s1 :&&&&: s2 :<===: nullVect)
        allUnordPairsOfPartSets s =
            let fff res [] = res
                fff res (h:t) = fff (res ++ (map (\x -> (h,x)) t)) t
            in  fff [] s
        partSetsDisjoint = map subsDisjoint (allUnordPairsOfPartSets vts)
        syntSetUnion = foldr (:||||:) nullVect vts
        partitionSetsExhaust = VF (syntSetUnion :====: (UnivV d))
        rEquivalenceClosure = Rela "rEquCl" s0 s0
        equivalenceTimesSetEqualsSet s =
            VF (RC rEquivalenceClosure :****: s :====: s)
    in  TH "StrongConnCompOntolEnhancedTheory" [s0] [] pl
        [r,rEquivalenceClosure] [] []
        (partitionSetsNonEmpty ++ partSetsDisjoint ++
        [partitionSetsExhaust] ++ map equivalenceTimesSetEqualsSet vts)
```

The piece of code schematically generates the formulae satisfied by strongly connected components, namely that they be nonempty, disjoint, and exhaust the set. In addition, they are closed with respect to the equivalence formed of the reflexive-transitive closure intersected with its transpose. We have not included the properties the equivalence closure enjoys, as these are standard.

This schema of a theory cannot be checked for well-formedness with checkTheory-WellDefined, as it is composed of lists of sets and formulae with lengths not yet determined. After applying the decomposition algorithm, there will be the result model where subset constants are filled in corresponding to the sets of the partition as long as there exist further strongly connected components.

```
strongConnCompResultModel ::
          Theory -> Model -> ([VectConst] -> Theory) -> (Theory,Model)
strongConnCompResultModel sparseTheory givenModel enhancedTheorySchema =
    let TH _ [co] _ _ [rc] _ _ _ = sparseTheory
        MO _ iC _ _ iR _ _ = givenModel
        eThS = enhancedTheorySchema
        r = interpreteRelaConst givenModel ([],[],[]) rc
        reflTransClos = reflTranClosure r
        eq = transpMat reflTransClos &&& reflTransClos
        rowTypesH     = sort $ nub reflTransClos
        rowTypesH1 [] = []
        rowTypesH1 (hh:tt) = hh : (rowTypesH1
                     (map (\ pp -> zipWith (&&) (map not hh) pp) tt))
        rowTypes   = reverse $ rowTypesH1 rowTypesH
        partitionSetsNamed =
            zipWith (\a b -> Vect ("partSet" ++ show a) co) [1..] rowTypes
        th@(TH _ _ _ _ [_,rEqCl] _ _ _) = eThS partitionSetsNamed
        partitionSetsInterpreted =
            zipWith (\a b -> InterVec a b) partitionSetsNamed rowTypes
        resModel = MO "StrongConnCompResultModel" iC []
            partitionSetsInterpreted (iR ++ [InterRel rEqCl eq]) [] []
    in  (th,resModel)
```

Here, the three parameters of the pushout are taken and the reflexive-transitive closure is formed, as well as the equivalence defined by it. Then the rows of the closure are considered. First, duplicates are eliminated, then the rows are sorted in order to have the matrix later with small elements first followed by greater ones. As now the number of strongly connected components is known, they may be named as $partSet_i$, followed by building the enhanced theory. To these names are then attached the resulting partition sets and delivered in the result model. Once gR is concretely given as the matrix on the left, a sophisticated TₑX-generating matrix printing algorithm will produce the subdivided matrix on the right showing the connected components.

	1	2	3	4	5	6	7	8	9	10	11	12	13
1	1	0	0	0	0	0	0	0	0	0	0	0	1
2	0	0	0	0	0	0	0	1	0	0	0	1	1
3	0	0	1	0	0	1	0	0	0	1	0	0	0
4	0	0	0	0	0	0	0	0	0	0	0	0	0
5	0	0	0	0	0	0	0	0	0	0	1	0	0
6	0	0	0	0	0	1	0	0	0	0	0	1	0
7	0	1	0	0	0	0	0	0	1	0	0	0	0
8	0	1	0	0	0	0	1	0	0	0	0	0	0
9	0	0	0	0	1	0	0	0	1	0	0	0	0
10	0	0	0	0	0	0	0	0	1	0	0	0	0
11	0	0	0	0	0	0	0	0	1	1	0	0	0
12	1	0	1	0	0	0	0	0	0	0	0	0	0
13	0	0	1	0	0	0	0	0	0	0	1	0	0

	2	7	8	1	3	6	12	13	4	5	9	10	11
2	0	0	1	0	0	1	1	0	0	0	0	0
7	1	0	0	0	0	0	0	0	0	1	0	0
8	1	1	0	0	0	0	0	0	0	0	0	0
1	0	0	0	1	0	0	0	1	0	0	0	0
3	0	0	0	0	1	1	0	0	0	0	1	0
6	0	0	0	0	0	1	1	0	0	0	0	0
12	0	0	0	1	1	0	0	0	0	0	0	0
13	0	0	0	0	1	0	0	0	0	0	0	1
4	0	0	0	0	0	0	0	0	0	0	0	0
5	0	0	0	0	0	0	0	0	0	0	0	1
9	0	0	0	0	0	0	0	0	1	1	0	0
10	0	0	0	0	0	0	0	0	0	1	0	0
11	0	0	0	0	0	0	0	0	0	1	1	0

An original relation and the rearranged relation

One will observe that all this is based on the following

Proposition. Any given finite homogeneous relation R can by simultaneously permuting rows and columns be transformed into a matrix of the following form: It has upper triangular pattern with square diagonal blocks

$$\begin{pmatrix} \square & * & * & * \\ \mathbb{L} & \square & * & * \\ \mathbb{L} & \mathbb{L} & \square & * \\ \mathbb{L} & \mathbb{L} & \mathbb{L} & \square \end{pmatrix}$$

where $* = \mathbb{L}$ unless the generated preorder R^* allows entries $\neq \mathbb{L}$. The reflexive-transitive closure of every diagonal block is the universal relation $\mathbb{T}$. $\square$

7 Galois Decompositions

We now present some more involved possibilities to decompose a relation. They are closely related, as all of them may be formulated using a Galois correspondence. Afterwards a schema for the decompositions will be given that enables us to handle them more or less simultaneously.

In all of these cases, we will need two antitone mappings between powersets, which we call $\sigma : \mathcal{P}(V) \to \mathcal{P}(W)$ and $\pi : \mathcal{P}(W) \to \mathcal{P}(V)$. These mappings are usually determined by a relational construct based on some relation $B : V \leftrightarrow W$. Nested iterations will then start with the empty subset of V on the left and the full subset of W on the right – or vice versa. While there is a lot of theory necessary for the infinite case, the finite case is rather simple. Consider the starting configuration with its trivial containments $\mathbb{L} \subseteq \pi(\mathbb{T})$ and $\sigma(\mathbb{L}) \subseteq \mathbb{T}$ which are perpetuated by the antitone mappings to $\mathbb{L} \subseteq \pi(\mathbb{T}) \subseteq \pi(\sigma(\mathbb{L})) \subseteq \dots$ and $\dots \subseteq \sigma(\pi(\mathbb{T})) \subseteq \sigma(\mathbb{L}) \subseteq \mathbb{T}$. In the finite case, these two sequences will eventually become stationary. The effect of the iteration is that the least fixed point a of $v \mapsto \pi(\sigma(v))$ on the side started with the empty set is related to the greatest fixed point b of $w \mapsto \sigma(\pi(w))$ on the side started from the full set. The final situation obtained will be characterized by $a = \pi(b)$ and $\sigma(a) = b$.

7.1 Termination

The set of all points of a graph, from which only paths of finite length emerge,

$$J(R) := \inf\{x \mid \overline{x} = R_i \overline{x}\}$$

is called the **initial part** $J(R)$ of the relation R underlying the graph. We are going to determine the initial part of that relation. A relation is *progressively finite* if $J(R) = \mathbb{T}$. A slightly different property is being *progressively bounded*, $\sup_{h \geq 0} \overline{B^h {}_i \mathbb{T}} = \mathbb{T}$. A difference between the two exists only for non-finite relations; it may, thus, be neglected here.

Looking at the definition of the initial part, the two antitone functionals $v \mapsto \sigma(v) := \overline{v}$ and $w \mapsto \pi(w) := \overline{R;\overline{w}}$ seem to play a major role. One will later easily identify them in Constituents.

The algorithm applied to the relation R will result in a pair (a, b) of vectors. The relational formulae valid for the final pair (a, b) of the iteration are $a = \pi(b) = \overline{R;\overline{b}}$ and $b = \sigma(a) = \overline{a}$. (In this case, it is uninteresting to start with the empty set and the full set exchanged from left to right.)

Here, b is the initial part belonging to R: There are no paths of infinite length from the vertices of b, which, however, do exist starting from vertices of a. This is based on the following

Proposition. Any finite homogeneous relation may by simultaneously permuting rows and columns be transformed into a matrix satisfying the following basic structure with square diagonal entries:

$$\begin{pmatrix} \text{progressively bounded} & \mathbb{\perp} \\ * & \text{total} \end{pmatrix} \qquad \square$$

This subdivision into groups "initial part/infinite path exists" is uniquely determined, and indeed

$$a = \begin{pmatrix} \mathbb{\perp} \\ \mathbb{T} \end{pmatrix} = \begin{pmatrix} \text{progressively bounded} & \mathbb{\perp} \\ * & \text{total} \end{pmatrix} ; \overline{\begin{pmatrix} \mathbb{T} \\ \mathbb{\perp} \end{pmatrix}}, \quad b = \begin{pmatrix} \mathbb{T} \\ \mathbb{\perp} \end{pmatrix} = \overline{\begin{pmatrix} \mathbb{\perp} \\ \mathbb{T} \end{pmatrix}}$$

The termination-oriented decomposition may prove useful in the following case: Assume a preference relation being given, where it is not clear from the beginning that this preference is circuit-free. There is a tendency of ranking equal all those who belong to a circuit. The initial part collects all items from which one will not run into a circuit at all, so that they are properly ranked by the given relation. The others should be treated with the strongly connected component ontology and then be ranked groupwise.

	1	2	3	4	5	6	7	8	9	10	11
1	0	1	0	0	0	0	0	0	0	0	0
2	0	0	0	0	0	0	0	0	0	1	0
3	0	1	1	1	0	1	0	0	0	0	0
4	0	0	0	1	1	0	0	0	0	0	0
5	0	1	0	0	0	1	1	0	0	1	0
6	1	0	1	0	0	0	0	0	0	0	0
7	0	0	1	1	0	0	0	0	0	0	0
8	0	0	0	0	1	0	0	0	0	0	0
9	0	0	0	0	0	1	0	1	0	0	0
10	0	0	0	0	0	0	0	0	0	0	0
11	1	1	0	1	0	0	0	0	0	0	0

	1	2	10	3	4	5	6	7	8	9	11
1	0	1	0	0	0	0	0	0	0	0	0
2	0	0	1	0	0	0	0	0	0	0	0
10	0	0	0	0	0	0	0	0	0	0	0
3	0	1	0	1	1	0	1	0	0	0	0
4	0	0	0	0	1	1	0	0	0	0	0
5	0	1	1	0	0	0	1	1	0	0	0
6	1	0	0	1	0	0	0	0	0	0	0
7	0	0	0	1	1	0	0	0	0	0	0
8	0	0	0	0	0	1	0	0	0	0	0
9	0	0	0	0	0	0	1	0	1	0	0
11	1	1	0	0	1	0	0	0	0	0	0

A relation, original and rearranged according to its initial part

7.2 Matching and Assignment

A second Galois decomposition is known to exist in connection with matchings and assignments. Here we will for the first time consider heterogeneous relations. Let two matrices $Q, \lambda : V \leftrightarrow W$ be given, where $\lambda \subseteq Q$ is univalent and injective, i.e. a matching – possibly not yet of maximum cardinality, for instance

$$
Q = \begin{array}{c} 1 \\ 2 \\ 3 \\ 4 \\ 5 \\ 6 \\ 7 \end{array}
\begin{pmatrix}
1 & 0 & 0 & 1 & 0 \\
0 & 0 & 0 & 0 & 0 \\
1 & 0 & 0 & 1 & 0 \\
0 & 0 & 0 & 1 & 0 \\
0 & 1 & 1 & 1 & 1 \\
1 & 0 & 0 & 1 & 0 \\
0 & 0 & 1 & 0 & 1
\end{pmatrix}
\supseteq
\lambda =
\begin{pmatrix}
1 & 0 & 0 & 0 & 0 \\
0 & 0 & 0 & 0 & 0 \\
0 & 0 & 0 & 1 & 0 \\
0 & 0 & 0 & 0 & 0 \\
0 & 0 & 0 & 0 & 1 \\
0 & 0 & 0 & 0 & 0 \\
0 & 0 & 1 & 0 & 0
\end{pmatrix}
$$

Sympathy and matching

We consider Q to be a relation of sympathy between a set of boys and a set of girls and λ the set of current dating assignments, assumed only to be established if sympathy holds. We now try to maximize the number of dating assignments.

Definition. i) Given a possibly heterogeneous relation Q, the relation λ will be called a Q-**matching** if it is univalent, injective, and contained in Q, i.e., if

$$\lambda \subseteq Q \qquad \lambda_{;}\lambda^{\mathsf{T}} \subseteq \mathbb{I}, \qquad \lambda^{\mathsf{T}}_{;}\lambda \subseteq \mathbb{I}.$$

ii) We say that a point set x can be **saturated** if there exists a matching λ with $\lambda_{;}\mathbb{T} = x$. □

The current matching λ may have its origin from a procedure like the following that assigns matchings as long as no backtracking is necessary. The second parameter of the encapsulated function serves for accounting purposes so that no matching row will afterwards contain more than one assignment.

```
trivialMatchAbove q lambda =
  let colsOccupied = map or (transpMat lambda)
      trivialMatchRow []        []       = []
      trivialMatchRow (True:t) (False:_ ) =
        True :(replicate (length t) False)
      trivialMatchRow (_   :t) (_     :tf) = False:(trivialMatchRow t tf)
      trivialMatchAboveH []          _     = []
      trivialMatchAboveH ((hq, hl) : t) f =
          let actRow = case or hl of
                         True  -> hl
                         False -> trivialMatchRow hq f
              fNEW   = zipWith (||) actRow f
          in  actRow : (trivialMatchAboveH t fNEW)
  in  trivialMatchAboveH (zip q lambda) colsOccupied
```

Given this setting, it is again wise to design two antitone mappings. The first shall relate a set of boys to those girls not sympathetic to anyone of them, $v \mapsto \sigma(v) = \overline{Q^{\mathsf{T}}{}_i v}$. The second shall present the set of boys not assigned to some set of girls, $w \mapsto \pi(w) = \overline{\lambda_i w}$.

The iteration will end with two vectors (a, b) satisfying $a = \pi(b)$ and $\sigma(a) = b$ as before. Here, this means $\overline{a} = \lambda_i b$ and $\overline{b} = Q^{\mathsf{T}}{}_i a$. In addition $\overline{a} = Q_i b$. This follows from the chain $\overline{a} = \lambda_i b \subseteq Q_i b \subseteq \overline{a}$, which implies equality at every intermediate state. Only the resulting equalities for a, b have been used together with monotony and the Schröder rule.

One may discuss whether we had been right in deciding for starting the iteration procedure with $\mathbb{\perp}$ on the left side and $\mathbb{T}$ on the right. Assume we had decided the other way round. This would obviously mean the same as starting as before, but with Q, λ transposed. Instead of $\overline{a} = \lambda_i b$, $\overline{b} = Q^{\mathsf{T}}{}_i a$, and $\overline{a} = Q_i b$ we would then obtain the three conditions with Q replaced by Q^{T}, λ by λ^{T}, and a, b exchanged. While the two equations with Q just exchange each other, the first is transferred to $\overline{b} = \lambda^{\mathsf{T}}{}_i a$. This means that the resulting decomposition of the matrices does *not* depend on the choice – if this fourth equation is also satisfied.

It is thus not uninteresting to concentrate on condition $\overline{b} = \lambda^{\mathsf{T}}{}_i a$. After having applied `trivialMatch` to some sympathy relation and applying the iteration, it may not yet be satisfied. So let us assume $\overline{b} = \lambda^{\mathsf{T}}{}_i a$ *not* to hold, which means that $\overline{b} = Q^{\mathsf{T}}{}_i a \overset{\supseteq}{\neq} \lambda^{\mathsf{T}}{}_i a$.

We make use of the formula $\lambda_i \overline{S} = \lambda_i \mathbb{T} \cap \overline{\lambda_i S}$, which holds since λ is univalent. The iteration ends with $\overline{b} = Q^{\mathsf{T}}{}_i a$ and $\overline{a} = \lambda_i b$. This easily expands to

$$\overline{b} = Q^{\mathsf{T}}{}_i a = Q^{\mathsf{T}}{}_i \overline{\lambda_i b} = Q^{\mathsf{T}}{}_i \overline{\lambda_i \overline{Q^{\mathsf{T}}{}_i a}} = Q^{\mathsf{T}}{}_i \overline{\lambda_i Q^{\mathsf{T}}{}_i \overline{\lambda_i \overline{Q^{\mathsf{T}}{}_i a}}} \ \dots$$

from which the last but one becomes

$$\overline{b} = Q^{\mathsf{T}}{}_i a = Q^{\mathsf{T}}{}_i \overline{\lambda_i b} = Q^{\mathsf{T}}{}_i \overline{\lambda_i \mathbb{T} \cap \overline{\lambda_i Q^{\mathsf{T}}{}_i a}} = Q^{\mathsf{T}}{}_i (\overline{\lambda_i \mathbb{T}} \cup \lambda_i Q^{\mathsf{T}}{}_i a)$$
$$= Q^{\mathsf{T}}{}_i (\overline{\lambda_i \mathbb{T}} \cup \lambda_i Q^{\mathsf{T}}{}_i (\overline{\lambda_i \mathbb{T}} \cup \lambda_i Q^{\mathsf{T}}{}_i a))$$

indicating how to prove that

$$\overline{b} = (Q^{\mathsf{T}} \cup Q^{\mathsf{T}}{}_i \lambda_i Q^{\mathsf{T}} \cup Q^{\mathsf{T}}{}_i \lambda_i Q^{\mathsf{T}}{}_i \lambda_i Q^{\mathsf{T}} \cup \ \dots)_i \overline{\lambda_i \mathbb{T}}$$

If $\lambda^{\mathsf{T}}{}_i a \overset{\subsetneq}{\not\supseteq} \overline{b}$, we may thus find a point in $\overline{\lambda^{\mathsf{T}}{}_i a} \cap (Q^{\mathsf{T}} \cup Q^{\mathsf{T}}{}_i \lambda Q^{\mathsf{T}} \cup Q^{\mathsf{T}}{}_i \lambda Q^{\mathsf{T}}{}_i \lambda Q^{\mathsf{T}} \cup \ \dots)_i$ $\overline{\lambda_i \mathbb{T}}$ which leads to the famous alternating chain algorithm. While `trivialMatch` didn't do any backtracking, the alternating chain algorithm does. It therefore delivers cardinality maximum matchings and not just matchings that cannot be increased by finding an enclosing one.

We now visualize the results of this matching iteration by concentrating on the subdivision of the matrices Q, λ initially considered by the resulting vectors $a = \{2, 6, 4, 1, 3\}$ and $b = \{5, 3, 2\}$. One easily proves that $\overline{b} = \lambda^{\mathsf{T}}{}_i a$ is already satisfied. Some additional care must be taken concerning empty rows or columns in Q. To obtain the subdivided relations neatly, these are placed at the beginning of the rows, respectively at the end of the columns. In addition, rows and columns may be permuted so as to let λ appear as a diagonal.

$$
\begin{array}{c}
\begin{array}{ccccc} 1 & 4 & 5 & 3 & 2 \end{array} \\
\begin{array}{c} 2 \\ 6 \\ 4 \\ 1 \\ 3 \\ 5 \\ 7 \end{array}
\left(\begin{array}{cc|ccc}
0 & 0 & 0 & 0 & 0 \\
1 & 1 & 0 & 0 & 0 \\
0 & 1 & 0 & 0 & 0 \\
1 & 1 & 0 & 0 & 0 \\
1 & 1 & 0 & 0 & 0 \\
0 & 1 & 1 & 1 & 1 \\
0 & 0 & 1 & 1 & 0
\end{array}\right)
\qquad
\begin{array}{ccccc} 1 & 4 & 5 & 3 & 2 \end{array} \\
\begin{array}{c} 2 \\ 6 \\ 4 \\ 1 \\ 3 \\ 5 \\ 7 \end{array}
\left(\begin{array}{cc|ccc}
0 & 0 & 0 & 0 & 0 \\
0 & 0 & 0 & 0 & 0 \\
0 & 0 & 0 & 0 & 0 \\
1 & 0 & 0 & 0 & 0 \\
0 & 1 & 0 & 0 & 0 \\
0 & 0 & 1 & 0 & 0 \\
0 & 0 & 0 & 1 & 0
\end{array}\right)
\end{array}
$$

Sympathy and matching rearranged

Proposition. Any given heterogeneous relation Q admits a cardinality maximum matching $\lambda \subseteq Q$. Both relations may then in addition be simultaneously transformed into matrices of the following form by independently permuting rows and columns: Principally, they have a 4 by 4 pattern with possibly empty zones and not necessarily square diagonal blocks.

$$
\left(\begin{array}{c|ccc}
\text{⫫} & \text{⫫} & \text{⫫} & \text{⫫} \\
\text{total} & \text{⫫} & \text{⫫} & \text{⫫} \\
\text{Hall}^{\mathsf{T}} + \text{square} & \text{⫫} & \text{⫫} & \text{⫫} \\
\hline
* & \text{Hall} + \text{square} & \text{surj} & \text{⫫}
\end{array}\right)
\qquad
\left(\begin{array}{c|ccc}
\text{⫫} & \text{⫫} & \text{⫫} & \text{⫫} \\
\text{⫫} & \text{⫫} & \text{⫫} & \text{⫫} \\
\text{perm.} & \text{⫫} & \text{⫫} & \text{⫫} \\
\hline
\text{⫫} & \text{perm.} & \text{⫫} & \text{⫫}
\end{array}\right)
$$

- The first zone of rows and the last zone of columns of both matrices are zero rows resp. columns.
- The upper right 3 by 3 zones are again empty.
- The zones $\lambda_{3,1}$ and $\lambda_{4,2}$ are identity matrices.
- Zone $Q_{2,1}$ is a total relation.
- Zone $Q_{4,3}$ is a surjective relation.

The cardinality maximum matching λ is not uniquely determined, only by cardinality. The decomposition is, thus, uniquely determined up to the so-called term-rank, defined below, which here shows up as the total length of the diagonal in λ. $\qquad\square$

We also provide the definition of term-rank and Hall-condition.

Definition. i) Given a relation Q, the **term rank** is defined as the minimum number of lines (i.e., rows or columns) necessary to cover all entries 1 in Q, i.e.

$$\min\{|s| + |t| \mid Q \,\overline{;t} \subseteq s\}.$$

ii) Given a relation Q and a set x, we say that x satisfies the **Hall condition**

$$\Longleftrightarrow \quad |z| \leq |Q^{\mathsf{T}} ; z| \text{ for every subset } z \subseteq x. \qquad\square$$

8   Galois Decomposition Ontologies

Appropriate ontologies for these Galois decompositions shall now be developed. They may serve to solve a diversity of application problems, such as matching, line-covering, assignment, games, etc.

This shall be done simultaneously, i.e., in a schema that may be instantiated later to cope with these variants. So we give the task in a schematic form also and introduce the following constituents as a parameter.

```
type Constituents = (String,[CatObject],[RelaConst],[VectFct])
gameConstituents :: Constituents
gameConstituents =
   let singleObject = OC (CstO "NodeSet")
       b  = Rela "B" singleObject singleObject
       vv = VarV "v" singleObject
       f  = VFCT vv (NegVect (RC b :****: (VV vv)))
   in ("Game",[singleObject],[b],[f,f])
terminationConstituents :: Constituents
terminationConstituents =
   let singleObject = OC (CstO "NodeSet")
       b = Rela "B" singleObject singleObject
       vv1 = VarV "v1" singleObject
       vv2 = VarV "v2" singleObject
       f1 = VFCT vv1 (NegVect (VV vv1))
       f2 = VFCT vv2 (RC b :****: (NegVect (VV vv2)))
   in ("Termination",[singleObject],[b],[f1,f2])
matchAssignConstituents :: Constituents
matchAssignConstituents =
   let firstObject = OC (CstO "NodeSet1")
       secndObject = OC (CstO "NodeSet2")
       q       = Rela "B"      firstObject secndObject
       lambda = Rela "Lambda" firstObject secndObject
       vv1 = VarV "v1" firstObject
       vv2 = VarV "v2" secndObject
       f1 = VFCT vv1 (NegVect (RC q      :****: (VV vv1)))
       f2 = VFCT vv2 (NegVect (RC lambda :****: (VV vv2)))
   in ("MatchAssign",[firstObject,secndObject],[q,lambda],[f1,f2])
```

As can be seen, we have provided for denotations for category objects, relation constants, and antitone functions relating vectors on the domain side to vectors on the codomain side and vice versa. From this, we get the sparse theory in a schematic way. We may afterwards instantiate with the respective constituents.

```
sparseTheorySchema :: Constituents -> Theory
sparseTheorySchema cs =
   let (s,os,rs,vfs) = cs
   in  TH ("SparseTheory" ++ s) os [] [] rs [] [] []
sparseGameTheory        = sparseTheorySchema gameConstituents
sparseTerminationTheory = sparseTheorySchema terminationConstituents
sparseMatchAssignTheory = sparseTheorySchema matchAssignConstituents
```

Also the given models may be presented schematically, providing the respective constituents first and then the list of given matrices.

```
givenModelSchema :: Constituents -> [ [[Bool]] ] -> Model
givenModelSchema cs gRs =
   let TH s os _ _ rs _ _ _ = sparseTheorySchema cs
       irs = zipWith (\ rc bm -> InterRel rc bm) rs gRs
       osM = map (\(a,b) -> Carrier a b) $ nub $ concat $
             map (\(InterRel rc bm) -> [(domRC rc,rows bm),
                                        (codRC rc,cols bm)]) irs
   in  MO ("GivenModel" ++ s) osM [] [] irs [] []
givenModelGame        gRs = givenModelSchema gameConstituents        gRs
givenModeTermination gRs = givenModelSchema terminationConstituents gRs
givenModeMatchAssign gRs = givenModelSchema matchAssignConstituents gRs
```

With, e.g.,

```
    givenModelGame = givenModelSchema gameConstituents listOfMat
```
this may be instantiated. We can immediately check

```
    checkIsModelForTheory givenModelGame sparseGameTheory,
```
for instance.

As there will always be a result which simply subdivides the domain as well as the range set into two subsets, it is a feasible task to find the schema of a Galois ontology-enhanced theory.

```
galoisOntolEnhancedTheorySchema :: Constituents -> Theory
galoisOntolEnhancedTheorySchema cs =
   let (_,_,_,vfs) = cs
       TH s os es vs rs _ _ fs = sparseTheorySchema cs
       rsTerm = map RC rs
       [lr,rl] = vfs
       (d,c)   = typeOfFV lr

       leftFixAboveNull  = Vect "LeftFixAboveNull"  d
       rightFixBelowUniv = Vect "RightFixBelowUniv" c

       leftFixAboveNullT  = VC leftFixAboveNull
       rightFixBelowUnivT = VC rightFixBelowUniv

       sigmaLeftNullEqualsRightUniv =
          VF $ VFctAppl lr leftFixAboveNullT  :====: rightFixBelowUnivT
       leftNullEqualsPiRightUniv    =
          VF $ VFctAppl rl rightFixBelowUnivT :====: leftFixAboveNullT

   in  TH ("GaloisEnhancedTheoryTo" ++ s) os es
          [leftFixAboveNull,rightFixBelowUniv]
          rs vfs [] (fs ++
             [leftNullEqualsPiRightUniv,sigmaLeftNullEqualsRightUniv])
```

There is not much computation in this piece of code. Two vector denotations are provided for and made to vector constants terms. Then the formula is built that says that appplying the left-right function to the left vector will result in the right vector. Finally the formula is generated that the right-left function applied to the right vector will result in the left vector. As before, instantiation is possible, e.g.,

```
gameOntologyEnhancedTheory =
    galoisOntolEnhancedTheorySchema gameConstituents
```

Then we develop the result model in a schematic form. The antitone functions have to be inserted as appropriate. We formulate the basic iteration for the antitone functions along the well-known until-construct of Haskell with lr for σ and rl for π.

```
untilGalois lr rl (v, w)
    = let lrv = lr v
          rlw = rl w
      in  if (w == lrv) && (v == rlw) then (v, w)
                                       else untilGalois lr rl (rlw, lrv)
```

This untilGalois is the main algorithmic part in generating the result model in a schematic form. All the rest is designed to administrative purposes of getting the left-right and right-left functions appropriately out of the enhanced theory, interpreting them, and applying them, e.g.

```
galoisResultModelSchema :: Constituents -> [ [[Bool]] ] -> (Theory,Model)
galoisResultModelSchema cs gRs =
    let th@(TH s os _ [leftFixAboveNull,rightFixBelowUniv]
            rs vfs _ [lNP,sLN]) = galoisOntolEnhancedTheorySchema cs
        mo@(MO _ osM _ _ irs _ _) = givenModelSchema cs gRs
        [lR,rL] = vfs
        (d,c) = typeOfFV lR
        argVectConstLeft  = VarV "argL" d
        argVectConstRight = VarV "argR" c
        dSize = getObjectCarrierSize osM d
        cSize = getObjectCarrierSize osM c
        lr v = interpreteVectTerm mo ([],[(argVectConstLeft ,v)],[])
                (VFctAppl lR (VV argVectConstLeft ))
        rl w = interpreteVectTerm mo ([],[(argVectConstRight,w)],[])
                (VFctAppl rL (VV argVectConstRight))
        (leNuPi,siLeNu) = untilGalois lr rl (replicate dSize False,
                                             replicate cSize True)
        vectInterpretations = [InterVec leftFixAboveNull  leNuPi,
                               InterVec rightFixBelowUniv siLeNu]
        resModel = MO ("GaloisResultPushoutOf" ++ s)
                    osM [] vectInterpretations irs [(InterVFc lR lr),
                                                    (InterVFc rL rl)] []

    in  (th,resModel)
```

Instantiation is possible to

```
resGameModel =
   galoisResultModelSchema gameConstituents listOfMat
resTermionationModel =
   galoisResultModelSchema terminationConstituents listOfMat
resMatchAssignModel =
   galoisResultModelSchema matchAssignConstituents listOfMat
```

and one may check

```
isResultModelGame =
   checkIsModelForTheory mod the where (the,mod) = resGameModel
```

9   Conclusion and Outlook

We have provided several ontologies in which to embed newly presented relations for handling them in a pre-formatted way. With the methods presented it is possible to analyze a given relation with regard to different concepts and to visualize the results. This paper is in some regard related to work such as [Kit93,DL01,BR96]. Ordering decompositions have been studied using a similar technique in [Win03].

We hope that this will lead to future research. We have scanned a diversity of topics for their algebraic properties. On several occasions, we have replaced counting arguments by algebraic ones. Our hope is that these algebraic properties will be of value in handling fuzzy relations in this way, which do not lend themselves readily to counting methods.

In the course of this research, a wide-spectrum relational reference language [Sch03] far beyond the hints given here has been and is still being developed. It is conceived as part of the research of Work Area 2 *Mechanization* of the European COST Action TARSKI (*Theory and Applications of Relational Structures as Knowledge Instruments*) which attempts jointly to find ways to mechanize relational reasoning. Colleagues are expressly invited to take part in this endeavor and to further contribute to the design of the language.

Acknowledgments

Discussions with Michael Ebert, Eric Offermann, and Michael Winter provided considerable help.

References

[BR96]   R. B. Bapat and T. E. S. Raghavan. *Nonnegative Matrices and Applications*, volume 64 of *Encyclopaedia of Mathematics and its Applications*. Cambridge University Press, 1996.

[DL01]   Sašo Džeroski and Nada Lavrač, editors. *Relational Data Mining*. Springer-Verlag, 2001.

[Kit93] Leonid Kitainik. *Fuzzy Decision Procedures With Binary Relations – Towards a Unified Theory*, volume 13 of *Theory and Decision Library, Series D: System Theory, Knowledge Engineering and Problem Solving*. Kluwer Academic Publishers, 1993.

[Sch02] Gunther Schmidt. Decomposing Relations – Data Analysis Techniques for Boolean Matrices. Technical Report 2002-09, Fakultät für Informatik, Universität der Bundeswehr München, 2002, 79 pages.
http://ist.unibw-muenchen.de/People/schmidt/DecompoHomePage.html

[Sch03] Gunther Schmidt. Relational Language. Technical Report 2003, Fakultät für Informatik, Universität der Bundeswehr München, 2003, in preparation.

[SS89]   Gunther Schmidt and Thomas Ströhlein. *Relationen und Graphen*. Mathematik für Informatiker. Springer-Verlag, 1989. ISBN 3-540-50304-8, ISBN 0-387-50304-8.

[SS93]   Gunther Schmidt and Thomas Ströhlein. *Relations and Graphs – Discrete Mathematics for Computer Scientists*. EATCS Monographs on Theoretical Computer Science. Springer-Verlag, 1993. ISBN 3-540-56254-0, ISBN 0-387-56254-0.

[Win03] Michael Winter. Decomposing Relations Into Orderings. In *Participants Proc. of the International Workshop RelMiCS '7 Relational Methods in Computer Science and 2nd International Workshop on Applications of Kleene Algebra, in combination with a workshop of the COST Action 274: TARSKI*, pages 190–196, 2003.

An Environment for Specifying Properties of Dyadic Relations and Reasoning about Them I: Language Extension Mechanisms[*]

Pasquale Caianiello, Stefania Costantini, and Eugenio G. Omodeo

Dipartimento di Informatica, Università degli Studi di L'Aquila
{caianiel,stefcost,omodeo}@di.univaq.it

Abstract. We show how to enhance a low-level logical language, such as the 'Schröder-Tarski' *calculus of dyadic relations*, so as to make it amenable to a friendly usage. An equational formalism of that kind can play a fundamental role in a two-level architecture of logic-based systems. Three forms of definitional extensions are supported: (1) introduction of new term constructors; (2) 'disguisement' of special equations under new sentence constructors; (3) *templates* for parametric lists of sentences that will be actualized in the formation of axiomatic theories. The power of these extension mechanisms, fully supported by a Prolog program, is illustrated through examples and case studies.

Keywords: Calculus of relations, relation algebras, algebraic specifications, computational logic, logic programming

Introduction

In the architecture of a computerized system, translation techniques have the role of bridging languages which cater for friendly interaction with man at one end, and formalisms which can best cope with machine exploitation issues at the opposite end. Source language and target language meet at some intermediate level, thanks to pre-processing stages which proceed from higher levels towards the machine level, and to definitional extension mechanisms which proceed in the opposite direction. Well-engineered systems related to applications of logic are no exception. Preprocessing stages normalize and simplify source-level sentences, and can be supported by term- and graph-rewriting techniques; definitional extension mechanisms (behaving like macros and procedures) enrich the dictions available at target level, enabling the construction of a hierarchy of increasingly abstract dictions.

[*] The research described in this paper benefited from the cooperation fostered by the European COST action 274 (TARSKI). It is partially supported by the MURST/MIUR 40% project *"Aggregate- and number-reasoning for computing: From decision algorithms to constraint programming with multisets, sets, and maps."* This research benefited from collaborations fostered by the European action COST n.274 (TARSKI, see http://www.tarski.org).

H. de Swart et al. (Eds.): TARSKI, LNCS 2929, pp. 87–106, 2003.

Examples of this are not hard to find, since we are referring to logical systems in a broad sense. In an environment for declarative programming, an abstract logical machine can underlie the user-oriented language, typically a fragment of predicate logic or of some first-order theory (cf. [3, 30]); in a relational DBMS, a high-level language such as SQL gets translated into relational algebra before query optimization (cf. [29, 1]); even translating a regular definition into a family of finite automata and then into a concrete lexical analyzer (cf. [2]) can be viewed as an activity of the kind we are discussing.

The focus of this paper is on definitional extension mechanisms as a way of overcoming expressive limitations of logical formalisms which are simple enough to act as machine-oriented languages. This paper does not intend to propose any formalism as the ultimate machine-language for logic; however, we cannot make our points clear unless we focus on a specific formalism. We therefore consider a framework akin to the 'Schröder-Tarski' *calculus of dyadic relations* (here quotes are meant to indicate that one can retain the general features without being committed to some standard formulation). We think that formalisms of that kind, *purely equational* and *devoid of individual variables*, can play an important role in a two-level architecture of logic-based systems.

In fact, proof assistants often perform better in equational reasoning – which we view as being more machine-oriented – than in unrestricted forms of first-order reasoning, and hence they push the user in restating his/her axioms and lemmas in equational terms whenever (s)he can (cf. [12, 16]).

Major drawbacks of the language underlying the proposed calculus are its poor readability, its lengthy wording, and its rigid syntax. As stated above, such drawbacks, which are typical of a machine-oriented language, can be alleviated by definitional extension mechanisms: in the ongoing we propose mechanisms that are to some extent able to cope with these limitations.

Our presentation relies upon a Prolog program, Anamorpho[1], which sets our ideas to work. In particular, we have designed and implemented in Prolog three basic mechanism:

- introduction of new term constructors;
- 'disguisement' of special equations under new sentence constructors;
- *templates* for parametric lists of sentences that will be actualized in the formation of axiomatic theories. (Plenty of illustrations of this point will be given throughout the paper.)

A 'stress-test' to which Anamorpho has been subjected relates to the translation of an Entity-Relationship model into the calculus of relations, along the lines discussed in [22, 10, 11].

This paper aims at illustrating the features of this definitional environment, mainly via examples. We will show how easily one can progress from very basic and simple constructs to quite significant dictions, provided that convenient

---

[1] The complete system, named Metamorpho, will encompass another major component, Katamorpho, based on rewriting. An initial version of Anamorpho, written in SWI-Prolog, is available at the URL http://costantini.dm.univaq.it/online.htm

support for definition-handling is made available. It is worth noticing that the layered organization of constructs, which is a key for enhancing the expressiveness of the language, is also expected to play an important role in connection with automated proofs, provided that it is integrated with suitable lemma management capabilities (cf. [16, 15]).

1   Minimality Assumptions about the Privileged Formalism

By way of first approach, we develop our extension mechanisms for a logical formalism which is *equational* and *devoid of individual variables*. 'Equational' means that every formula can ultimately be reduced to an equation; the absence of variables implies that our formalism will have no quantifiers or binding constructs of any kind (descriptors, lambda-abstractors, etc.). Even a formalism subject to such syntactic restrictions can span from applications (e.g., ER-modeling and knowledge representation) to pure mathematics (e.g. number theories and set theories), offering adequate support to specifications and reasoning. We choose in fact as our 'drosophila' (not a niche language, though!) the historical Schröder-Tarski formalism of dyadic relations, which we call RELATION CALCULUS. On it, one can erect such full-fledged theories as the Zermelo-Fraenkel set theory (cf. [28, 12] and Sec.4.3), which gets much more often developed within first-order logic. A major drawback will be poor readability; on the other hand, thanks to its simplicity, relation calculus will not clutter with inessential details the nature of the design issues which here we address.

Moving from the same minimalist attitude, we will describe the syntax of our logical language simply by a *signature*, optionally equipped with a table of operator-precedences and associativity rules which enable and facilitate a prefix- infix- and postfix-usage of some operators. This signature will progressively grow, as new constructs will be brought into play via the extension mechanisms which we will propose. A Prolog parser can, hence, be exploited (in combination with a simple filtering recognizer) to analyze our logical expressions. Conversely, in order to 'pretty-print' the well-formed expressions of the extended logical language, a ready-made Prolog program translating them into LATEX will suffice[2].

To make the discussion even simpler, we could have set to work our definitional mechanisms within the equational theory of regular languages [9] (and in fact our Prolog program would offer support to that, cf. Figure 4); but, by choosing an overly limited framework as our '*drosophila*' into which to carve our examples, we might convey the wrong impression that the expressive power of a language endowed with only those features that we will indicate cannot lead very far.

---

[2] By and large, the tables and figures of this paper have been generated by means of this pretty-printer.

2   Logical Framework

Very much like any logical formalism, *relation calculus* consists of a symbolic language, an intended semantics, a collection of logical axiom schemata (which, according to the intended semantics, are valid, i.e. true in any legal interpretation), and a collection of inference rules.

Contexts and Theories.   Normally, a *derivation* gets performed within a *context* composed by a *calculus* and a bunch of *theories*. Calculus and theories consist of *axioms* and *inference rules*, the only difference being that: The axioms and inference rules of the calculus reflect some very general semantics associated with the formalism at work (they are, in a precise sense, *logically valid*); the ones of a theory, instead, describe specific assumptions concerning the domain(s) of an application (they are sometimes called *proper* axioms and rules). A context can comprise infinitely many axioms and rules, but in a computational setting we must insist that the collection of all axioms and rules be encompassed by finitely many *schemata* ('multiplied' so-to-speak, as explained below, by meta-variables occurring in them).

General Features of Relation Calculus

Our positive expectations on relation calculus arise from three orders of considerations:

- The algebraic (mostly equational) nature of this formalism.
- The simplicity of graphs and matrices as conceptual tools to support translation techniques as well as diagrammatic reasoning in this calculus [13], in a way similar to the way Karnaugh maps help in connection with switching algebra.
- The absence of variable-binding constructs (or, even, of individual variables) which, as argued above, in this framework facilitates one in the design of meta-level tools – such as definitional extension mechanisms of the kind which we are about to discuss.

Relation calculus was designed to ease reasoning about dyadic relations – MAPS, as they are sometimes called – over an unspecified, yet fixed, UNIVERSE $\mathcal{U}$ OF DISCOURSE. Its language fulfills the minimality assumptions stated in Sec.1; we feel therefore authorized to concentrate mainly on syntax and intended semantics.

Definition.   (RELATIONAL) EXPRESSIONS *are all terms of the following signature:*

symbol :	∅	𝟙	ι	p_i	∩	△	;	⌣	‾	−	∪	†
degree :	0	0	0	0	2	2	2	1	1	2	2	2
priority :					5	3	6	7		2	2	4

Of these, $\cap, \triangle, \,;, \cup, -, \dagger$ will be used as left-associative infix operators, $\smile$ as a postfix operator, and $^-$ as a line topping its argument. Larger priority numbers indicate higher 'cohesive power' (number 1 is not used because it is the priority number reserved to the relator $=$). We assume a countable infinity $p_1, p_2, p_3, \ldots$ of RELATION LETTERS to be available.

The language of relation calculus consists of EQUALITIES $Q=R$, where Q and R are relational expressions. □

The logical axioms of relational calculus are displayed in Figure 1, and we are taking the substitution law for equals as our only inference rule[3].

$$
\begin{array}{ll}
P \cup Q = Q \cup P & P \cup Q \cup R = P \cup (Q \cup R) \\
\overline{\overline{P \cup Q} \cup \overline{P \cup \overline{Q}}} = P & P \,\sharp\, Q \,\sharp\, R = P \,\sharp\, (Q \,\sharp\, R) \\
(P \cup Q) \,\sharp\, R = P \,\sharp\, R \cup Q \,\sharp\, R & P \,\sharp\, \iota = P \\
P^{\smile\smile} = P & (P \cup Q)^{\smile} = P^{\smile} \cup Q^{\smile} \\
(P \,\sharp\, Q)^{\smile} = Q^{\smile} \,\sharp\, P^{\smile} & P^{\smile} \,\sharp\, \overline{P \,\sharp\, Q} \cup \overline{Q} = \overline{Q}
\end{array}
$$

Fig. 1. Logical axioms of a version of relation calculus

Of the operators and constants in the above signature, only a few deserve being regarded as *primitive* constructs: all others, including the ones that will be added to the signature from time to time, will be regarded as *derived* constructs. For definiteness, we will treat as being primitive (apart from the p_is) only $\cup$, $\overline{}$, $\sharp$, $\smile$, and ι; but warn the reader that a complete basis of constructs can be chosen in many other ways (e.g., we could have adopted $\cap$, $\triangle$, $\mathbb{1}$, $\dagger$, ι).

For an *interpretation* of relation calculus one must indicate a nonempty $\mathcal{U}$ and a collection $\mathfrak{R}$ of sub-collections of the Cartesian square $\mathcal{U}^2 =_{\text{Def}} \mathcal{U} \times \mathcal{U}$, meeting the following closure properties:

- The *diagonal* relation $\{\langle a, a \rangle \mid a \in \mathcal{U}\}$ belongs to $\mathfrak{R}$.
- When Q belongs to $\mathfrak{R}$, the following *complement* and *converse* of Q also belong to $\mathfrak{R}$:

$$
\{\, \langle a, b \rangle \in \bigcup \mathfrak{R} \mid \langle a, b \rangle \notin Q \,\};
$$
$$
\{\, \langle b, a \rangle \in \mathcal{U}^2 \mid \langle a, b \rangle \in Q \,\}.
$$

(Here $\bigcup \mathfrak{R}$ designates the collection of of all pairs belonging to at least one of the P in $\mathfrak{R}$.)
- When Q and R belong to $\mathfrak{R}$, the following *join* and *composition* of Q, R also belong to $\mathfrak{R}$:

$$
\{\, \langle a, b \rangle \in \mathcal{U}^2 \mid \text{either } \langle a, b \rangle \in Q \text{ or } \langle a, b \rangle \in R \,\};
$$
$$
\{\, \langle a, b \rangle \in \mathcal{U}^2 \mid \text{there is a } c \in \mathcal{U} \text{ for which } \langle a, c \rangle \in Q \text{ and } \langle c, b \rangle \in R \,\}.
$$

Then one must assign a $p_i^{\mathfrak{R}}$ drawn from $\mathfrak{R}$ to each relation letter p_i, so that each expression P comes to designate, thanks to the rules below, a specific relation

---

[3] Instead of the last logical axiom in our list, various authors adopt the so-called *Schröder's law* shown at the bottom-right of Figure 3, cf. [27].

$P^{\Im}$ on $\mathcal{U}$ (any equality $Q=R$ between expressions turns out, accordingly, to be either true or false)[4]:

$$\iota^{\Im} =_{\text{Def}} \{\langle a,a\rangle \mid a \in \mathcal{U}\}; \quad \overline{Q}^{\Im} =_{\text{Def}} \{\langle a,b\rangle \in \mathcal{U}^2 \mid \langle a,b\rangle \notin Q^{\Im}\};$$
$$(Q \cup R)^{\Im} =_{\text{Def}} \{\langle a,b\rangle \in \mathcal{U}^2 \mid \text{either } \langle a,b\rangle \in Q^{\Im} \text{ or } \langle a,b\rangle \in R^{\Im}\};$$
$$(Q \mathbin{:} R)^{\Im} =_{\text{Def}} \{\langle a,b\rangle \in \mathcal{U}^2 \mid \text{there is a } c \in \mathcal{U} \text{ for which } \langle a,c\rangle \in Q^{\Im} \text{ and } \langle c,b\rangle \in R^{\Im}\};$$
$$(Q^{\smile})^{\Im} =_{\text{Def}} \{\langle b,a\rangle \in \mathcal{U}^2 \mid \langle a,b\rangle \in Q^{\Im}\}.$$

The interpretation of relation calculus obviously extends to any derived construct. E.g., we will state below that $\mathbb{1}$ and $P{\dagger}Q$ are shorts for $\iota \cup \bar{\iota}$ and for $\overline{\overline{P} \mathbin{:} \overline{Q}}$, respectively; hence it will ensue that $\mathbb{1}^{\Im} =_{\text{Def}} \bigcup \Re$ and that

$$(Q{\dagger}R)^{\Im} =_{\text{Def}} \{\langle a,b\rangle \in \mathcal{U}^2 \mid \text{for all } c \in \mathcal{U}, \text{ either } \langle a,c\rangle \in Q^{\Im} \text{ or } \langle c,b\rangle \in R^{\Im}\}.$$

Usually $\Re$ consists of all sub-collections of $\mathcal{U}^2$; accordingly, $\bigcup \Re = \mathcal{U}^2$.

Metavariables. *Meta-variables* are needed in the statement both of abbreviating definitions and of axiom schemata, inference rules, and templates. Thanks to our minimality assumptions, we can almost entirely avoid having to treat meta-variables of different types. We do not need meta-variables for formulas since, as seen above, a fully generic formula can be designated by $P=Q$.

It will turn out, however, that meta-variables representing lists of terms or formulas forcibly enter into play if we want to introduce variadic constructs, i.e. functors and relators whose numbers of arguments are not fixed, and if we want to keep the size of single formulas reasonably small (a 'granularity' issue which has some importance in the development of derivations). Moreover, meta-variables in templates sometimes stand for 'names'; i.e., they represent identifiers or special symbols which one will exploit within theories whose constructions depends on the template. Finally, one occasionally wants to represent a generic (say dyadic) construct by a metavariable.

Luckily, we will be able to cope with these accessory meta-variables without the burden of associating explicitly a type to each meta-variable. We will, in fact, exploit directly Prolog's logical variables in the role of meta-variables; and will rely on valuable features of Prolog to avoid certain otherwise necessary distinctions when working at the meta-level. To escape ambiguities, we will implicitly relate the type of language expressions which are represented by a meta-variable to the *positions* which the latter occupies in a (meta-)formula.

3   Uses and Formats of Definitions

Definitions extend in a bottom-up fashion the basic language, enabling one to specify a context more concisely and readably. We will now see a few introductory examples of their use.

We introduce a first kind of definitions by means of the $=\mathbin{:}$ sign, which enriches the term sublanguage. Having assumed union, complement, composition, and converse operators $\cup, \bar{}, \mathbin{:}, \smile$ to be available from the outset, we can put

---

[4] In the light of this semantics, notice that the operation designated by $\mathbin{:}$ is a special case of the popular *equi-join* (cf. [1]); moreover, it is related to classical function composition $\circ$ as follows: $G \circ F = F \mathbin{:} G$.

$$\delta \ =: \ \bar{\iota}, \qquad\qquad 1 \ =: \ \iota \cup \delta, \qquad\qquad \emptyset \ =: \ \bar{1},$$

$$P \cap Q \ =: \ \overline{\bar{P} \cup \bar{Q}}, \qquad P - Q \ =: \ P \cap \bar{Q}, \qquad P \triangle Q \ =: \ (P-Q) \cup (Q-P),$$

$$P \dagger Q \ =: \ \overline{\bar{P} \,\textrm{i}\, \bar{Q}}, \qquad \mathrm{mult}(P) \ =: \ P \cap P \,\textrm{i}\, \delta, \qquad \mathrm{bros}(P,Q) \ =: \ P^{\smile} \,\textrm{i}\, Q,$$

thus stating in particular that any term of either the form $P \cap Q$ or the form $P \triangle Q$ stands for the corresponding right-hand-side term. For example, $r_1 \triangle r_2$ stands for $r_1 \cap \bar{r_2} \cup r_2 \cap \bar{r_1}$, and ultimately (unless we simplify) for $\overline{\bar{r_1} \cup \bar{\bar{r_2}}} \cup \overline{\bar{r_2} \cup \bar{\bar{r_1}}}$.

Likewise, by means of the ↔: sign, one can introduce new forms of sentences. For example,

$$P \subseteq Q \leftrightarrow: P - Q = \emptyset, \qquad P = Q \& R = S \leftrightarrow: P \triangle Q \cup R \triangle S = \emptyset,$$

$$f \leftrightarrow: \iota = \emptyset, \qquad P = Q \rightarrow R = S \leftrightarrow: \overline{1 \,\textrm{i}\, (P \triangle Q) \,\textrm{i}\, 1} \,\textrm{i}\, (R \triangle S) = \emptyset,$$

disguise special equalities under the new inclusion relator and various connectives (which retain their standard meanings of conjunction, falsehood, and material implication). One can again interchange notation based exclusively on the primitive constructs with customized notation exploiting the derived ones as well.

One can continue by putting, for example,

$$\mathrm{Disj}(P,Q) \leftrightarrow: P \cap Q = \emptyset, \quad \mathrm{isFunc}(P) \leftrightarrow: \mathrm{bros}(P,P) \subseteq \iota, \quad \mathrm{RUniq}(P) \leftrightarrow: \mathrm{mult}(P) = \emptyset,$$

$$\mathrm{rA}(P) \ =: \ P \,\textrm{i}\, 1, \qquad \mathrm{diag}(P) \ =: \ P \cap \iota, \qquad\qquad \mathrm{Coll}(P) \leftrightarrow: \mathrm{is\_diag}(P),$$

$$\mathrm{Total}(P) \leftrightarrow: \mathrm{rA}(P) = 1, \quad \mathrm{dom}(P) \ =: \ \mathrm{diag}(\mathrm{rA}(P)), \quad \mathrm{img}(P) \ =: \ \mathrm{dom}(P^{\smile}).$$

It goes without saying – at least for the implemented Prolog program – that is_diag(P) stands for diag(P)=P. One hence easily recognizes that isFunc(P) and RUniq(P) are equivalent ways of stating that P designates a single-valued map; that is, a function partially defined on the universe $\mathcal{U}$. Coll(P), which requires P to be a sub-diagonal map, in a sense states that P is monadic, namely that P can be regarded as the representation of a sub-collection of $\mathcal{U}$. Another way of representing collections, is by means of those P for which is_rA(P) holds. Throughout, we will generally name operators by identifiers with a lowercase initial, and will name relators by identifiers which have either an uppercase initial or one of the forms "is...", "are...", "has...".

The above two forms of definitions act like macros: in fact, whatever construct matches something appearing on the left of =: or of ↔: could in principle disappear from the formulation of a theory, being reducible to what appears on the right. As we have seen, definitions can be nested; namely, the right-hand-side expressions of definitions may involve, along with constructs of the basic endowment, the additional constructs introduced by earlier definitions.

Recursion can be exploited in these kinds of 'macro' definitions, mainly in order to introduce variadic operators, i.e. operators with an unrestricted number of arguments. Tail-recursion (in essence, simple iteration) suffices to this aim. In this connection, we adopt the Prolog notation for lists, where $[\,]$, $[E_1, \ldots, E_n]$, and $[E_1, \ldots, E_{n+1} \,|\, T]$ represent, respectively: the void list; a list of length n whose i^{th} component is E_i; and a list whose length is at least $n+1$, whose i^{th} component is E_i for $i = 1, \ldots, n+1$, and whose components from the $(n+2)^{\mathrm{th}}$

on form the suffix list T. The following definition 'implements' iterated relation difference[5]:

$$-([P]) \;=:\; P \qquad -([P, Q \,|\, T]) \;=:\; -([P \cap \overline{Q} \,|\, T]).$$

Plainly, the first element P of the argument list represents the relation from which all other relations in the list will be subtracted. Among others, this definition yields

$$-([P, Q]) =: P \cap \overline{Q}, \quad -([P, Q, R]) =: P \cap \overline{Q} \cap \overline{R}, \quad -([P, Q, R, S]) =: P \cap \overline{Q} \cap \overline{R} \cap \overline{S}.$$

A substantially different kind of recursion is exemplified by the following definitions (cf also those in Figure 7)[6]:

$$
\begin{aligned}
\mathsf{th}(L, R \,\|\, 1) &=: L & \mathsf{th}(L, R \,\|\, i+1) &=: R \,\mathord{\scriptstyle\vdots}\, \mathsf{th}(L, R, i) \\
\mathsf{succth}(L, R \,\|\, N-1) &=: \mathsf{th}(L, R, N) \\
\mathsf{tuples}(R \,\|\, N) &=: \Big(\mathsf{img}(R) \cap \mathsf{dom}(\mathsf{th}(R, R, N)) \Big) - \mathsf{dom}(\mathsf{succth}(R, R, N))
\end{aligned}
$$

These introduce a construct $\mathsf{th}(L, R, N)$, and related constructs $\mathsf{succth}(L, R, M)$ and $\mathsf{tuples}(R, N)$, where the parameters N and M ($N = 1, 2, 3, \ldots$ and $M = 0, 1, 2, \ldots$) act as 'outer' parameters. To clarify what is understood there, let us try a more perspicuous printing of those definitions:

$$N^{\mathsf{th}}(L, R) =: \underbrace{R \,\mathord{\scriptstyle\vdots}\, \cdots \,\mathord{\scriptstyle\vdots}\, R}_{N-1 \text{ times}} \,\mathord{\scriptstyle\vdots}\, L, \qquad (N-1)^{\mathsf{succth}}(L, R) =: N^{\mathsf{th}}(L, R),$$

$$N\text{-tuples}(R) =: \big(\mathsf{img}(R) \cap \mathsf{dom}(N^{\mathsf{th}}(R, R))\big) - \mathsf{dom}(N^{\mathsf{succth}}(R, R)).$$

The intended meaning of L ad R is that they represent functions which extract the left part (=the first component) and the right part (=the sub-tuple consisting of all components but the first) from any non-void tuple in $\mathcal{U}$. Thus, roughly speaking, in order to extract the Nth component from a tuple, we must move $N-1$ times right, and then move left. Tuples (including the void tuple) can be thought of as being those elements of $\mathcal{U}$ which are R-images. An N-tuple then is any tuple within which we can move N times right (ending, presumably, in the void tuple), but which do not enable $N+1$ consecutive moves to the right.

As an application, let us consider the notion of *key* pertaining to relational databases: this is a tuple of attributes which uniquely characterizes an entity. One way of specifying keys, which exploits the th operator introduced above, is by the definition

$$\mathsf{Key}([L, R, A_0, \ldots, A_n]) \;\leftrightarrow:\; \mathsf{isFunc}\left(\bigcap_{j=0}^{n} (j+1)^{\mathsf{th}}(L, R) \,\mathord{\scriptstyle\vdots}\, A_j^{\smallsmile} \right),$$

whose level is, however, too high w.r.t. our current treatment of definitions, not catering for the very popular "$\ldots$" construct. We hence resort to the following specification:

$$
\begin{aligned}
\mathsf{keyFunc}([A], L, R \,\|\, I) &=: \mathsf{succth}(L, R, I) \,\mathord{\scriptstyle\vdots}\, A^{\smallsmile}, \\
\mathsf{keyFunc}([A, B|T], L, R \,\|\, J-1) &=: \mathsf{th}(L, R, J) \,\mathord{\scriptstyle\vdots}\, A^{\smallsmile} \cap \mathsf{keyFunc}([B|T], L, R, J),
\end{aligned}
$$

$$\mathsf{Key}([L, R|S]) \;\leftrightarrow:\; \mathsf{isFunc}(\mathsf{keyFunc}(S, L, R, 0)).$$

---

[5] Note, incidentally, that we are allowing use of the same name for symbols of different degrees: e.g., $-$ is being used in both a dyadic and a variadic way.

[6] As shown here, to separate the inner from the outer parameters in a *definiens*, we will use "$\|$" instead of "$,$". We avoid doing the same within *definienda*.

The reader is now invited to give a glance at Figure 2, where various customary properties which relations can meet are associated with newly generated constructs (cf. [8, pp. 34, 44–49]). The defined construct is sometimes a new relator (e.g., isSymmetric), and we proceed as before; but in other cases, specifying the property by a single equation would seem unnatural to us. For example, should we put

$$\text{isEquivalence}(P) \ \leftrightarrow: \ \text{isSymmetric}(P) \ \& \ \text{isTransitive}(P),$$

then (by the definition of $\&$ given above) isEquivalence(P) would reduce to $P^{\smile}\triangle P \cup (P \mathbin{\mathsf{i}} P - P)\triangle\emptyset=\emptyset$, where the constituent conditions would loose their features.

This is why we introduce definitions of another kind, called *templates*, which contain the Θ: sign. These will act like procedures in the construction of theories and contexts, during which they will be invoked with actual terms in place of the formal parameters (which are the meta-variables occurring to the left of Θ:). The "," separator appearing in the body of Θ-definitions behaves as a primitive and soft conjunction which seems preferable and more natural to us than $\&$ in most cases.

As illustrated by the definition of InductClosed in Figure 2, templates can be nested one inside another (although the list of syntactic element which they will generate upon invocation will always be flat); moreover they may contain, in addition to sentence schemata which will become axioms when templates will be invoked during the formation of a theory, also context-specific inference rules. In the case at hand, the rule $[\text{Coll}(S), G \subseteq S, S \mathbin{\mathsf{i}} R \subseteq \mathbb{1} \mathbin{\mathsf{i}} S] \Rightarrow D \subseteq S$ is meant to indicate that when D is the inductive R-closure of a set G of generators, then D will be included in any superset S of G which is closed with respect to the relation R (in the sense that any R-image of an element of S belongs to S in its turn).

Recursion helps in definitions of this kind too. In templates, however, simple tail-recursion does not always suffice: Forms of recursion more unwieldy than in macros are sometimes needed, as was shown in [22] in specifying the role of place-holders in ER-modeling – a quick recollection of this can be found in Sec.4.4 below. As a simple tail-recursive example, let us characterize an IsA chain among collections of 'individuals' of some sort. Assuming that the parameters Y, P designate the collection of all individuals (an unspecified sub-collection of $\mathcal{U}$), and the one (included in Y) of all 'place-holders', which none of the collections in an IsA chain is allowed to intersect, we can put

$$\text{IsA}([Y, P, F]) \quad \Theta: \quad [\iota] \ [\ F \subseteq Y, \ \text{Disj}(F, P) \ ],$$
$$\text{IsA}([Y, P, E, F \,|\, T]) \quad \Theta: \quad [\iota] \ [\ E \subseteq F \,|\, \text{IsA}([Y, P, F \,|\, T]) \ ].$$

Thus IsA$([Y, P, E_0, \ldots, E_n])$ states that $E_0 \subseteq \cdots \subseteq E_n \subseteq Y$ and $E_n \cap P=\emptyset$, where it is understood that $Y \subseteq \iota$ and $P \subseteq Y$. The term ι which appears in the default-list after Θ: will become the actual value of Y, should Y still be uninstantiated at invocation time.

isTransitive(P)	$\leftrightarrow$:	$P \,\mathring{,}\, P \subseteq P$
isSymmetric(P)	$\leftrightarrow$:	$P^{\smile}=P$
isReflexive(P)	$\leftrightarrow$:	$P \cup P^{\smile} \subseteq \mathrm{rA}(\iota \cap P)$
isStrict(P)	$\leftrightarrow$:	$\mathrm{diag}(P)=\emptyset$
isAntisymmetric(P)	$\leftrightarrow$:	$P \cap P^{\smile} \subseteq \iota$
isTrichotomic(P)	$\leftrightarrow$:	$\mathbb{1}=P \cup \iota \cup P^{\smile}$
isAsymmetric(P)	$\leftrightarrow$:	$P \cap P^{\smile}=\emptyset$
isTotallyReflexive(P)	$\leftrightarrow$:	$\iota \subseteq P$
isConnex(P)	$\leftrightarrow$:	$P \cup P^{\smile}=\mathbb{1}$
isPreorder(P)	Θ:	$[\ \text{isReflexive}(P),\ \text{isTransitive}(P)\ ]$
isEquivalence(P)	Θ:	$[\ \text{isSymmetric}(P),\ \text{isTransitive}(P)\ ]$
isEquivalence(P, Ch)	Θ:	$[\ \text{isFunc}(Ch),\ \text{is\_}\mathring{,}(Ch, Ch),\ Ch \,\mathring{,}\, Ch^{\smile}=P\ ]$
isGaloisCoresp(G)	Θ:	$[\ G \,\mathring{,}\, G \subseteq \iota,\ \text{isStrict}(G),\ G^{\smile} \subseteq \mathrm{rA}(G)\ ]$
isDense(Le)	$\leftrightarrow$:	$Le-\iota \subseteq (Le-\iota) \,\mathring{,}\, (Le-\iota)$
hasNoEndPoints(Le)	$\leftrightarrow$:	$\iota \subseteq (Le-\iota) \,\mathring{,}\, \mathbb{1} \,\mathring{,}\, (Le-\iota)^{\smile}$
isNDMonotonic(F, Le)	$\leftrightarrow$:	$Le \,\mathring{,}\, F \cap F \,\mathring{,}\, \overline{Le}=\emptyset$
Bisimulation(B, Oss)	$\leftrightarrow$:	$\mathbb{1} \,\mathring{,}\, (B-B^{\smile}) \cup (Oss \,\mathring{,}\, B - B \,\mathring{,}\, Oss\,)=\emptyset$
NonVoid(P)	$\leftrightarrow$:	$\mathbb{1} \,\mathring{,}\, P \,\mathring{,}\, \mathbb{1}=\mathbb{1}$
Const(P)	Θ:	$[\ \text{Coll}(P \,\mathring{,}\, \mathbb{1} \,\mathring{,}\, P),\ \text{NonVoid}(P)\ ]$
Point(P)	Θ:	$[\ \text{is\_rA}(P),\ \text{Coll}(P \,\mathring{,}\, P^{\smile}),\ \text{NonVoid}(P)\ ]$
Between(D, R, C)	$\leftrightarrow$:	$R \subseteq D \,\mathring{,}\, \mathbb{1} \,\mathring{,}\, C$
Maps(R, D, C)	Θ:	$[\ \text{Coll}(D),\ \text{Coll}(C),\ D \,\mathring{,}\, R \subseteq \mathbb{1} \,\mathring{,}\, C\ ]$
InductClosed(D, R, G)	Θ:	$[\ G \subseteq D,\ \text{Maps}(R, D, D-G),$ $[\text{Coll}(S), G \subseteq S, S \,\mathring{,}\, R \subseteq \mathbb{1} \,\mathring{,}\, S] \Rightarrow D \subseteq S\ ]$
semiGroup(P)	Θ:	$[\ P(P(Q, R), S)=P(Q, P(R, S))]$
monoid(P, U)	Θ:	$[\ \text{semiGroup}(P),\ P(U, R)=R,\ P(R, U)=R]$
convolution(C, P)	Θ:	$[\ C(C(Q))=Q,\ C(P(Q, R))=P(C(R), C(Q))]$
rightDistrib(P, Q)	Θ:	$[\ P(Q(R, S), T) = Q(P(R, T), P(S, T))]$
leftDistrib(P, Q)	Θ:	$[\ P(T, Q(R, S)) = Q(P(T, R), P(T, S))]$
Skolem(P, Q, N)	Θ:	$[\ N =:\ Q,\ N \subseteq P,\ \text{isFunc}(N),\ \text{rA}(N)=\text{rA}(P)]$

Fig. 2. Widespread properties of dyadic relations

semiGroup$(\cup)$	convolution$(\smile, \cup)$	$P \cup Q=Q \cup P$
monoid$(\mathring{,}, \iota)$	convolution$(\smile, \mathring{,})$	rightDistrib$(\mathring{,}, \cup)$
$\overline{\overline{P} \cup \overline{Q} \cup \overline{P} \cup \overline{Q}}=P$	$[\ P \,\mathring{,}\, Q \cap R=\emptyset\ ] \Rightarrow P^{\smile} \,\mathring{,}\, R \cap Q=\emptyset$	

Fig. 3. Variant version of the logical axioms for relation calculus

$P^* =:$ P^*	is_$\cup(P, P)$	$P \cup Q=Q \cup P$	
$P \cup Q =:$ $P \cup Q$	monoid$(\cup, \emptyset)$	$(\iota \cup P)^*=\iota \cup P^+=P^*$	
$P \,\mathring{,}\, Q =:$ $P \,\mathring{,}\, Q$	leftDistrib$(\mathring{,}, \cup)$	rightDistrib$(\mathring{,}, \cup)$	
$\emptyset =:$ $\emptyset$	monoid$(\mathring{,}, \iota)$	$X \,\mathring{,}\, \emptyset = \emptyset \,\mathring{,}\, X = \emptyset$	
$\iota =:$ $\emptyset^*$		$[\ P \cup Q \,\mathring{,}\, R=Q\ ] \Rightarrow P \,\mathring{,}\, R^*=Q$	
$P^+ =:$ $P \,\mathring{,}\, P^*$			

Fig. 4. Primitive and derived symbols, and logical axioms, for regular expressions

Tense($T1, T2, Tid$) Θ:	[Tid =: $T1$,	[is_rA(P)] $\Rightarrow P \subseteq \overline{T1} \dagger T2 : P$,
		$T1 : \overline{Q-T1} : \overline{P} \subseteq T1 : (P-Q)$]
Tense($\mathsf{p_2, p_1, future}$)	Tense($\mathsf{p_1, p_2, past}$)	

Fig. 5. Proper axioms of a formulation of minimal tense logic based on relation calculus

The following example hints at a totally different use of parameters in templates, by which one can associate mnemonic identifiers or symbols to relation letters within a theory:

nameLets([ ]) Θ: [ ],   nameLets([$P, Q|R$]) Θ: [P=:Q | nameLets(R)].

The very useful definitions of Skolem (which is meant to introduce a new name for an inclusion-maximal function contained in a given relation) and of semiGroup, monoid, etc., at the bottom of Figure 2, take advantage, similarly but with a different purpose (cf. Figures 3 and 4), of the allowed usage of a metavariable in the role of a constructor.

To see some of the above machinery at work, consider the file in Figure 5. Loading this will lead to a theory [5, 7] consisting of two axioms (which happen to be valid, and as such are redundant) and two inference rules. The Tense template, which serves a local purpose, will be discarded when the loading of the axiom file ends. The meta-variables in the body of the template which do not occur among parameters (viz. P and Q) represent propositional sentences.

4   Case Studies

4.1   Templates on Graph Isomorphism

A graph devoid of isolated nodes can be represented simply by the set of its edges, which we can designate by a relation letter. Then, in order to describe an isomorphism f between two graphs g, h, we can simply resort to the following theory:

g =: $\mathsf{p_1}$	h =: $\mathsf{p_2}$	f =: $\mathsf{p_3}$
isFunc(f)	isFunc(f$^\smile$)	g=f : h : f$^\smile$
rA(f)=rA(g$\cup$g$^\smile$)	rA(f$^\smile$)=rA(h$\cup$h$^\smile$)	

Here the first three items introduce aliases for three relation letters, the next three items state that f is a function, that f is injective, and that f is a morphism between g and h; the last two items state that the domain of f consists of all nodes of g, and that its image consists of all nodes of h.

Since the notion of graph isomorphism is an important one, it may be worthwhile to characterize it by templates, which is doable as follows:

graphIsom(G, F, H) Θ: [isFunc(F),   isFunc($F^\smile$),

dom(F)=nodes(G),   img(F)=nodes(H),

$G=F : H : F^\smile$],

graphIsom(G', F', H', G, F, H) Θ: [nameLets([G, G', H, H', F, F']),

nodes(G) =: dom($G \cup G^\smile$),

graphIsom(G, F, H)].

link(P,Q) =: $P \mathbin{;} \mathbb{1} \mathbin{;} Q$ sibs(P) =: bros$(P^\smile, P^\smile)$
areQProj(L,R,Y,T) Θ: $[\_,\_,\iota,\iota]$ [isFunc(L), isFunc(R), link$(Y,T) \subseteq$ bros(L,R)]
areQProj(L,R) Θ: areQProj$(L,R,\_,\_)$
areProj(L,R,Y,T) Θ: $[\_,\_,\iota,\iota]$ [areQProj(L,R,Y,T), Coll(sibs$(L) \cap$ sibs(R)), rA(L)=rA(R)]
areProj(L,R) Θ: areProj$(L,R,\_,\_)$
HdTlPure(L,R,E) Θ: [areProj(L,R), Const(E), rA(L)=rA$(\iota-E)$]
HdTl(L,R,Y,E) Θ: [areProj$(L,R,Y,\iota-Y)$, Coll(Y), Const(E), Disj(E,Y), Disj$(Y,\mathbb{1} \mathbin{;} R)$, rA$(L)$=rA$(E \cup Y)$]
HdTlFlat(P,L,R,Y,E) Θ: $[\emptyset]$ [$P \subseteq Y$, HdTl(L,R,Y,E), NonVoid(Y), rA$(L^\smile)$=rA$(Y^\smile)$]

Fig. 6. Quasi-projections, projections, and head–tail operations

At this point, one can easily construct a theory of the same kind of the theory seen at the beginning of this section, e.g. by the following series of invocations:

graphIsom(p_{110}, p_7, p_{89}),   graphIsom(p_1, p_2, p_3, g, f, h),
graphIsom(p_4, p_5, p_6, i, j, k),   graphIsom(p_3, p_7, p_4, h, l, i).

Referring to this example, let us notice that the first invocation, which is graphIsom(p_{110}, p_7, p_{89}), should be regarded as ill-formed unless a definition of the construct nodes were already available at the level of the calculus. Anyway, each one of the subsequent three invocations will *not* refer to the global definition of nodes, but to the one (which appears in the second template) that is local to the theory. This overriding mechanism may at first look confusing, because each invocation calls again into play the definition of nodes. Normally, it is not legal to define a construct twice in the same context; but in a case such as the one at hand, the system will easily recognize that the three local definitions are in fact the same, and therefore it will only store the first of the three. Similar considerations can be made concerning i and h. These aliases get in fact defined repeatedly, through the invocation of nameLets: if the definitions were inconsistent (e.g., if i were an alias both for p_4 and for p_{110}), they would cause an error during the loading of the theory. As a last remark concerning the scope of names, note that the definition of nodes within the above template of graphIsom contains a meta-variable G, which clearly is not the same G which appears among formal parameters in the header of the template.

4.2   Templates for Tuple Theories

Figure 6 illustrates how one can organize a hierarchical family of templates in sight of modeling divergent but akin situations which will be met frequently in significant application contexts. For example, many theories will describe a universe $\mathcal{U}$ where a pairing operation is available; but in some cases (such as those

$$
\begin{array}{lcl}
\mathsf{th}(L,R \,\|\, 1) \;=:\; L & & \mathsf{th}(L,R \,\|\, i+1) \;\;=:\; R \,\mathbf{;}\, \mathsf{th}(L,R,i) \\
\mathsf{sibs}(L,R \,\|\, [\,]) & =: & \mathbb{1} \\
\mathsf{sibs}(L,R \,\|\, [\mathsf{v}_i | \vec{V}]) & =: & \mathsf{th}(L,R,i) \,\mathbf{;}\, \mathsf{th}^{\smile}(L,R,i) \cap \mathsf{sibs}(L,R,\vec{V}) \\
\mathsf{mXpr}\!\left(L,R \,\|\, p(\mathsf{v}_i,\mathsf{v}_j)\right) & =: & \left(\mathsf{th}(L,R,i) \,\mathbf{;}\, p \cap \mathsf{th}(L,R,j)\right) \,\mathbf{;}\, \mathbb{1} \\
\mathsf{mXpr}(L,R \,\|\, \neg\varphi) & =: & \overline{\mathsf{mXpr}}(L,R,\varphi) \\
\mathsf{mXpr}(L,R \,\|\, \varphi\,\&\,\psi) & =: & \mathsf{mXpr}(L,R,\varphi) \cap \mathsf{mXpr}(L,R,\psi) \\
\mathsf{mXpr}(L,R \,\|\, \exists\,\vec{V}\,\varphi) & =: & \mathsf{sibs}\!\left(L,R,\mathsf{freeVars}(\exists\,\vec{V}\,\varphi)\right) \,\mathbf{;}\, \mathsf{mXpr}(L,R,\varphi) \\
\mathsf{Maddux}(L,R \,\|\, \chi) & \leftrightarrow: & \mathsf{mXpr}(L,R,\chi) = \mathbb{1} \\
\mathsf{areTotProj}(L,R,Tr) & \boldsymbol{\Theta:} & [\; \mathsf{areQProj}(L,R), \quad \mathsf{Total}(L), \quad \mathsf{Total}(R), \\
& & \qquad\qquad Tr(\chi) \,\leftrightarrow\!: \mathsf{Maddux}(L,R,\chi) \;]
\end{array}
$$

$i,j = 1,2,\dots,\;$ p relation letter, $\vec{V}$ variable-list, and φ,ψ,χ first-order formulas

Fig. 7. Translation of first-order formulas/sentences into relational expressions/equations

which arise in relational database applications) one can distinguish 'individuals' of some sort from 'tuples' (namely those objects which, with the only exception of a void tuple, result from pairing), whereas in other cases one cannot sharply draw such a distinction (e.g., Cantor's historical pairing function operates on natural numbers, and it also produces natural numbers as results). Rather than on the pairing operation, we will focus on the 'left' and 'right' operations which in a sense invert it (cf. the discussion in Sec.3). In this connection, divergent situations may arise again; in some cases these operations will turn out to have the same domain (namely, the collection of non-void tuples proper), and, moreover, distinct tuples will be guaranteed to differ either in their left elements or in their right sub-tuples: when this happens, the 'left' and 'right' functions are called CONJUGATED PROJECTIONS; in the contrary case, one speaks of CONJUGATED QUASI-PROJECTIONS.

Figure 7 specifies a classical method for translating any dyadic first-order formula φ devoid of constants and function symbols into a relational expression $E_\varphi = \mathsf{mXpr}(L,R,\varphi)$, in a context where conjugated quasi-projections L,R, both total, are available (cf. [28, pp. 95–145]). To explain what is meant, let us refer to an enumeration $\mathsf{v}_1, \mathsf{v}_2, \dots$ of all individual variables, and to an interpretation $\mathfrak{I}$; for all a in the universe $\mathcal{U}$, and for all positive integer i, let a_i be the value for which $\langle a, a_i \rangle \in \mathsf{th}(L,R,i)^{\mathfrak{I}}$ holds. The definitions are so given as to ensure that

$$
E_\varphi^{\mathfrak{I}} \;=\; \{\, \langle a,b \rangle \in \mathcal{U}^2 \mid \mathfrak{I} \models \varphi(a_1,\dots,a_{i-1}) \,\}
$$

holds provided that no variable v_j with $i \leqslant j$ occurs free in φ. It should hence be clear that the equation $E_\varphi = \mathbb{1}$, viz. $\mathsf{Maddux}(L,R,\varphi)$, has the same truth-value as φ when φ is a sentence. In the literature one finds similar algorithms which can translate all sentences of special first-order theories by taking advantage of the availability of a fork operator [31, 19, 17] instead of conjugated quasi-projections.

$$
\begin{aligned}
\mathsf{syq}(P,Q) &=: \ \overline{\mathsf{bros}}(P,\overline{Q}) - \mathsf{bros}(\overline{P},Q) \\
\mathsf{valve}(P,Q) &=: \ P - \boldsymbol{\delta} \mathbin{\vdots} (P-Q) \\
\mathsf{mkTotal}(P) &=: \ P \cup (\boldsymbol{\iota} - \mathsf{rA}(P))
\end{aligned}
$$

$$
\begin{array}{ll}
\in \ =: \ \mathsf{p}_1 & \in\!\in \ =: \ \in \mathbin{\vdots} \in \\
\ni \ =: \ \in^{\smile} & \notin\!\in \ =: \ \overline{\in} \mathbin{\vdots} \in \\
\ni\!\in \ =: \ \mathsf{bros}(\in,\in) & \mathsf{mix} \ =: \ \in\!\in \cap \notin\!\in \\
\lambda \ =: \ \mathsf{valve}(\mathsf{mix},\emptyset) & \varrho \ =: \ \lambda \mathbin{\vdots} (\in \cap \ni \mathbin{\vdots} \overline{\boldsymbol{\delta} \mathbin{\vdots} \mathsf{mix}}) \\
\end{array}
$$

$$
\mathsf{SetMaddux}(\chi) \leftrightarrow: \ \mathsf{Maddux}\Big(\mathsf{mkTotal}(\lambda^{\smile}),\ \mathsf{mkTotal}(\varrho^{\smile}),\ \chi\Big)
$$

(E)	$\mathsf{Coll}\big(\mathsf{syq}(\in,\in)\big)$
(NWL)	$\mathbb{1} = \lambda \mathbin{\vdots} \ni$
(R)	$\in \ \subseteq \mathbb{1} \mathbin{\vdots} (\in - \ni\!\in)$
	$\mathsf{Skolem}(\ni \cup \boldsymbol{\iota} - \ni\!\in, \mathsf{p}_2, \mathsf{arb})$

Fig. 8. A weak set theory

4.3 Weak Set Theory

The following five first-order sentences,

$$
\begin{aligned}
&\textbf{(E)} &&\forall x\,\forall y\,\exists d\,\big(\,x \neq y \to (d \in x \leftrightarrow d \notin y)\,\big), \\
&\textbf{(N)} &&\exists z\,\forall v\ \ v \notin z, \\
&\textbf{(W)} &&\forall x\,\forall y\,\exists w\,\forall v\,\big(\,v \in w \leftrightarrow (v \in x \vee v = y)\,\big), \\
&\textbf{(L)} &&\forall x\,\forall y\,\exists \ell\,\forall v\,\big(\,v \in \ell \leftrightarrow (v \in x \,\&\, v \neq y)\,\big), \\
&\textbf{(R)} &&\forall x\,\exists r\,\forall v\,\big(\,v \in x \to (r \in x \,\&\, v \notin r)\,\big),
\end{aligned}
$$

form the system of axioms of a very weak set theory: extensionality, null set, single-element addition ('with' operation $x, y \mapsto x \cup \{y\}$), single-element removal ('less' operation $x, y \mapsto x \setminus \{y\}$), and regularity. By dropping **(R)** – whose role is to forbid membership cycles – one gets an even weaker theory of sets.

Discussing this theory gives us the opportunity to compare the language of relation calculus with the one of first-order predicate calculus. The translation of **(E)**, **(N)**, and **(R)** can be carried out straightforwardly, because each one of these sentences involves three variables at most. On the other hand, it can be proved that neither **(W)** nor **(L)**, separately taken, nor the conjunction of **(W)** with **(L)** and **(E)**, can be translated into relation calculus. Somewhat surprisingly, the conjunction of **(W)** with **(L)** and **(N)** can be restated as the single statement

$$
\textbf{(NWL)} \ \forall x\,\forall y\,\exists p\,\Big(\,y \in p \,\&\, \forall u\,\big(u = x \leftrightarrow (\ \ \exists v\,(u \in v \,\&\, v \in p)
$$
$$
\&\, \exists w\,(u \notin w \,\&\, w \in p)\,)\,\big)\,\Big),
$$

which gets compactly translated into relation calculus, in the way displayed in Figure 8. As a matter of fact, it can be shown easily – even automatically – that in first-order logic the following equivalence holds:

$$
\textbf{(E)} \vdash \big(\,\textbf{(N)} \,\&\, \textbf{(W)} \,\&\, \textbf{(L)}\,\big) \leftrightarrow \textbf{(NWL)}.
$$

The intuitive idea is that **(NWL)** entails in a roundabout fashion that sets $\emptyset$, $x \cup \{y\}$ and $x \setminus \{y\}$ can be obtained from given sets x and y, by stating that one can form the somewhat more formidable doubleton set

$$x @ y =_{\text{Def}} \{\, x \setminus \{y\},\ x \cup \{y\}\, \}.$$

As we have recalled above, if one succeeds in deriving $\mathsf{areQproj}(L, R)$ for suitable L, R in a theory formalized within relation calculus, this indicates that the theory has the same power, both in means of expression and in means of proof, as the corresponding theory formalized in full first-order logic. Here we can adopt the set $(x @ y) @ x$ as the encoding of the ordered pair x, y of sets, and this is the rationale behind the definitions of λ and ϱ shown in Figure 8. It thus turns out that the theory in Figure 8 and the first-order theory with which we have started are equipollent, because all three sentences in $\mathsf{areQproj}(\lambda^{\smile}, \varrho^{\smile})$ can be derived from the logical axioms in Figure 1 taken together with the proper axioms in Figure 8. A key step in these derivations is the intermediate lemma that $\mathsf{isFunc}(Q^{\smile})$ entails $\mathsf{isFunc}(\mathsf{valve}(P, Q)^{\smile})$ for all P, Q, a general fact which can be proved in the relation calculus, without contribution of any proper axioms. This is why the definition of the valve construct should be introduced at the global level, instead of being kept local to set theory (as are the definitions of $\in, \in\in, \ldots, \lambda, \varrho$). Similar considerations advise us that syq (the symmetric quotient construct, cf. [27, pp. 18–20]), and $\mathsf{mkTotal}$ (a construct which functionally prolongs any relation, without affecting the image of any element in its previous domain), should be made available globally.

Once we have found quasi-projections in a theory based on relation calculus, and have made them total, we can import all first-order notation, as we do here by means of the $\mathsf{SetMaddux}$ translator. Another definition reuse shown in Figure 8 is a conservative Skolem extension, by which we introduce an operator arb meeting the condition

$$\forall x \big(\, (\mathsf{arb}\ x \in x \lor \mathsf{arb}\ x = x)\ \&\ \neg\exists y (y \in x\ \&\ y \in \mathsf{arb}\ x)\, \big).$$

4.4   Keys and Placeholders

As discussed in [22, 11], the notions of *key* and *place-holder* are crucial for the semantics of ER-models. While showing how such notions can be specified in relation calculus, we get a chance to illustrate some additional features (a bit more esoteric than what we have seen so far) of the definitional apparatus we have made concrete in Prolog. In relational databases, a *key* is a tuple of attributes which uniquely characterizes an entity. One way of specifying this, which exploits the th operator introduced in Sec.3, is by the definition

$$\mathsf{Key}([L, R, A_0, \ldots, A_n])\ \leftrightarrow:\ \mathsf{isFunc}\left(\bigcap_{j=0}^{n}(j+1)^{\mathsf{th}}(L, R) \,\dot{\,}\, A_j^{\smile}\right),$$

whose level is, however, too high w.r.t. our current treatment of definitions. Since our system cannot deal directly, as yet, with the very important "$\ldots$" construct, we shall resort to the following specification:

$$\mathsf{keyFunc}([A], L, R \parallel I)\ =:\ \mathsf{succth}(L, R, I) \,\dot{\,}\, A^{\smile},$$
$$\mathsf{keyFunc}([A, B|T], L, R \parallel J - 1)\ =:\ \mathsf{th}(L, R, J) \,\dot{\,}\, A^{\smile} \cap \mathsf{keyFunc}([B|T], L, R, J),$$

$$\mathsf{Key}([L, R|S])\ \leftrightarrow:\ \mathsf{isFunc}(\mathsf{keyFunc}(S, L, R, 0)).$$

Place-holders form a collection P of individuals such that for every relation Q in an ER-model:

$$IA(P) =: \mathbb{1} ; P$$

$RUniq(P,Q) \leftrightarrow: mult(Q) \cap rA(P) = \emptyset$	$LUniq(P,Q) \leftrightarrow: RUniq(P,Q^{\smile})$
$RXcl(P,Q) \leftrightarrow: mult(Q) \cap IA(P) = \emptyset$	$LXcl(P,Q) \leftrightarrow: RXcl(P,Q^{\smile})$

$$
\begin{aligned}
IBoth(S,Q) &=: & rA(S) \cap rA(Q) \\
NoLLBoth(P,Q,S) &\leftrightarrow: & IBoth(Q,S) \cap P = \emptyset \\
NoLRBoth(P,Q,S) &\leftrightarrow: & NoLLBoth(P,Q,S^{\smile}) \\
NoTogether(P,Q) &\leftrightarrow: & rA(P) \cap IA(P) \cap Q = \emptyset \\
NoTwice([P]) &\Theta: & true(P) \\
NoTwice([P,Q|T]) &\Theta: & [\, RUniq(P,Q), \quad LUniq(P,Q), \\
& & \quad NoTogether(P,Q), \\
& & \quad RXcl(P,Q), \quad LXcl(P,Q) \\
& & \,|\, NoTwice([P|T]) \,] \\
NoLRBoth([P,Q]) &\Theta: & true([P,Q]) \\
NoLRBoth([P,Q,S|T]) &\Theta: & [\, NoLLBoth(P,Q,S), \quad NoLRBoth(P,Q,S), \\
& & \quad NoLRBoth(P,S,Q) \\
& & \,|\, NoLRBoth([P,Q|T])] \\
NoBoth([P]) &\Theta: & true(P) \\
NoBoth([P,Q|T]) &\Theta: & [\, NoLRBoth(P,Q,Q) \\
& & \,|\, \{NoLRBoth([P,Q|T]), \quad NoBoth([P|T])\} \,] \\
PlaceHolders([P,Q|T]) &\Theta: & \{NoTwice([P,Q|T]), \quad NoBoth([P,Q|T])\}
\end{aligned}
$$

Fig. 9. Characterization of place-holders

- each place-holder occurs at most once in the domain (respectively, in the image) of Q;
- no pair of Q has place-holders on both sides;
- place-holders are left- and right-exclusive in Q: e.g., if a pair $\langle *, b \rangle$ with $*$ in P belongs to Q, then no pair $\langle a, b \rangle$ with $a \neq *$ can belong to P;
- a place-holder cannot occur sometimes on the left and sometimes on the right of pairs in Q;
- a place-holder cannot occur both in Q and in some other relation S of the same ER-model.

We can specify all of the above conditions on place-holders as shown in Figure 9.

This PlaceHolders notion can be combined with that of HdT|Flat introduced in Sec.4.2. This leads to the template

$$
\begin{aligned}
PlaceHolders(P', L', R', S', E', T, P, L, R, S, E) \quad \Theta: \quad [\\
P =: P', \quad L =: L', \quad R =: R', \quad S =: S', \quad E =: E' \quad |\, \{HdT|Flat(P,L,R,S,E), \\
PlaceHolders([P|T])\} \,],
\end{aligned}
$$

which can be invoked, e.g., as follows:

$$PlaceHolders(p_5, p_1, p_2, p_3, p_4, [p_6, p_7, p_8, p_9, p_{10}], \pi, \lambda, \varrho, \upsilon, \epsilon).$$

Notice the use of braces in some of the above definitions, which is aimed at protecting their recursive core against off-line template expansion (which would lead to endless loops). Also notice that true (with any number of arguments) stands for a void template: It is used to ensure that all parameters in the left-hand sides of some template definitions do also appear in the corresponding right-hand sides.

5   Concluding Remarks

Definitional extension mechanisms of the kind discussed in this paper are rarely bestowed in logic textbooks the attention they deserve (one noticeable exception is [21]). This is surprising, because many issues which are regarded as fundamental in the design of any programming language (e.g., scope of declarations, implementation of recursion, inheritance, overriding, loop-detection, treatment of defaults, etc.) enter into the design of logical systems as well. Here we have treated definitional mechanisms which can be useful near the level of a logical machine; elsewhere [23], we have addressed modularization issues regarding large-scale proof development in terms of what one might call 'proof-engineering'.

Supplied features which make the mechanisms proposed in this paper really flexible are the ones enabling the user to:

- introduce *variadic* constructs, to wit, constructs with an unrestricted number of arguments (e.g. equality chains);
- exploit, in the *definienda*, parameters which do not belong to the formalism: typically, natural numbers, lists, and even quantified first-order formulas (cf. Figure 7);
- make some definitions local to specific theories;
- make some templates for parametric theories 'ephemeral', so that they are available throughout the loading of the context where they are created, but then they automatically vanish;
- pass identifiers as parameters to templates: a typical use will be Skolemization;
- fix default expressions for some template parameters.

Moreover, the system checks the overall consistency of definitions: the same construct cannot be defined twice in the same context, all parameters of a *definiendum* must appear in the *definiens*, etc. The process of definition expansion should be guaranteed to converge by such checks, although this fact has not been proved formally so far.

This article is meant to be the first of a series, whose second paper will treat extensible inference mechanisms for equational languages of the kind seen above, and whose third paper will complete the picture of how to move from front-end languages to back-end formalisms. In each case we will describe tools implemented in Prolog, illustrating their usefulness through significant case-studies: in their final integrated form, these tools will form a broad-spectrum translation system named Metamorpho.

As regards the inferential apparatus, we have in mind to investigate two very different, and in a way complementary, approaches. On the one hand, we are experimenting with entirely algebraic methods such as Knuth-Bendix (cf. [20]), with which theorem-provers show a large autonomy, although at a very

low expressive level. On the other hand, we are getting acquainted with the
Rasiowa-Sikorski approach (cf. [26]), which has considerable appeal for interactive automated deduction. The latter method proceeds analytically in a way
similar to the very popular tableau-based systems, although it constructs proofs
instead of refutations (mainly in logics involving only dyadic predicates).

The next significant goal in which we feel engaged is the one of reaching a
satisfactory unified design for the main front-end component of the envisaged
Metamorpho system. To tackle this design issue, we will first carry out the
detailed analysis of a number of translation algorithms and techniques. Various
such algorithms are known, e.g. some which link nonclassical logics with relation
algebras [24, 25, 18], and others which operate on 3-variable sentences of first-
order logic [28, 14]; some others are still under study [7]. Indeed, translation
techniques to switch between man-oriented and machine-oriented formalisms
need to be unceasingly developed and improved.

References

1. S. Abiteboul, R. Hull, and V. Vianu. *Foundations of Databases*. Addison-Wesley,
   1995.
2. A. V. Aho, R. Sethi, and J. D. Ullman. *Compilers - Principles, techniques and
   tools*. Addison-Wesley, 1986. Reprinted.
3. H. Aït Kaci. *Warren's Abstract Machine - A Tutorial Reconstruction*. The MIT
   Press, Cambridge, Mass., 1991.
4. Dines Bjørner, Manfred Broy, and Igor V. Pottosin, editors. *Formal Methods in
   Programming and Their Applications, International Conference, Akademgorodok,
   Novosibirsk, Russia, June 28 - July 2, 1993, Proceedings*, volume 735 of *Lecture
   Notes in Computer Science*. Springer, 1993.
5. J.P. Burgess. Basic tense logic. In D. M. Gabbay and F. Guenthner, editors,
   Handbook of Philosophical Logic, volume II, pages 89–133. D. Reidel, Dordrecht-
   Holland, 1984.
6. R. Caferra and G. Salzer, editors. *Automated Deduction in Classical and Non-
   Classical Logics*, LNCS 1761 (LNAI). Springer-Verlag, January 2000.
7. D. Cantone, A. Formisano, E.G. Omodeo, and C.G. Zarba. Compiling dyadic first-
   order specifications into map algebra. *Theoretical Computer Science*, 303, 2002.
8. D. Cantone, E.G. Omodeo, and A. Policriti. *Set Theory for Computing - From
   decision procedures to declarative programming with sets*. Texts and Monographs
   in Computer Science. Springer-Verlag, Berlin, 2001.
9. F. Corradini, R. De Nicola, and A. Labella. An equational axiomatization of
   bisimulation over regular expressions. *J. Logic and Comput.*, 12(2):89–108, 2002.
10. E.-E. Doberkat and E.G. Omodeo. Algebraic semantics of ER-models in the con-
    text of the calculus of relations. II: Dynamic view. In H. de Swart, editor, *Re-
    lational methods in computer science*, volume 2561 of *Lecture Notes in Computer
    Science*, pages 50–65. Springer-Verlag, December 2002. (6th International Confer-
    ence RelMiCS 2001, and 1st Workshop of COST Action 274 TARSKI, Oisterwijk,
    The Netherlands, Oct.16-21, 2001 Revised Papers).
11. E.-E. Doberkat and E.G. Omodeo. ER modelling from first relational principles.
    Theoretical Computer Science, To appear.

12. A. Formisano and E. Omodeo. An equational re-engineering of set theories. In Caferra and Salzer [6], pages 175–190.

13. A. Formisano, E.G. Omodeo, and M. Simeoni. A graphical approach to relational reasoning. In W. Kahl, D. L. Parnas, and Schmidt G., editors, *Proc. of Relational Methods in Software, RelMiS 2001*, Bericht No.2001-02. Fakultät für Informatik, Universität der Bundeswehr Muenchen, april 2001. To appear on Electronic Notes in Theoretical Computer Science 44(3).

14. A. Formisano, E.G. Omodeo, and M. Temperini. Goals and benchmarks for automated map reasoning. *J. Symb. Computation*, 29(2):259–297, 2000. (Special issue on Advances in First-order Theorem Proving, M.-P. Bonacina and U. Furbach eds).

15. A. Formisano, E.G. Omodeo, and M. Temperini. Instructing equational set-reasoning with Otter. In R. Gore, A. Leitsch, and T. Nipkow, editors, *Automated Reasoning. Proc. of First International Joint Conference, IJCAR 2001– (CADE+FTP+TABLEAUX)*, number 2083 in Lecture Notes in Computer Science, pages 152–167, Berlin, 2001. Springer-Verlag.

16. A. Formisano, E.G. Omodeo, and M. Temperini. Layered map reasoning: An experimental approach put to trial on sets. In A. Dovier, M.-C. Meo, and A. Omicini, editors, *Declarative Programming – Selected Papers from AGP 2000*, number 48 in Electronic Notes in Theoretical Computer Science, pages 1–28. Elsevier Science B. V., 2001.

17. M. F. Frias. *Fork algebras in Algebra, Logic, and Computer Science*. World Scientific Publishing Co., 2002.

18. M. F. Frias and E. Orłowska. Equational reasoning in non cassical logics. *Journal of Applied Non Classical Logics*, 8(1-2):27–66, 1998.

19. A. M. Haeberer, G. A. Baum, and G. Schmidt. On the smooth calculation of relational recursive expressions out of first-order non-constructive specifications involving quantifiers. In Bjørner et al. [4], pages 281–298.

20. D.E. Knuth and P.B. Bendix. Simple word problems for universal algebras. In J. Leech, editor, *Computational problems in abstract algebra*, pages 263–297. 1970.

21. A. P. Morse. *A Theory of Sets*. Pure and Applied Mathematics. Academic Press, New York, 1965.

22. E.G. Omodeo and E.-E. Doberkat. Algebraic semantics of ER-models in the context of the calculus of relations. I: Static view. In W. Kahl, D. L. Parnas, and Schmidt G., editors, *Proc. of Relational Methods in Software, RelMiS 2001*, Bericht No.2001-02. Fakultät für Informatik, Universität der Bundeswehr Muenchen, april 2001. To appear on Electronic Notes in Theoretical Computer Science 44(3).

23. E.G. Omodeo and J. T. Schwartz. A 'theory' mechanism for a proof-verifier based on first-order set theory. In A. Kakas and F. Sadri, editors, *Computational Logic: Logic Programming and Beyond – Essays in honour of Bob Kowalski*, Part II, volume 2408 of *Lecture Notes in Artificial Intelligence*, pages 214–230. Springer-Verlag, Berlin, 2002.

24. E. Orłowska. Relational interpretation of modal logics. In H. Andreka, D. Monk, and I. Nemeti, editors, *Algebraic Logic. Colloquia Mathematica Societatis Janos Bolyai 54*, pages 443–471. North-Holland, Amsterdam, 1988.

25. E. Orłowska. Relational semantics for nonclassical logics: Formulas are relations. In J. Wolenski, editor, *Philosophical Logic in Poland*, pages 167–186. 1994.

26. H. Rasiowa and R. Sikorski. *The mathematics of metamathematics*, volume 12 of *PWN*. Polish Scientific Publishers, Warsaw, 1963.

27. G. Schmidt and T. Ströhlein. *Relations and graphs*. Monographs on Theoretical Computer Science. Springer-Verlag, Berlin, 1993.

28. A. Tarski and S. Givant. *A formalization of Set Theory without variables*, volume 41 of *Colloquium Publications*. American Mathematical Society, 1987.

29. J. D. Ullman. *Database and Knowledge-base Systems, vol. 1*, volume 49 of *Principles of Computer Science*. Computer Science Press, Stanford University, 1988.

30. P. L. Van Roy. *Can Logic Programming Execute as Fast as Imperative Programming?* Ph.D. thesis, Univ. of California at Berkeley, 1990.

31. P. A. S. Veloso and A. M. Haeberer. A finitary relational algebra for classical first-order logic. *Bulletin of the Section of Logic*, 20:52–62, 1991.

Consistent Representation of Rankings

Kim Cao-Van and Bernard De Baets

Department of Applied Mathematics, Biometrics and Process Control
Ghent University, Coupure links 653, B-9000 Gent, Belgium
{Kim.CaoVan,Bernard.DeBaets}@UGent.be

Abstract. In this paper, we discuss how a proper definition of a ranking can be introduced in the framework of supervised learning. We elaborate on its practical representation, and show how we can deal in a sound way with reversed preferences by transforming them into uncertainties within the representation.

Keywords: MCDA, Monotonicity, Ordinal Classification, Ranking, Reversed preference, Supervised learning.

1   Introduction

Supervised learning has been studied by many research groups, largely coming from statistics, machine learning and information systems science. In these studies, the problems of classification (discrete) and regression (continuous) have received a lot of attention. More recently the problem of ranking gained in interest because of the wide variety of applications it can be used for.

Ranking can be interpreted as monotone classification or monotone regression. The addition of the word "monotone" to the definition is, however, less trivial than it seems. And the problems this addition to the definition entails in the mathematical model used to deal with classification or regression are even more persistent. In this paper, we will mainly focus on discrete models, in other words, we will discuss how we can deal with monotone classification, which is equivalent to monotone ordinal regression.

Compared to classification, methods for solving ranking problems are only beginning to emerge. The reason for this slow progress lies partly in the fact that a solid framework for dealing with rankings in the supervised learning context has not yet been developed. For example, not all methods can guarantee that the generated classifier behaves monotonically, e.g. [3, 14]. Developing such a framework is the main goal of this paper.

Another bare terrain in machine learning is that of working with ordinal data, e.g. [7, 10, 17], even though "In measurement theory the raw data are typically postulated to be ordinal in nature" [12]. Mostly "...the question addressed is the conditions under which the data structure exhibits numerical measures having certain properties". However, we prefer to work directly on the ordinal nature of the data itself, clearing ourselves of any conditions to be met.

H. de Swart et al. (Eds.): TARSKI, LNCS 2929, pp. 107–123, 2003.

2 Problems with Earlier Proposals

The aim of supervised learning is to discover a function $\lambda : \Omega \to \mathcal{L}$ from a set of objects Ω to a set of labels $\mathcal{L}$, based on a finite set of example pairs $(a, \lambda(a))$ with $a \in \Omega$. If $\mathcal{L}$ is finite, then λ is referred to as a classification. Generally, the objects $a \in \Omega$ are described by means of a finite set $Q = \{q_1, \ldots, q_n\}$ of attributes $q : \Omega \to \mathcal{X}_q$. Therefore, to each $a \in \Omega$ corresponds a vector $\mathbf{a} = (q_1(a), \ldots, q_n(a)) \in \mathcal{X} = \prod_{q \in Q} \mathcal{X}_q$ (called the **data space**, a.k.a. measurement space), and the problem is then restated as learning the function λ based on examples $(\mathbf{a}, \lambda(a))$ with $a \in \Omega$. Although this new definition is not less restrictive if handled with care, it does tend to encourage a more narrow view, where $\lambda(a)$ is interpreted as $\lambda(\mathbf{a})$. This may lead to conflicting situations, since it is possible that $a, b \in \Omega$, $\mathbf{a} = \mathbf{b}$ but $\lambda(a) \neq \lambda(b)$. We will use the term *doubt* to refer to such a situation.

The problem of ranking is generally formulated as a classification problem in the narrow view, with the additional restriction that it has to be monotone, i.e. for all $\mathbf{x}, \mathbf{y} \in \mathcal{X}$ we must have that $\mathbf{x} \leq_{\mathcal{X}} \mathbf{y}$ implies $\lambda(\mathbf{x}) \leq_{\mathcal{L}} \lambda(\mathbf{y})$, where $(x_1, \ldots, x_n) \leq_{\mathcal{X}} (y_1, \ldots, y_n)$ if and only if $x_i \leq_{\mathcal{X}_{q_i}} y_i$ for $i = 1, \ldots, n$, and the relations $\leq_{\mathcal{X}_q}$ on $\mathcal{X}_q$ and $\leq_{\mathcal{L}}$ on $\mathcal{L}$ are complete orders. Again, conflicting situations may arise, which we will refer to as *reversed preference* (see Section 6). Some authors [13, 15] impose some additional restrictions, such as demanding the training data to fulfill the monotonicity requirement, to ensure that these conflicts do not occur. Others [2] propose a form of naive conflict resolution.

However, a fundamental flaw in this definition is that it is formulated as a restriction not on the original definition $\lambda : \Omega \to \mathcal{L}$ of a classification, but on its operational counterpart $\lambda : \mathcal{X} \to \mathcal{L}$, which was introduced in function of the description of the objects. Yet another problem is that ranking is not merely a restriction, but can also be seen as a generalisation of classification, in which the equality relation is replaced by an order relation. For example, in the formulation of a ranking given above, we see that "$\mathbf{x} \leq_{\mathcal{X}} \mathbf{y}$ implies $\lambda(\mathbf{x}) \leq_{\mathcal{L}} \lambda(\mathbf{y})$" is an extension of "$\mathbf{x} = \mathbf{y}$ implies $\lambda(\mathbf{x}) = \lambda(\mathbf{y})$". It is well known that different points of view on a basic definition may lead to completely different extensions, so, if possible, the most intrinsic definition should be chosen. In our case, this means that $\lambda : \Omega \to \mathcal{L}$ is preferred.

It should be remarked that, while we focus in the firs part of this paper on a representation based on sets of λ, there exists another strategy using cumulative models coming from statistics (e.g. cumulative logit model [1]) and later on applied in other learning schemes [7–9]. Instead of λ, it considers a family $(\lambda_i)_{i \in \mathcal{L}}$, where the $\lambda_i : \Omega \to \{0, 1\}$ are defined by $\lambda_i(a) = 0$ if $\lambda(a) \leq_{\mathcal{L}} i$, and 1 otherwise. In the last two sections of this paper, we will incorporate this cumulative model (a probabilistic one) to construct a probabilistic representation of λ.

3 Classification and Ranking

Classification. Thus, following the preceding discussion, we define a **classification** λ in Ω as the assignment of the objects belonging to Ω, to some element, called a **class label**, in a universe $\mathcal{L}$ of labels. If $\mathcal{L}$ is a continuum, λ is usually referred to as a **regression**. The class labels can be identified with their inverse image in the object space Ω, where they constitute a partition. We will call these inverse images (**object**) **classes**. For any class label $i \in \mathcal{L}$, we denote the corresponding class by $C_i := \lambda^{-1}(i)$, and $\mathrm{Cl} := \{C_i \mid i \in \mathcal{L}\}$. So, the set of all classifications in Ω stands in one-to-one correspondence to the set of all partitions of Ω (which is equivalent to the set of all equivalence relations on Ω). Remark that we may assume that λ is surjective by constraining $\mathcal{L}$ to the image of λ (this can be done without loss of generality since by definition, we are only interested in objects from Ω, which may well be infinite).

Ranking: Preliminaries. If we want to define a ranking based on this definition, it becomes clear that there is no room for a concept such as monotonicity since Ω has no inherent structure such as $\mathcal{X}$. Still, we have to plant the seeds for it, such that monotonicity will appear naturally when the data space $\mathcal{X}$ is introduced as a representation of Ω. This can be done, following the ideas from [9, 16], by returning to the semantics behind ranking, which declares that the higher an object's rank, the more it is preferred.

We can model this preferential information by a complete preorder. A *preorder* R on Ω is a binary relation R on Ω that is reflexive and transitive. It is called *complete* if for all $a, b \in \Omega$ we have aRb or bRa. A preorder can be seen as a special kind of *weak* (also called *large*) *preference relation* [19] S (which is only reflexive), and where the expression aSb stands for "a is at least as good as b" (it is also said that "a *outranks* b"). Given a complete preorder S, we can define a *strict preference relation* P by aPb if and only if aSb and not bSa, and an *indifference relation* I by aIb if and only if aSb and bSa. We clearly have that $S = P \cup I$.

Ranking: Definition. In the MCDA literature [16], the term "sorting" is used to refer to a classification into a pre-defined finite set of ordered classes. The best way of to enlighten the meaning of words is to turn to a dictionary. According to Webster's Encyclopedia Unabridged Dictionary [20], the meaning of "sort" is: "n. a particular kind, species, variety, class, or group, distinguished by a common character or nature; v.t. to arrange according to sort, kind or class". On the other hand, looking up the word "rank" results in: "n. a number of persons forming a separate class in a social hierarchy or in any graded body; relative position or standing; v.t. to arrange in ranks or in regular formation". Therefore, we feel the term "ranking" would be better suited. More general, we define a ranking as:

Definition 1. *A* **ranking** *in* Ω *is a classification/regression* $\lambda : \Omega \to \mathcal{L}$, *together with an order* $\geq_\mathcal{L}$ *on* $\mathcal{L}$. *We denote this ranking by* $(\lambda, \geq_\mathcal{L})$. *Moreover, the order* $\geq_\mathcal{L}$ *defines a weak preference relation* S *on* Ω *as follows:*

$$aSb \iff \lambda(a) \geq_\mathcal{L} \lambda(b),$$

or stated differently, aSb *if and only if* $s \geq_\mathcal{L} r$, *for any* $a \in C_s$ *and* $b \in C_r$, *with* $C_i = \lambda^{-1}(i)$ *denoting the class associated with the label* $i \in \mathcal{L}$.

In this paper, we will only consider **complete** rankings, where $\geq_{\mathcal{L}}$ is a complete order on $\mathcal{L}$, i.e. $(\mathcal{L}, \geq_{\mathcal{L}})$ is a chain. This is in line with most of the current problems considered in supervised learning. In this case, we have a specific preference structure on Ω linked with the classes. For $a \in C_s$ and $b \in C_r$ it holds that

$$aPb \iff s >_{\mathcal{L}} r \qquad \text{and} \qquad aIb \iff s = r.$$

So, the set of all rankings in Ω stands in one-to-one correspondence to the set of all complete preorders S on Ω. The classes are formed by the indifference relation I which is an equivalence relation (transitivity holds since S is a complete preorder). Hence, the indifference relation I determines the classification.

Remarks. The foregoing definition of a ranking consists of two parts: firstly a classification/regression with an ordered image; secondly, the associated semantics expressed by a weak preference relation. It is the second condition that ensures that a (finite) ranking is not simply an ordinal classification. A second point worth noting in the definition is that the set $\mathcal{L}$ is not necessarily finite (the same is true in the definition of a classification). Still, in the remainder of this paper, we will assume that $\mathcal{L}$ is finite. We will also assume that $\mathcal{X}$ is finite to obtain a pure ordinal setting.

4   Representing a Classification: Sets

The above definitions are not really useful in practice since they relate to a universe Ω that is in essence just an enumeration of all the objects. To access some of the interesting properties of the objects, we fall back on a set of attributes Q. In this way, we can represent each object $a \in \Omega$ by a vector $\mathbf{a} = (q_1(a), \ldots, q_n(a)) \in \widehat{\Omega} \subseteq \mathcal{X}$, where $\widehat{\Omega}$ is the set of all measurement vectors corresponding to objects in Ω. This also leads to a representation $\hat{\lambda}$ of the classification $\lambda : \Omega \to \mathcal{L}$ in the following way:

$$\hat{\lambda} : \widehat{\Omega} \to 2^{\mathcal{L}},$$
$$\mathbf{x} \mapsto \hat{\lambda}(\mathbf{x}) = \{\lambda(a) \mid a \in \Omega \wedge \mathbf{a} = \mathbf{x}\},$$

where $2^{\mathcal{L}}$ is the power set of $\mathcal{L}$, i.e. the set of all subsets of $\mathcal{L}$. Thus, the representation of a classification is again a classification, but now in the space $\widehat{\Omega} \subseteq \mathcal{X}$, with classes $\mathbf{C}_I := \hat{\lambda}^{-1}(I)$, where $I \subseteq \mathcal{L}$ (we may define a classification $\hat{\lambda}^* : \Omega \to 2^{\mathcal{L}}$ by setting $\hat{\lambda}^*(a) = \hat{\lambda}(\mathbf{a})$) [1]. Moreover, if Ω is isomorphic to $\mathcal{X}$, then $\hat{\lambda}$ is isomorphic to λ.

This property of isomorphism states that the representation $\hat{\lambda}$ is a very natural one. We certainly want to keep this property in the case of rankings. Moreover, the more general observation that representing a classification results again into a classification is also a very desirable property. We know that the real problem

---

[1] Since $\hat{\lambda}$ and $\hat{\lambda}^*$ both express the same idea, we will not restrain ourselves from mixing their usage.

is one of classification, but we will work with a representation, so it would be against our intuition that this representation would become something different from a classification. In the same line of thinking, we would like this property, if possible, to hold for rankings as well.

5   Representing a Ranking: Intervals

General Notation. As mentioned above, we would like to have that if $\Omega \cong \mathcal{X}$, then a representation of a ranking $(\lambda, \geq_{\mathcal{L}})$ should be a ranking once more. therefore, we will denote a **representation of a ranking** by

$$\boxed{(\lambda_{\mathrm{repr}}, \unrhd)}$$

where $\unrhd$ is some relation on the image of λ_{repr} that should be isomorphic to $\geq_{\mathcal{L}}$ if $\Omega \cong \mathcal{X}$.

In the remainder if this paper, the term "order" is used in its strict mathematical sense (i.e. a reflexive, antisymmetric, transitive binary relation). The term "ordering", however, is used more freely, conveying a semantical idea rather than a mathematical one.

Conditions on the Ordering $\unrhd$. There are some intuitive conditions we want to impose on the relation $\unrhd$. It should be (i) reflexive, to ensure that equal elements of the image of λ_{repr} are also treated equally, (ii) an extension of $\geq_{\mathcal{L}}$, as discussed previously, and (iii) meaningful, just as $\geq_{\mathcal{L}}$ has a meaning in terms of a preference relation. So, we would like to have an interpretation such as

$$\boxed{\lambda_{\mathrm{repr}}(\mathbf{a}) \unrhd \lambda_{\mathrm{repr}}(\mathbf{b}) \iff a\widehat{S}b} \tag{1}$$

where $a\widehat{S}b$ means: *based on the information derived from Q and λ_{repr}, we conclude that a is at least as good as b.* Lastly, closely related to the previous, we would like that (iv) $\unrhd$ does not depend on λ. Remark that condition (i) has an additional advantage since it enables us to interpret $\unrhd$ as a weak preference relation.

Next, we study a representation continuing with $\lambda_{\mathrm{repr}} = \hat{\lambda}$ in the spirit of the previous section. Later on, in Section 8, we will consider another possibility for λ_{repr}.

The Image of a Ranking. Let $(\lambda, \geq_{\mathcal{L}})$ be a (complete) ranking. A first remark concerns the range of $\hat{\lambda}$ when we are dealing with rankings. Because of the ordinal nature of the class labels, it is only meaningful to associate intervals to objects. Therefore, we define

$$\boxed{\begin{aligned} \hat{\lambda} : \widehat{\Omega} &\to \mathcal{L}^{[2]} = \{[r, r'] \mid (r, r') \in \mathcal{L}^2 \wedge r \leq_{\mathcal{L}} r'\}, \\ \mathbf{x} &\mapsto [\hat{\lambda}_\ell(\mathbf{x}), \hat{\lambda}_r(\mathbf{x})], \end{aligned}} \tag{2}$$

where

$$\hat{\lambda}_\ell(\mathbf{x}) = \min\{\lambda(a) \mid a \in \Omega \land \mathbf{a} = \mathbf{x}\},$$
$$\hat{\lambda}_r(\mathbf{x}) = \max\{\lambda(a) \mid a \in \Omega \land \mathbf{a} = \mathbf{x}\}.$$

Remark that if we write $I \in \mathcal{L}^{[2]}$ as an interval $[r, s]$, it holds by definition that

$$\mathbf{C}_I \neq \emptyset \implies (\exists a \in C_r)(\exists b \in C_s)(\mathbf{a} \in \mathbf{C}_I \land \mathbf{b} \in \mathbf{C}_I), \qquad (3)$$

Finding an Ordering of $\mathcal{L}^{[2]}$ (1). Starting from (ii) and (iii), we will try to generalise $\geq_\mathcal{L}$ while keeping in mind its underlying semantics. Recall that we have by definition of S derived from a ranking that

$$s \geq_\mathcal{L} r \iff (\forall a \in C_s)(\forall b \in C_r)(aSb)$$
$$\iff (\exists a \in C_s)(\exists b \in C_r)(aSb),$$

where the second equivalence is due to the fact that the objects inside one class are all considered to be indifferent to each other. We can reformulate these expressions as follows:

$$s \geq_\mathcal{L} r \iff (\forall(a, b) \in \Omega^2)((a, b) \in C_s \times C_r \Rightarrow aSb) \qquad (4)$$
$$\iff (\exists(a, b) \in \Omega^2)((a, b) \in C_s \times C_r \Rightarrow aSb). \qquad (5)$$

These expressions can easily be generalised by replacing $s \geq_\mathcal{L} r$ by $I \trianglerighteq I'$, where $I, I' \in \mathcal{L}^{[2]}$, and C_s (resp. C_r) by $\mathbf{C}_I \neq \emptyset$ (resp. by $\mathbf{C}_{I'} \neq \emptyset$). If we impose reflexivity, and write $I = [s_1, s_2]$, $I' = [r_1, r_2]$, this finally results in (using Expression (3))

$$[s_1, s_2] \trianglerighteq^1 [r_1, r_2] \iff s_1 \geq_\mathcal{L} r_2,$$

as a generalisation of (4), and

$$[s_1, s_2] \trianglerighteq^2 [r_1, r_2] \iff s_2 \geq_\mathcal{L} r_1,$$

as a generalisation of (5). Relation $\trianglerighteq^1$ is an order, and relation $\trianglerighteq^2$ is an interval order[2].

It is clear that these two relations fulfill conditions (i), (ii) and (iv). However, in both cases there are some problems with condition (iii). We have for instance that $[1, 3]$ and $[1, 2]$ are incomparable w.r.t. $\trianglerighteq^1$. However, we would prefer an object with label $[1, 3]$ over another with label $[1, 2]$ if that is all we know about these objects. Indeed, if an object is assigned to $[1, 2]$, it may belong to either of the classes C_1 and C_2, whereas an object assigned to $[1, 3]$ may also belong to the better class C_3. So, for two objects $a, b \in \Omega$, knowing only $\hat{\lambda}(\mathbf{a}) = [1, 3]$ and $\hat{\lambda}(\mathbf{b}) = [1, 2]$, we would prefer a over b, implying that we would like to have $[1, 3] \trianglerighteq [1, 2]$. This reasoning is in line with the semantics (1) we are pursuing. Whereas $\trianglerighteq^1$ proves to be too restrictive, the relation $\trianglerighteq^2$ seems to be too permissive, being indifferent between $[1, 2]$ and $[1, 3]$. This means that we need to find something in between the generalisations of (4) and (5).

---

[2] An **interval order** is a reflexive Ferrers (whence complete) relation. A relation R on X is called **Ferrers** if $(aRb \land cRd) \Rightarrow (aRd \lor cRb)$, for any $a, b, c, d \in X$. The Ferrers property can be seen as a relaxation of transitivity.

Finding an Ordering of $\mathcal{L}^{[2]}$ (2). There exist quite some different kinds of orders that could be possible candidates for $\trianglerighteq$, see e.g. [6]. Unfortunately, their known characterisations are not particularly helpful in pinpointing one or more of them in the present context. We could simply check them all out and see whether they suit our purpose, but at this point, it is more interesting to let us guide by the reasoning demonstrated in the previous paragraph, and to first state the desired semantics and translate it afterwards into a suitable expression.

In that way we define $I \trianglerighteq I'$ if and only if I is an improvement over I' or I' is a deterioration compared to I. We will now translate this into mathematical expressions. Assuming that $\mathbf{C}_I$ and $\mathbf{C}_{I'}$ are non-empty, we say that I is an **improvement** of I' if and only if

$$\begin{cases} (\exists a \in \Omega)(\forall b \in \Omega)((\mathbf{a},\mathbf{b}) \in \mathbf{C}_I \times \mathbf{C}_{I'} \Rightarrow aSb) \\ (\forall a \in \Omega)(\exists b \in \Omega)((\mathbf{a},\mathbf{b}) \in \mathbf{C}_I \times \mathbf{C}_{I'} \Rightarrow aSb) . \end{cases} \quad (6)$$

Likewise, I' is a **deterioration** of I if and only if

$$\begin{cases} (\exists b \in \Omega)(\forall a \in \Omega)((\mathbf{a},\mathbf{b}) \in \mathbf{C}_I \times \mathbf{C}_{I'} \Rightarrow aSb) \\ (\forall b \in \Omega)(\exists a \in \Omega)((\mathbf{a},\mathbf{b}) \in \mathbf{C}_I \times \mathbf{C}_{I'} \Rightarrow aSb) . \end{cases} \quad (7)$$

It immediately strikes that all of these expressions can be seen as intermediate to the generalisations of (4) and (5). If we write $I = [s_1, s_2]$ and $I' = [r_1, r_2]$, it can be shown quite easily that (remember that $\mathbf{C}_I$ and $\mathbf{C}_{I'}$ are assumed to be non-empty)

$$(6) \iff (7) \iff ((r_1 \leq_{\mathcal{L}} s_1) \wedge (r_2 \leq_{\mathcal{L}} s_2)) .$$

Hence, we find that $\trianglerighteq$ is an order, which we will denote by $\geq^{[2]}$. If $I, I' \in \mathcal{L}^{[2]}$ and we write $I = [r_1, r_2]$ and $I' = [s_1, s_2]$, then we have

$$[r_1, r_2] \leq^{[2]} [s_1, s_2] \iff ((r_1 \leq_{\mathcal{L}} s_1) \wedge (r_2 \leq_{\mathcal{L}} s_2)) .$$

So, based on a semantical discourse, we arrived at the order $\leq^{[2]}$, which is already extensively studied and whose properties are well known [5], for example, it turns $(\mathcal{L}^{[2]}, \leq^{[2]})$ into a complete lattice[3].

Some Afterthoughts. It is clear that now all four conditions are met. It should be noted that the order $\leq^{[2]}$ was derived from the premise that we only have access to intervals of values to reach a decision. If other information would be available, other orderings might prevail. For example, distributional information might lead to a stochastic ordering (see Section 8), or if risk aversion underlies the decision, then we could consider the leximin order $\leq_1$. Consider for example $\mathcal{L} = \{1, 2, 3, 4\}$, then we have

$$1 \leq_1 [1,2] \leq_1 [1,3] \leq_1 [1,4] \leq_1 2 \leq_1 [2,3] \leq_1 [2,4] \leq_1 3 \leq_1 [3,4] \leq_1 4.$$

---

[3] Potharst [15] also introduced this order in the setting of rankings, however, without being aware of its semantics in the context of ranking.

If risk would be favoured, the following order $\leq_2$ might be suitable:

$$1 \leq_2 [1,2] \leq_2 2 \leq_2 [1,3] \leq_2 [2,3] \leq_2 3 \leq_2 [1,4] \leq_2 [2,4] \leq_2 [3,4] \leq_2 4.$$

Note that both situations are in line with the order $\leq^{[2]}$ just defined ($\leq^{[2]} \subseteq \leq_i$). Even more, the order $\leq^{[2]}$ is nothing else but the intersection of $\leq_1$ and $\leq_2$. This also pleads for the non-invasive character of the order $\leq^{[2]}$ (no presuppositions about the preferences are imposed).

6   The Monotonicity Constraint

Introduction. Up to now, we have given a definition of a (complete) ranking $(\lambda, \geq_{\mathcal{L}})$ and have shown how it can be represented by the (not necessarily complete) ranking $(\hat{\lambda}, \geq^{[2]})$. Note that we did not need any form of monotonicity for the definition or the representation. Monotonicity will arise in a natural way when taking into account the attributes used to describe the properties of objects.

Preliminaries. Let us first turn back to classifications. Even if we assume a classification λ to be deterministic in the sense that any object $a \in \Omega$ is assigned to exactly one class with label in $\mathcal{L}$, we still cannot guarantee that we have $|\hat{\lambda}^*(a)| = 1$ for all $a \in \Omega$. This is a consequence of the possible occurrence of *doubt* (called "inconsistency" in rough set theory [11, 18]).

Definition 2.

(i) *There is **doubt** between the classification λ and the set of attributes Q if*

$$(\exists (a,b) \in \Omega^2)(\mathbf{a} = \mathbf{b} \wedge \lambda(a) \neq \lambda(b)).$$

(ii) *There is **doubt** inside the representation $\hat{\lambda}$ if*

$$(\exists \mathbf{x} \in \widehat{\Omega})(|\hat{\lambda}(\mathbf{x})| > 1).$$

It is clear that these two notions of doubt coincide. In the case of doubt, the function $\hat{\lambda}^*$ can assign a set of labels to an object, indicating that it is not possible to label the object with one specific class label based on the associated vector. Remark that the first definition emphasises the conflict between the vector representations of the objects and the classification λ (as discussed in Section 2), the second definition stresses the ensuing idea of uncertainty. Also note that if $\hat{\lambda}$ is interpreted as a classification, we have that there is never doubt between the classification $\hat{\lambda}$ and the (corresponding) set of attributes Q.

Multicriteria Decision Aid (MCDA). In the context of ranking, the attributes have a specific interpretation, and are usually referred to as criteria. A **criterion** [16] is defined as a mapping $c : \Omega \to (\mathcal{X}_c, \geq_c)$, where $(\mathcal{X}_c, \geq_c)$ is a chain, such that it appears meaningful to compare two objects a and b, according to a particular point of view, on the sole basis of their evaluations $c(a)$ and $c(b)$. In this paper, we will only consider **true criteria** [4], where the induced weak preference relation is a complete preorder defined by $aS_c b \Leftrightarrow c(a) \geq_c c(b)$. We assume to have a finite set $C = \{c_1, \ldots, c_n\}$ at our disposal.

The **dominance relation**[4] $\triangleright_C$ on Ω w.r.t. C is defined by

$$a \triangleright_C b \iff \begin{cases} (\forall c \in C)(aS_c b) \\ (\exists c \in C)(aP_c b) \end{cases}$$

for any $a, b \in \Omega$. It is said that a **dominates** b. We may also write $b \triangleleft_C a$, saying that b **is dominated by** a. We say that a **weakly** dominates b, $a \trianglerighteq b$, if only $(\forall c \in C)(aS_c b)$. Since we are working with true criteria we have that $a \trianglerighteq_C b$ is equivalent with $\mathbf{a} \geq_\chi \mathbf{b}$.

Monotonicity. A basic principle stemming from MCDA [16] is that $a \trianglerighteq_C b \Rightarrow aSb$. On the other hand we have that $aSb \iff \lambda(a) \geq_\mathcal{L} \lambda(b)$. Merging all these expressions we find in a natural way the monotonicity constraint

$$\mathbf{a} \geq_\chi \mathbf{b} \Rightarrow \lambda(a) \geq_\mathcal{L} \lambda(b).$$

Remark that this constraint does not tolerate the presence of doubt since it advocates $\mathbf{a} = \mathbf{b} \Rightarrow \lambda(a) = \lambda(b)$. Thus, it is too restrictive for applications in supervised learning. The reason for this lies in the fact that we have adopted a principle from MCDA without considering its context: build a ranking based on the set C of criteria. This is a different setting than for supervised learning where we try to reconstruct a ranking based on the set C. In the former, the set C is a framework, in the latter, this same set C is a restriction. We can solve this problem by applying the same principle but with the additional demand that we restrict our knowledge to the information we can retrieve from C. In that case, we define the dominance relation on $\widehat{\Omega} \subseteq \mathcal{X}$, resulting in $\mathbf{x} \trianglerighteq \mathbf{y}$ if and only if $\mathbf{x} \geq_\chi \mathbf{y}$, and the principle becomes $\mathbf{x} \trianglerighteq \mathbf{y} \Rightarrow \mathbf{x}\hat{S}\mathbf{y}$. Together with (1), this finally leads to the monotonicity constraint

$$\mathbf{x} \geq_\chi \mathbf{y} \Rightarrow \lambda_{\mathrm{repr}}(\mathbf{x}) \trianglerighteq \lambda_{\mathrm{repr}}(\mathbf{y}).$$

In this case, doubt is tolerated since $\mathbf{x} = \mathbf{y} \Rightarrow \lambda_{\mathrm{repr}}(\mathbf{x}) = \lambda_{\mathrm{repr}}(\mathbf{y})$ is a trivial demand. This also means that the monotonicity constraint reduces to

$$\boxed{\mathbf{x} >_\chi \mathbf{y} \Rightarrow \lambda_{\mathrm{repr}}(\mathbf{x}) \trianglerighteq \lambda_{\mathrm{repr}}(\mathbf{y}).} \tag{8}$$

We can now adapt the two equivalent definitions of Definition 2, which leads us to two different notions:

---

[4] In the literature, the dominance relation is usually denoted by Δ_C. Because of the symmetrical nature of the symbol Δ_C, we feel it does not clearly denote its meaning and prefer to use the notation $\triangleright_C$.

Definition 3.

(i) *There is* **reversed preference** *between the ranking* $(\lambda, \geq_{\mathcal{L}})$ *and the set of criteria* C *if*
$$(\exists(a,b) \in \Omega^2)(\mathbf{a} >_{\mathcal{X}} \mathbf{b} \wedge \lambda(a) \not\geq_{\mathcal{L}} \lambda(b)).$$

(ii) *There is* **reversed preference** *inside the representation* $(\lambda_{repr}, \trianglerighteq)$ *if*
$$(\exists(\mathbf{x},\mathbf{y}) \in \widehat{\Omega}^2)(\mathbf{x} >_{\mathcal{X}} \mathbf{y} \wedge \lambda_{repr}(\mathbf{x}) \not\trianglerighteq \lambda_{repr}(\mathbf{y})).$$

A representation of a ranking is said to be **consistent** *if there is no reversed preference inside it.*

These two notions do not coincide, we must make a clear distinction between the first definition that considers inconsistencies in a ranking, and the second definition that considers inconsistencies in a representation of a ranking.

About Doubt and Reversed Preference. There is a major difference between the existence of reversed preference in a ranking, and the existence of doubt. People can accept doubt in a classification, but they will not accept reversed preference in a ranking. For example, you might accept that it is difficult to choose between two candidates, either because you feel you don't have enough information about them, or you feel they are too similar to differentiate, e.g. athlete A wins the first 5 challenges in a decathlon, and ends second in the other 5, while athlete B ends second in the first 5, but wins the remaining 5. On the other hand, it is never tolerated that the candidate with the lower marks ends up in a higher rank than the candidate with the better marks, e.g. athlete A always ends in the fourth or fifth place, while athlete B is always among the first three, then it is unacceptable that A would end up in a higher position on the league table than B.

This is nicely reflected in the definitions: there is never doubt between the classification $\hat{\lambda}$ and the corresponding set of attributes Q, but there can exist reversed preference between the ranking $(\hat{\lambda}, \leq^{[2]})$ and the corresponding set of criteria C. It is also the reason why we introduced the term **consistency** instead of **monotonicity**. While these two notions are clearly equivalent, "consistency" conveys the semantical idea behind the property of monotonicity of the representation.

7 Transforming Reversed Preference into Doubt: The Consistent Representation

Introduction. As just mentioned, the occurrence of reversed preference in $\hat{\lambda}$ is not satisfactory (even unacceptable). This can be solved by redefining λ (or in terms of supervised learning, altering the training data) in such a way that the reversed preferences disappear. A very drastic solution could be to demand that $\mathbf{a} <_{\mathcal{X}} \mathbf{b} \Rightarrow \lambda(a) \leq_{\mathcal{L}} \lambda(b)$ for the training data. A less drastic one, is to find a maximally consistent subset by eliminating some of the data. Another possibility (see below) is to redefine C until the resulting $\hat{\lambda}$ behaves monotonically according to (8). All these proposals have an invasive character, and might even be unfeasible in certain circumstances. We therefore propose another, non-invasive method, that uses all available information, and results in the closest possible consistent representation by defining a mapping $\tilde{\lambda}$ such that $(\tilde{\lambda}, \leq^{[2]})$ does no longer contain reversed preferences.

Sources of Reversed Preference. We begin this section by an enumeration of how reversed preferences can arise in the ranking problem. In this case, the classification λ on Ω (or rather on a finite sample $S \subseteq \Omega$) and the order $\leq_{\mathcal{L}}$ on $\mathcal{L}$ are furnished by a Decision Maker (DM). The DM also gives the set of criteria C to be considered. We first focus on reversed preference between the ranking $(\lambda, \leq_{\mathcal{L}})$ in Ω and the set of criteria C, i.e. there exist objects $a, b \in S$ such that

$$\mathbf{a} <_{\mathcal{X}} \mathbf{b} \wedge \lambda(a) \not\leq_{\mathcal{L}} \lambda(b).$$

Now assuming that reversed preference occurs, there are several scenarios possible:

1. The DM has based his decisions on the set C:
   (a) The DM has used some additional information not present in C. A solution could be to find the criteria $c_1, \ldots, c_r \notin C$ the DM uses and add them to C.
   (b) The DM has made an inconsistent decision. In this case, the mapping λ might be redefined in a consistent way by the DM. If this is not a possibility (e.g. the DM has no time, the samples are taken from past decisions, $\ldots$), both C and λ must remain unchanged. We will show next how we can deal with this case.
   (c) It is agreed that C is the final set of criteria to be considered. Again we have to keep C and $(\lambda, \leq_{\mathcal{L}})$, including the conflict between them.
2. The DM has made his decisions before the set C was defined:
   (a) Some meaningful criteria $c_1, \ldots, c_r \notin C$ were missed in the first attempt to construct the set C, such that $(\exists c \in \{c_1, \ldots, c_r\})(a \neg S_c b)$. This means that a does no longer dominate b w.r.t. $C \cup \{c_1, \ldots, c_r\}$.
   (b) It is agreed that C is the final set of criteria to be considered. We are in the same situation as case (1c).

Since for practical purposes we use the representation of the ranking, it is not really necessary to solve the problems just stated. Indeed, we must only take care of reversed preference inside the representation $(\hat{\lambda}, \leq^{[2]})$ because of the following proposition:

Proposition 1. *Substituting the ranking* $(\lambda, \leq_{\mathcal{L}})$ *by its representation* $(\hat{\lambda}, \leq^{[2]})$

(i) *might eliminate existing reversed preference (i.e. if* $\mathbf{a} <_{\mathcal{X}} \mathbf{b}$ *and* $\lambda(a) \not\leq_{\mathcal{L}} \lambda(b)$, *then it can happen that* $\hat{\lambda}(\mathbf{a}) \trianglelefteq \hat{\lambda}(\mathbf{b})$.)
(ii) *will never introduce new reversed preferences.*

Proof.
We will illustrate the first assertion with a little example. Consider a ranking $(\lambda, \leq_{\mathcal{L}})$ with $\mathcal{L} = \{1, 2, 3\}$, $1 <_{\mathcal{L}} 2 <_{\mathcal{L}} 3$, $\{a, b, c, d\} \subseteq \Omega$ and $\mathbf{a} = \mathbf{b} = \mathbf{x} <_{\mathcal{X}} \mathbf{c} = \mathbf{d} = \mathbf{y}$. Furthermore, assume that $\lambda(a) = 1, \lambda(b) = 3, \lambda(c) = 2, \lambda(d) = 3$. This means b and c give rise to reversed preference. However, we have $\hat{\lambda}(\mathbf{x}) = [1, 3] \leq^{[2]}$ $\hat{\lambda}(\mathbf{y}) = [2, 3]$ and thus $\mathbf{x}$ and $\mathbf{y}$ do not give rise to any reversed preferences.

To prove the second assertion, consider a ranking $(\lambda, \leq_{\mathcal{L}})$. For $a_1, a_2, b_1, b_2 \in \Omega$, we assume there is doubt: $\hat{a}_1 = \hat{a}_2 = \mathbf{x}$ and $\hat{b}_1 = \hat{b}_2 = \mathbf{y}$. We put $\lambda(a_1) = r_1, \lambda(a_2) = r_2, \lambda(b_1) = s_1, \lambda(b_2) = s_2$. Without restrictions we may assume that $r_1 \leq_{\mathcal{L}} r_2$, $s_1 \leq_{\mathcal{L}} s_2$, $\mathbf{x} \leq_{\mathcal{X}} \mathbf{y}$, $\hat{\lambda}(\mathbf{x}) = [r_1, r_2]$ and $\hat{\lambda}(\mathbf{y}) = [s_1, s_2]$. Now suppose there is no reversed preference between $\mathbf{x}$ and $\mathbf{y}$ w.r.t. $(\lambda, \leq_{\mathcal{L}})$ and C, i.e. $(\forall a \in \Omega)(\hat{a} = \mathbf{x})$ and $(\forall b \in \Omega)(\hat{b} = \mathbf{y})$ we have $\lambda(a) \leq_{\mathcal{L}} \lambda(b)$. This means we must have that $r_2 \leq_{\mathcal{L}} s_1$, which automatically leads to $\hat{\lambda}(\mathbf{x}) = [r_1, r_2] \leq^{[2]} \hat{\lambda}(\mathbf{y}) = [s_1, s_2]$.
$\square$

Dealing with Reversed Preference. The above proposition implies that if there is reversed preference inside the representation $(\hat{\lambda}, \leq^{[2]})$, it must have its origins in one of the situations described above. If both λ and C should remain unchanged, as in cases (1b), (1c) and (2b), we may transform $\hat{\lambda}$ into a mapping $\tilde{\lambda}$ such that $(\tilde{\lambda}, \leq^{[2]})$ does no longer contain reversed preference. This transformation should stay as close as possible to the original $\hat{\lambda}$, and if there is no reversed preference in $(\hat{\lambda}, \leq^{[2]})$, then $\tilde{\lambda}$ should be equal to $\hat{\lambda}$.

We will now show how this transformation can be done. Assume there is reversed preference inside the representation $(\hat{\lambda}, \leq_I)$, meaning we can find $\mathbf{x}, \mathbf{y} \in \widehat{\Omega}$ such that

$$\mathbf{x} \leq_{\mathcal{X}} \mathbf{y} \text{ and } \hat{\lambda}(\mathbf{x}) = [r_1, r_2] \not\leq^{[2]} \hat{\lambda}(\mathbf{y}) = [s_1, s_2].$$

Because of expression (3), we can never make the interval $[r_1, r_2]$ or $[s_1, s_2]$ smaller without removing an object from the sample space. This would result in neglecting the information that does not fit our formalisation. Thus, removing objects cannot be defended. As a consequence, we may only enlarge the intervals in order to remove the inconsistency. To stay as close as possible to the original (inconsistent) information, we will enlarge the intervals in a minimal way to eliminate the reversed preference. We have

$$[r_1, r_2] \not\leq^{[2]} [s_1, s_2] \iff (r_1 >_{\mathcal{L}} s_1) \vee (r_2 >_{\mathcal{L}} s_2),$$

and we must try to find intervals $[\tilde{r}_1, \tilde{r}_2] \supseteq [r_1, r_2]$ and $[\tilde{s}_1, \tilde{s}_2] \supseteq [s_1, s_2]$ such that $[\tilde{r}_1, \tilde{r}_2] \leq^{[2]} [\tilde{s}_1, \tilde{s}_2]$ or still $(\tilde{r}_1 \leq_{\mathcal{L}} \tilde{s}_1) \wedge (\tilde{r}_2 \leq_{\mathcal{L}} \tilde{s}_2)$.

(i) $s_1 <_{\mathcal{L}} r_1$. We may only resolve this by reducing s_1 to some $\tilde{s}_1$ and r_1 to some $\tilde{r}_1$. It is obvious that we can turn the inequality into an equality if we put $\tilde{r}_1 := s_1$ and keep $\tilde{s}_1 := s_1$. Moreover, this is the smallest[5] change possible in order to obtain $\tilde{r}_1 \leq_{\mathcal{L}} \tilde{s}_1$ with $\tilde{r}_1 \leq_{\mathcal{L}} r_1$ and $\tilde{s}_1 \leq_{\mathcal{L}} s_1$.

---

[5] There is only one acceptable distance measure on an ordinal scale, namely the distance in the underlying graph.

(ii) $s_2 <_{\mathcal{L}} r_2$. The smallest possible change to obtain $\tilde{r}_2 \leq_{\mathcal{L}} \tilde{s}_2$ with $\tilde{r}_2 \geq_{\mathcal{L}} r_2$ and $\tilde{s}_2 \geq_{\mathcal{L}} s_2$, is putting $\tilde{r}_2 := r_2$ and $\tilde{s}_2 := r_2$.

So, we may eliminate this reversed preference by subjecting $\hat{\lambda}$ to the transformation t:

$$t(\hat{\lambda}(\mathbf{x})) = t([\hat{\lambda}_\ell(\mathbf{x}), \hat{\lambda}_r(\mathbf{x})]) := [\min(\hat{\lambda}_\ell(\mathbf{x}), \hat{\lambda}_\ell(\mathbf{y})), \hat{\lambda}_r(\mathbf{x})],$$

$$t(\hat{\lambda}(\mathbf{y})) = t([\hat{\lambda}_\ell(\mathbf{y}), \hat{\lambda}_r(\mathbf{y})]) := [\hat{\lambda}_\ell(\mathbf{y}), \max(\hat{\lambda}_r(\mathbf{x}), \hat{\lambda}_r(\mathbf{y}))].$$

Moreover, if there was no reversed preference, this transformation would just yield

$$t(\hat{\lambda}(\mathbf{x})) = \hat{\lambda}(\mathbf{x}) \quad \text{and} \quad t(\hat{\lambda}(\mathbf{y})) = \hat{\lambda}(\mathbf{y}).$$

It should be noted that this transformation may create new reversed preferences. The previous procedure must therefore be repeated until no more reversed preferences exist. It is clear that in the present finite setting, there is convergence, in the worst case we would end up with the constant mapping to $[\inf \mathcal{L}, \sup \mathcal{L}]$.

Proposition 2. *Let* $(\lambda, \leq_{\mathcal{L}})$ *be a ranking in* Ω, *and let* C *be a set of criteria. Now define*

$$\tilde{\lambda} : \widehat{\Omega} \to \mathcal{L}^{[2]},$$
$$\mathbf{x} \mapsto [\min_{\mathbf{y} \in [\mathbf{x}]} \hat{\lambda}_\ell(\mathbf{y}), \max_{\mathbf{y} \in (\mathbf{x}]} \hat{\lambda}_r(\mathbf{y})],$$

where $[\mathbf{x}) = \{\mathbf{x}' \in \widehat{\Omega} \mid \mathbf{x} \leq_{\mathcal{X}} \mathbf{x}'\}$ *and* $(\mathbf{x}] = \{\mathbf{x}' \in \widehat{\Omega} \mid \mathbf{x}' \leq_{\mathcal{X}} \mathbf{x}\}$. *We have that*

(i) *The representation* $(\tilde{\lambda}, \leq^{[2]})$ *is consistent with* C.
(ii) *For all* $\mathbf{x} \in \mathcal{X}$, *it holds that* $\hat{\lambda}(\mathbf{x}) \subseteq \tilde{\lambda}(\mathbf{x})$, *and if* $(\hat{\lambda}, \leq^{[2]})$ *is consistent, then* $\tilde{\lambda} = \hat{\lambda}$.
(iii) *There exists no other representation* $(\overline{\lambda}, \leq^{[2]})$ *consistent with* C *such that for all* $\mathbf{x} \in \mathcal{X}$, *it holds that* $\hat{\lambda}(\mathbf{x}) \subseteq \overline{\lambda}(\mathbf{x}) \subseteq \tilde{\lambda}(\mathbf{x})$.

Proof.
(i) Consider $\mathbf{x}, \mathbf{y} \in \widehat{\Omega}$ such that $\mathbf{x} \leq_{\mathcal{X}} \mathbf{y}$. This implies that $\mathbf{x} \in (\mathbf{y}]$ and $\mathbf{y} \in [\mathbf{x})$, and so $(\mathbf{x}] \subseteq (\mathbf{y}]$ and $[\mathbf{y}) \subseteq [\mathbf{x})$. From this it immediately follows that

$$\tilde{\lambda}(\mathbf{x}) = [\min_{\mathbf{z} \in [\mathbf{x})} \hat{l}(\mathbf{z}), \max_{\mathbf{z} \in (\mathbf{x}]} \hat{r}(\mathbf{z})] \leq^{[2]} \tilde{\lambda}(\mathbf{y}) = [\min_{\mathbf{z} \in [\mathbf{y})} \hat{l}(\mathbf{z}), \max_{\mathbf{z} \in (\mathbf{y}]} \hat{r}(\mathbf{z})].$$

(ii) Self-evident.
(iii) Let $\overline{\lambda}(\neq \tilde{\lambda})$ be such that for all $\mathbf{x} \in \mathcal{X}$, it holds that $\hat{\lambda}(\mathbf{x}) \subseteq \overline{\lambda}(\mathbf{x}) \subseteq \tilde{\lambda}(\mathbf{x})$. Remark that, because of (ii), this is only possible if $(\hat{\lambda}, \leq^{[2]})$ is not consistent. We have that there exists at least one $\mathbf{x} \in \mathcal{X}$ such that $\overline{\lambda}(\mathbf{x}) = [l, r] \subset \tilde{\lambda}(\mathbf{x})$. Assume that $l > \min_{\mathbf{y} \in [\mathbf{x})} \hat{\lambda}_\ell(\mathbf{y})$. Evidently, this implies there exist a non-empty set Y of $\mathbf{y} >_{\mathcal{X}} \mathbf{x}$ such that $\hat{\lambda}_\ell(\mathbf{y}) = i <_{\mathcal{L}} \hat{\lambda}_\ell(\mathbf{x}) = j$. Let $\mathbf{y}^* \in \arg\min_{\mathbf{y} \in Y} \hat{\lambda}_\ell(\mathbf{y})$, and $i^* = \hat{\lambda}_\ell(\mathbf{y}^*)$. Since $(\forall \mathbf{z} \in \mathcal{X})(\hat{\lambda}(\mathbf{z}) \subseteq \overline{\lambda}(\mathbf{z}))$, we have that $j \in \overline{\lambda}(\mathbf{x})$ and $i^* \in \overline{\lambda}(\mathbf{y}^*)$. If $(\hat{\lambda}, \leq^{[2]})$ is to be consistent, it must hold that $i^* \in \overline{\lambda}(\mathbf{x})$. By construction, we have that $i^* = \min_{\mathbf{y} \in [\mathbf{x})} \hat{\lambda}_\ell(\mathbf{y}) < l$, a contradiction. $\square$

Essentially, we have enlarged the intervals in a minimal way such that there are no more violations against the monotonicity requirement (8). In other words, we transform the unacceptable reversed preferences into acceptable doubt.

Example 1. Let us demonstrate this on a small example taken from [8]. Assume we have $\Omega = \{a_1, \ldots, a_6\}$, a single criterion $c : \Omega \rightarrow (\{1, \ldots, 6\}, \leq)$ and a ranking $(\lambda, \leq)$ with $f : \Omega \rightarrow \{1, \ldots 6\}$, as shown in Table 1. There is no doubt, so $\Omega \cong \widehat{\Omega}$. Table 2 lists the consistent representation of this ranking.

Table 1. A simple ranking $(\lambda, \leq)$.

	a_1	a_2	a_3	a_4	a_5	a_6
c	2	1	4	3	6	5
λ	1	2	3	4	5	6

Table 2. The consistent representation $(\tilde{\lambda}, \leq^{[2]})$ of $(\lambda, \leq)$.

	a_2 $\leq_X$	a_1 $\leq_X$	a_4 $\leq_X$	a_3 $\leq_X$	a_6 $\leq_X$	a_5
$\hat{\lambda}$	2	1	4	3	6	5
$\tilde{\lambda}$	$[1,2] \leq^{[2]}$	$[1,2] \leq^{[2]}$	$[3,4] \leq^{[2]}$	$[3,4] \leq^{[2]}$	$[5,6] \leq^{[2]}$	$[5,6]$

8   Representing a Ranking: Distributions

Introduction. Up to now, we have focussed on classifiers (functions with domain $\widehat{\Omega}$) that return the possible values that can be assigned to each object. In this section, we go one step further and consider probabilistic classifiers $\hat{\lambda}_{\text{prob}}$ assigning to each object a probability distribution over the labels. (In the case Ω is finite, this is easily achieved by normalising the frequency distributions associated with the elements of $\widehat{\Omega}$.)

Stochastic Dominance. As before, we need to establish an order on the set $\mathcal{F}(\mathcal{L})$ of all possible distributions[6] over $\mathcal{L}$. In the present context, the ordering that comes immediately to mind is the stochastic dominance ordering. Let $f_1, f_2 \in \mathcal{F}(\mathcal{L})$, and denote their cumulative distribution functions as F_1 and F_2. **Weak (first order) stochastic dominance** $\unrhd_{(1)}$ is defined by

$$f_1 \unrhd_{(1)} f_2 \iff (\forall i \in \mathcal{L})(F_1(i) \leq F_2(i)).$$

Remark that if we consider the support $\text{Supp}(f) = \{\ell \in \mathcal{L} \mid f(\ell) > 0\}$, then

$$f_1 \unrhd_{(1)} f_2$$
$$\Downarrow$$
$$[\min \text{Supp}(f_1), \max \text{Supp}(f_1)] \geq^{[2]} [\min \text{Supp}(f_2), \max \text{Supp}(f_2)].$$

---

[6] If $\mathcal{L} = \{1, \ldots, k\}$, then we denote a distribution $f_X \in \mathcal{F}(\mathcal{L})$ as a vector of dimension k: $f_X = (\mathcal{P}(X = 1), \ldots, \mathcal{P}(X = k))$.

Of course, the converse does not hold, e.g. for $\mathcal{L} = \{1, 2, 3\}$, $f_1 = (0.4, 0.4, 0.2)$ and $f_2 = (0.2, 0.8, 0)$.

Along the same line, we have the following lemma:

Lemma 1. *If we interpret the intervals of $\mathcal{L}^{[2]}$ in the previous sections as uniform distributions, the order $\leq^{[2]}$ just coincides with $\unlhd_{(1)}$.*

Meaningful Representations. For a representation to be meaningful, it should at least make the second assertion of Proposition 1 true. Because of its importance, we will look for a class of representations that can guarantee this assertion. This can easily be done by imposing a kind of minimal consistency on the representation $(\lambda_{\mathrm{repr}}, \unlhd)$.

Definition 4. *We say that the representation $(\lambda_{\mathrm{repr}}, \unlhd)$ of $(\lambda, \leq_{\mathcal{L}})$ is **minimally consistent** in $U \subseteq \mathcal{X}$, if for all $\mathbf{x}, \mathbf{y} \in U$ with $\mathbf{x} \leq_{\mathcal{X}} \mathbf{y}$, it holds that*

$$\hat{\lambda}_r(\mathbf{x}) \leq_{\mathcal{L}} \hat{\lambda}_\ell(\mathbf{y}) \implies \lambda_{\mathrm{repr}}(\mathbf{x}) \unlhd \lambda_{\mathrm{repr}}(\mathbf{y}).$$

Proposition 3. *Substituting the ranking $(\lambda, \leq_{\mathcal{L}})$ by a representation $(\lambda_{\mathrm{repr}}, \unlhd)$ that is minimally consistent in $\widehat{\Omega}$ will never introduce new reversed preferences.*

Proof.
Similar to Proposition 1. □

It is clear that $(\hat{\lambda}_{\mathrm{prob}}, \unlhd_{(1)})$ satisfies Definition 4. Moreover, also the first part of Proposition 1 can be kept using the same proof, but this cannot be generalised for all $(\lambda_{\mathrm{repr}}, \unlhd)$.

A Consistent Representation. In this new setting, we aim once more at finding a representation $(\hat{\lambda}_{\mathrm{prob}}, \unlhd_{(1)})$ without reversed preferences, and without reducing the support of the distributions (which is equivalent to not removing any objects as advocated in Section 7). Consider an $\mathbf{x} \in \widehat{\Omega}$, with $\hat{\lambda}_{\mathrm{prob}}(\mathbf{x}) = f_{\mathbf{x}}$. We know that for all $\mathbf{y} \leq_{\mathcal{X}} \mathbf{x}$ (and of course $\mathbf{y} \in \widehat{\Omega}$), it should hold that $f_{\mathbf{y}} \unlhd_{(1)} f_{\mathbf{x}}$ or $F_{\mathbf{y}}(i) \geq F_{\mathbf{x}}(i)$, and therefore

$$F_{\mathbf{x}}(i) \leq \min_{\mathbf{y} \in (\mathbf{x}]} F_{\mathbf{y}}(i).$$

At the same time, for all $\mathbf{y} \geq_{\mathcal{X}} \mathbf{x}$, it should hold that $f_{\mathbf{x}} \unlhd_{(1)} f_{\mathbf{y}}$, or

$$F_{\mathbf{x}}(i) \geq \max_{\mathbf{y} \in [\mathbf{x})} F_{\mathbf{y}}(i).$$

However, it may very well happen (if $(\hat{\lambda}_{\mathrm{prob}}, \unlhd_{(1)})$ contains reversed preferences) that $\min_{\mathbf{y} \in (\mathbf{x}]} F_{\mathbf{y}}(i) < \max_{\mathbf{y} \in [\mathbf{x})} F_{\mathbf{y}}(i)$, for example: with $\widehat{\Omega} = \{\mathbf{x}, \mathbf{y}\}$, $\mathbf{x} \leq_{\mathcal{X}} \mathbf{y}$, $\mathcal{L} = \{1, 2\}$, $f_{\mathbf{x}} = (0, 1)$ and $f_{\mathbf{y}} = (1, 0)$, we find the constraints $F_{\mathbf{x}}(1) \leq 0$ and $F_{\mathbf{x}}(1) \geq 1$.

Proposition 4. *For all* $\mathbf{y} \in \widehat{\Omega}$, *we denote* $\hat{\lambda}_{prob}(\mathbf{y}) = f_{\mathbf{y}}$. *Let* $s \in [0, 1]$. *For all* $\mathbf{x} \in \widehat{\Omega}$, *for all* $i \in \mathcal{L}$, *we define:*

$$\tilde{F}_{\mathbf{x}}(i) := (1 - s) \min_{\mathbf{y} \in (\mathbf{x}]} F_{\mathbf{y}}(i) + s \max_{\mathbf{y} \in [\mathbf{x})} F_{\mathbf{y}}(i).$$

If we set $\tilde{\lambda}_{prob}(\mathbf{x}) = \tilde{f}_{\mathbf{x}}$, *then* $(\tilde{\lambda}_{prob}, \trianglelefteq_{(1)})$ *is consistent.*

Proof.
Consider $\mathbf{x}, \mathbf{y} \in \widehat{\Omega}$, with $\mathbf{x} \leq_{\chi} \mathbf{y}$. We need to prove that $\tilde{f}_{\mathbf{x}} \trianglelefteq_{(1)} \tilde{f}_{\mathbf{y}}$.

We know that $(\mathbf{x}] \subseteq (\mathbf{y}]$ and $[\mathbf{x}) \supseteq [\mathbf{y})$. Let $i \in \mathcal{L}$. We have

$$\min_{\mathbf{z} \in (\mathbf{x}]} F_{\mathbf{z}}(i) \geq \min_{\mathbf{z} \in (\mathbf{y}]} F_{\mathbf{z}}(i),$$

$$\max_{\mathbf{z} \in [\mathbf{x})} F_{\mathbf{z}}(i) \geq \max_{\mathbf{z} \in [\mathbf{y})} F_{\mathbf{z}}(i).$$

Hence $\tilde{F}_{\mathbf{x}}(i) \geq \tilde{F}_{\mathbf{y}}(i)$. □

The proposed family of solutions to the problem is certainly not the sole one. This in severe contrast with the unique solution put forward in Proposition 2.

Example 2. A small example demonstrating the above proposition is given in Table 3.

Table 3. The consistent representation $(\tilde{\lambda}_{prob}, \trianglelefteq_{(1)})$ of $(\lambda, \leq)$ using $s = \frac{1}{2}$.

	$\mathbf{x}_1$	$\leq_{\chi}$	$\mathbf{x}_2$	$\leq_{\chi}$	$\mathbf{x}_3$
$\hat{\lambda}_{prob}(\mathbf{x}) = f_{\mathbf{x}}$	$(0.4, 0.4, 0.2)$		$(0.2, 0.8, 0.0)$		$(0.2, 0.3, 0.5)$
$F_{\mathbf{x}}$	$(0.4, 0.8, 1.0)$		$(0.2, 1.0, 1.0)$		$(0.2, 0.5, 1.0)$
$\tilde{F}_{\mathbf{x}}$	$(0.4, 0.9, 1.0)$		$(0.2, 0.9, 1.0)$		$(0.2, 0.5, 1.0)$
$\tilde{\lambda}_{prob}(x)$	$(0.4, 0.5, 0.1)$	$\trianglelefteq_{(1)}$	$(0.2, 0.7, 0.1)$	$\trianglelefteq_{(1)}$	$(0.2, 0.3, 0.5)$

9   Conclusion and Future Research

We have paved the path to deal in a mathematically and semantically sound way with rankings in the context of supervised learning. We have given a definition, a representation when dealing with attributes (in which case we are on the territory of ordinal regression), and a consistent interval and stochastic representation when dealing with criteria.

Although our main interest was in trying to find acceptable representations that could be presented to a DM, it appears that Proposition 4 is also a good basis for a lazy learning method, providing an alternative to OLM, the Ordinal Learning Method [2]. We are currently experimenting with several variations of such a lazy learner, undertaking extensive comparisons with other learning methods.

References

1. A. Agresti, **An introduction to categorical data analysis**, John Wiley & Sons, New York, 1996.
2. A. Ben-David, **Automatic generation of symbolic multiattribute ordinal knowledgebased dss: Methodology and applications**, Decision Sciences **23** (1992), 1357–1372.
3. _____, **Monotonicity maintenance in information-theoretic machine learning algorithms**, Machine Learning **9** (1995), 29–43.
4. D. Bouyssou, **Building criteria: a prerequisite for MCDA**, Readings in Multiple Criteria Decision Aid (C. Bana e Costa, ed.), Springer-Verlag, Heidelberg, 1990, pp. 58–80.
5. P.C. Fishburn, **Interval orders and interval graphs**, John Wiley, 1985.
6. _____, **Generalizations of semiorders: A review note**, Journal of Mathematical Psychology **41** (1997), 357–366.
7. E. Frank and M. Hall, **A simple approach to ordinal classification**, Lecture Notes in Computer Science **2167** (2001), 145–56.
8. G. Gediga and I. Düntsch, **Approximation quality for sorting rules**, Computational Statistics and Data Analysis **40** (2002), 499–526.
9. S. Greco, B. Mattarazo, and S. Słowiński, **Rough set theory for multicriteria decision analysis**, European Journal of Operational Research **129** (2001), 1–47.
10. R. Herbrich, T. Graepel, and K. Obermayer, **Large margin rank boundaries for ordinal regression**, Advances in Large Margin Classifiers (A. Smola, P. Bartlett, B. Schölkopf, and D. Schuurmans, eds.), MIT Press, Cambridge, MA, 2000, pp. 115–132.
11. J. Komorowski, L. Polkowski, and A. Skowron, **Rough sets: a tutorial**, Rough-Fuzzy Hybridization: A New Method for Decision Making (S.K. Pal and A. Skowron, eds.), Springer-Verlag, Singapore, 1998, pp. 3–98.
12. R.D. Luce, **Several unresolved conceptual problems of mathematical psychology**, Journal of Mathematical Psychology **41** (1997), 79–87.
13. K. Makino, T. Suda, H. Ono, and T. Ibaraki, **Data analysis by positive decision trees**, IEICE Trans. Inf. and Syst. **E82-D** (1999), 76–88.
14. R. Potharst, J.C. Bioch, and R. van Dordregt, **Quasi-monotone decision trees for ordinal classification**, Tech. Report EUR-FEW-CS-98-01, Erasmus University Rotterdam, 1998.
15. R. Potharst and J.C. Bioch, **Decision trees for ordinal classification**, Intelligent Data Analysis **4** (2000), 97–112.
16. B. Roy, **Multicriteria methodology for decision aiding**, Kluwer Academic Publishers, Dordrecht, 1996.
17. A. Shashua and A. Levin, **Taxonomy of large margin principle algorithms for ordinal regression problems**, Tech. Report 2002-39, Leibniz Center for Research, School of Computer Science and Eng., the Hebrew University of Jerusalem, 2002.
18. J. Stefanowski, **On rough set based approaches to induction of decicion rules**, Rough Sets in Data Mining and Knowledge Discovery (Skowron S. Polkowski L., ed.), Physica-Verlag, Heidelberg, 1998, pp. 500–529.
19. P. Vincke, **Basic concepts of preference modelling**, Readings in Multiple Criteria Decision Aid (C. Bana e Costa, ed.), Springer-Verlag, Heidelberg, 1990, pp. 101–118.
20. **Webster's Encyclopedia Unabridged Dictionary of the English Language**, Gramercy Books, New York, 1989.

Axiomatic and Strategic Approaches to Bargaining Problems

Agnieszka Rusinowska[1,2]

[1] Tilburg University, Department of Philosophy, P.O. Box 90153,
5000 LE Tilburg, The Netherlands
a.rusinowska@uvt.nl
[2] Warsaw School of Economics
Department of Mathematical Economics
Al. Niepodleglosci 162, 02-554 Warsaw, Poland
arusin@sgh.waw.pl

Abstract. In this paper, a survey of bargaining theory is presented. First, an axiomatic approach to bargaining problems is considered. I describe the Nash solution to bargaining problems, and present an example of an application of this solution. Some other solutions to bargaining problems, such as the egalitarian solution, the utilitarian solution, and the solution proposed by Kalai and Smorodinsky, are also mentioned. Next, I describe the strategic approach to the bargaining problem. Rubinstein's bargaining game of alternating offers, the form of Nash equilibria, and the subgame perfect equilibrium of this game are presented. I also describe two special bargaining models with stationary preferences of the players. Moreover, several generalizations of Rubinstein's model, including models with non-stationary preferences of the players, are presented. Finally, I mention some applications of dynamic bargaining models.

1   Introduction

Bargaining problems concern situations in which some parties, referred to as players, bargain over the division of certain goods. A solution to a bargaining problem means the determination of such a division. In simple examples of bargaining problems, one often refers to an amount of money, a piece of cake, or a portion of ice-cream. In more complex applications, one may analyze, for instance, a division of the profits of a company, negotiated between the management and the trade unions, a division of the budget, or negotiations between two countries on fishing quotas. One may apply an axiomatic approach derived from John Nash ([9]) to bargaining problems, i.e., one can postulate some axioms concerning a potential solution, and then investigate its existence and properties resulting from the adopted axioms. Besides the Nash solution to a bargaining problem, some other solutions may be found in the literature: the egalitarian solution, the utilitarian solution, and the solution proposed by Kalai and Smorodinsky ([5]).

A different approach to bargaining problems, however, called the dynamic (or strategic) approach, may also be applied. It involves the representation of a

H. de Swart et al. (Eds.): TARSKI, LNCS 2929, pp. 124–146, 2003.

bargain as a non-cooperative game and the investigation of solutions from among the equilibria of the game. As one of the first to do this, Ariel Rubinstein considered the strategic model ([3], [10], [11]), in which two players bargain over the division of a certain good. In the bargaining literature, many generalizations of Rubinstein's original model have been analyzed. Several were presented in [10], for instance, a three-player model, models in which a player may withdraw from a bargain, a model with a risk of breaking the bargain, models with incomplete information. A bargaining model concerning the division of two goods simultaneously was presented in [2]. Ariel Rubinstein analyzed a bargaining game, in which the preferences of the players were stationary and constant in time. In particular, he considered a model with constant discount rates, and a model with constant bargaining costs. One of the generalizations of Rubinstein's model concerns bargaining models with non-stationary preferences of the players [14]), in particular, with preferences varying in time ([12], [14]). One may analyze, for instance, the bargaining model in which the preferences of each player are expressed not by a constant discount rate, but by a sequence of discount rates varying in time ([14], [15]). A model in which the preferences of each player are expressed not by a constant bargaining cost, but by a sequence of bargaining costs varying in time, has also been analyzed ([14], [16]). Finally, a 'mixed' model may be considered, in which the preferences of each player are expressed simultaneously by a sequence of discount rates and a sequence of bargaining costs varying in time ([17]).

In this paper, I present a survey of bargaining theory. In Sect. 2, an axiomatic approach to bargaining problems is described in detail. The notions of bargaining problem and of bargaining solution are first defined in Sect. 2.1. Next, I recapitulate Nash's axioms and Nash's theorem. In Sect. 2.2, an application of Nash's solution is presented. In Sect. 2.3, some other bargaining solutions, different from the Nash solution, are mentioned. In Sect. 3, Rubinstein's bargaining model is presented. In Sect. 3.1, his bargaining game of alternating offers is described. I describe the bargaining procedure, define the strategies of the players, and introduce Rubinstein's axioms, including the stationarity axiom. Sect. 3.2 and Sect. 3.3 are concerned with the Nash equilibria and the subgame perfect equilibrium of this game, respectively. In Sect. 3.4 and Sect. 3.5, I present Rubinstein's models with stationary preferences, described by constant discount rates, and constant bargaining costs, respectively. Several generalizations of Rubinstein's bargaining model are presented in this paper. One is described in Sect. 4, and concerns bargaining models with non-stationary preferences of the players. In Sect. 4.1, Nash equilibria and subgame perfect equilibria for the non-stationary case are presented. A model in which the players' preferences are expressed by sequences of discount rates varying in time is presented in Sect. 4.2. In Sect. 4.3, a model with preferences defined by sequences of bargaining costs is analyzed. In Sect. 4.4, I mention 'mixed' models in which the players' preferences are described simultaneously by discount rates and bargaining costs. In Sect. 4.5, a delay in reaching an agreement is considered. I present a subgame perfect equilibrium, in which an agreement is reached with a delay of one period. Some other generalizations of Rubinstein's model, different from those described in Sect. 4, are presented in Sect. 5. In Sect. 5.1, two models in which player 2 has an option of leaving player 1 and terminating the game are presented. In Sect. 5.2, I mention the three-player bargaining model. In Sect. 5.3, a bargaining game with short periods is introduced. In Sect. 5.4, a bargaining model with a risk of breakdown is presented. Finally, in Sect. 6, some applications of dynamic bargaining models to economic and social situations are mentioned.

2   Axiomatic Approach

One of the approaches that can be applied to bargaining problems is the axiomatic (also called static) approach. In this approach, instead of taking into account the strategic aspects of bargaining (such as rules, and course of negotiating), one concentrates on the properties of a potential solution to the problem. In this section, we recapitulate some bargaining solutions of the axiomatic approach. We start with the solution derived from John Nash ([9]).

2.1   The Nash Solution

Let N be the set of all bargainers, also called players. The players either manage to reach an agreement or they fail to reach one. This means that a bargain may result either in an agreement in a certain set X or in disagreement D. Each player $i \in N$ has a preference ordering $\succeq_i$ (that is, a complete, transitive, and reflexive binary relation) over the set $X \cup \{D\}$, where for any $x, y \in X$, $x \succeq_i y$ if and only if player i either prefers x to y or is indifferent to them. The objects N, X, D, and $\succeq_i$ for each $i \in N$, define a *bargaining situation*. For each player $i \in N$, there is a function $u_i : X \cup \{D\} \to \mathbb{R}$, called a *utility function*.

Let us assume that $N = \{1, 2\}$. Let S be the set of all utility pairs $(u_1(x), u_2(x))$ for $x \in X$. The point $d = (d_1, d_2) = (u_1(D), u_2(D))$ determines the payoffs the players get when they fail to reach an agreement. It is assumed that $d \in S$, which means that the players can agree to disagree, and that there is an agreement preferred by both players to the disagreement outcome. The latter means that the players are interested in reaching an agreement. Formally, one can write the following definition:

Definition 1 *Let X be the set of possible agreements, D be the disagreement event, and $u_i : X \cup \{D\} \to \mathbb{R}$ be the utility function of player $i = 1, 2$. A bargaining problem is a pair (S, d), where*

$$S = \{(u_1(x), u_2(x)) \mid x \in X\} \subseteq \mathbb{R}^2, \quad d = (d_1, d_2) = (u_1(D), u_2(D)), \quad (1)$$

such that

- *S is compact (i.e., closed and bounded), and convex*
- *$d \in S$*
- *there exists $(s_1, s_2) \in S$ such that $s_1 > d_1$ and $s_2 > d_2$.*

Definition 2 *A bargaining solution is a function* $\phi : BP \to \mathbb{R}^2$, *where* BP *is the set of all bargaining problems, which assigns a unique element* $(s_1^*, s_2^*) \in S$ *to each bargaining problem* $(S, d) \in BP$, *i.e.,*

$$\phi(S, d) = (\phi_1(S, d), \phi_2(S, d)) = (s_1^*, s_2^*) \in S. \tag{2}$$

The pair (s_1^*, s_2^*) *satisfies individual rationality, that is,*

$$s_1^* \geq d_1 \quad and \quad s_2^* \geq d_2. \tag{3}$$

Individual rationality means that, for each player, reaching an agreement is at least as attractive as disagreement.

Nash imposed four axioms on a bargaining solution $\phi : BP \to \mathbb{R}^2$. Let us first introduce some definitions.

Definition 3 *The bargaining problem* (S, d) *is symmetric if*

$$d_1 = d_2 \quad and \quad (s_1, s_2) \in S \quad if \quad and \quad only \quad if \quad (s_2, s_1) \in S. \tag{4}$$

Definition 4 *We say that* (S', d') *is obtained from the bargaining problem* (S, d) *by the transformation* $s_i \mapsto \alpha_i s_i + \beta_i$ *for* $i = 1, 2$ *if*

$$d_i' = \alpha_i d_i + \beta_i \quad for \quad i = 1, 2, \quad and \tag{5}$$

$$S' = \{(\alpha_1 s_1 + \beta_1, \alpha_2 s_2 + \beta_2) \in \mathbb{R}^2 \mid (s_1, s_2) \in S\}. \tag{6}$$

It may be shown that, if $\alpha_i > 0$ for $i = 1, 2$, then (S', d') is itself a bargaining problem.

Nash's Axioms

(N-1) *Pareto Efficiency:*
   Suppose (S, d) is a bargaining problem, $s = (s_1, s_2) \in S$, $s' = (s_1', s_2') \in S$, and $s_i' > s_i$ for $i = 1, 2$. Then, $\phi(S, d) \neq s$.
(N-2) *Symmetry:*
   If the bargaining problem (S, d) is symmetric, then $\phi_1(S, d) = \phi_2(S, d)$.
(N-3) *Invariance to Equivalent Utility Representations:*
   Suppose that the bargaining problem (S', d') is obtained from (S, d) by the transformation $s_i \mapsto \alpha_i s_i + \beta_i$ for $i = 1, 2$, where $\alpha_i > 0$ for $i = 1, 2$. Then, $\phi_i(S', d') = \alpha_i \phi_i(S, d) + \beta_i$ for $i = 1, 2$.
(N-4) *Independence of Irrelevant Alternatives:*
   If (S, d) and (W, d) are bargaining problems with $W \subset S$ and $\phi(S, d) \in W$, then $\phi(W, d) = \phi(S, d)$.

Note that axioms (N-1) and (N-2) concern the solution of single bargaining problems, while axioms (N-3) and (N-4) require the solution to satisfy some consistency among bargaining problems. According to (N-1), the players never agree on an outcome s if another outcome s' is available, in which they are both better off. If the players agreed on the non-efficient outcome s, then there would be good reason to re-negotiate. Axiom (N-1) also implies that the players never disagree (see Definition 1). By virtue of axiom (N-2), both players are treated equally. If the players are interchangeable, then the bargaining solution must assign the same utility to each player. Axiom (N-3) formalizes the assumption that the players' preferences, not the specific utility functions representing the preferences, are basic. The bargaining solution does not depend on equivalent utility representations. Axiom (N-4) means that the solution should not depend on 'irrelevant' alternatives. If we remove some alternatives from the set S, but the solution $\phi(S, d)$ is still available, then $\phi(S, d)$ should remain the solution also in the smaller set. Axiom (N-4) may be illustrated by the following story. Two friends make an order in a restaurant. There are three menu options: A, B, and C. After choosing dinner B, they are informed that, in fact, dinner A is not available. If, after receiving this information, the friends decline dinner B and choose dinner C, they violate the independence of irrelevant alternatives. The choice between dinners B and C should be independent of the 'irrelevant' option A.

Nash proved in [9] that there exists precisely one bargaining solution satisfying the four axioms. This solution assigns to each bargaining problem the utility pair that maximizes the product of the players' gains in utility over the disagreement outcome.

Theorem 1 *There is a unique bargaining solution $\phi^N : BP \to \mathbb{R}^2$ satisfying axioms (N-1), (N-2), (N-3), and (N-4), and it is given by*

$$\phi^N(S, d) = \arg \max_{(d_1, d_2) \leq (s_1, s_2) \in S} (s_1 - d_1)(s_2 - d_2). \tag{7}$$

The bargaining solution ϕ^N is called the *Nash solution* to bargaining problems. If we consider n-person bargaining problem (S, d), where

$$S = \{(u_1(x), ..., u_n(x)) \mid x \in X\} \subseteq \mathbb{R}^n, \quad d = (d_1, ..., d_n) = (u_1(D), ..., u_n(D)), \tag{8}$$

then the Nash solution is given by

$$\phi^N(S, d) = \arg \max_{d \leq s \in S} \prod_{i \in N} (s_i - d_i). \tag{9}$$

2.2 Applying the Nash Solution

In this section, an application of the Nash solution is presented. Let us consider an example in which a firm and a labor union negotiate a wage-employment package ([8], [10], [13]). Suppose that the union represents L workers. Each worker can obtain a wage w_0 outside the firm. If the firm hires l workers, then it produces $k(l)$ units of output. We assume that the price of the output is 1, k is strictly concave, $k(0) = 0$, and $k(l) > lw_0$ for some l. An agreement is a wage-employment pair (w, l). The utility to the firm of the agreement (w, l) is its profit $k(l) - lw$. The utility to the union of (w, l) is the total amount of money received by its members, i.e., $lw + (L - l)w_0$. We restrict agreements to pairs (w, l) for which the profit of the firm is non-negative (i.e., $k(l) \geq lw$) and the wage is at least w_0. The set of utility pairs that can result from agreement is

$$S = \{(k(l) - lw, lw + (L - l)w_0) \mid k(l) \geq lw, \ 0 \leq l \leq L, \ w \geq w_0\}. \quad (10)$$

If the two parties fail to reach an agreement, then the firm receives a profit of 0, and the union obtains Lw_0. Hence, the disagreement utility pair is equal to $d = (0, Lw_0)$. Let l^* be the unique maximizer of the sum of the utilities of the firm and the union, $k(l) + (L - l)w_0$. Then, we have

$$S = \{(s_1, s_2) \in \mathbb{R}^2 \mid s_1 + s_2 \leq k(l^*) + (L - l^*)w_0, \ s_1 \geq 0, \ s_2 \geq Lw_0\}. \quad (11)$$

This is a compact and convex set, and $d = (0, Lw_0) \in S$. Therefore, (S, d) is a bargaining problem. The Nash solution, being Pareto efficient, gives the labor l^* which maximizes the profit $k(l) - lw$. Note that the difference between the union's payoff at the agreement (w, l) and its disagreement payoff is equal to $lw + (L - l)w_0 - Lw_0 = l(w - w_0)$. Hence, the predicted wage w^* is equal to

$$w^* = \arg \max_{w \geq w_0} (k(l^*) - l^*w)l^*(w - w_0), \quad (12)$$

and by virtue of the necessary condition for the existence of a maximum of a continuous function, we receive

$$w^* = \frac{w_0 + \frac{k(l^*)}{l^*}}{2}. \quad (13)$$

The wage w^* is the average wage outside the firm and the average product of labor.

2.3 Other Solutions to the Bargaining Problems

Apart from the Nash solution to bargaining problems, some other solutions have been proposed. Among them, the egalitarian solution, the utilitarian solution, and the Kalai-Smorodinsky solution are worth mentioning. For an extensive analysis of the bargaining solutions, see, for instance, [6].

In the *egalitarian solution* $\phi^E(S, d)$ of the bargaining problem (S, d), the gains arising from an agreement are shared equally between the players, that is, $\phi_1^E(S, d) = \phi_2^E(S, d)$.

For each bargaining problem (S, d), the *utilitarian solution* $\phi^U(S, d)$ maximizes the sum of the players' utilities, that is, $\sum_{i=1}^2 (s_i - d_i)$, in the set $S \cap \mathbb{R}_+^2$, where $\mathbb{R}_+^2 = \{(x_1, x_2) \in \mathbb{R}^2 \mid x_i > 0 \ for \ i = 1, 2\}$.

Another bargaining solution has been proposed by Kalai and Smorodinsky ([5]). For each bargaining problem (S, d), let $u^i(S) \in \mathbb{R}$ (for $i = 1, 2$) be the maximal utility which player i can reach in $S \cap \mathbb{R}_+^2$. In the *Kalai-Smorodinsky solution* $\phi^{KS}(S, d) = (\phi_1^{KS}(S, d), \phi_2^{KS}(S, d))$, $\phi_1^{KS}(S, d)$ and $\phi_2^{KS}(S, d)$ are proportional to $u^1(S)$ and $u^2(S)$, respectively.

All these bargaining solutions satisfy individual rationality (i.e., equation (3)), axioms (N-1), and (N-2). Moreover, both the egalitarian and the utilitarian solutions satisfy (N-4), but none satisfies axiom (N-3). The Kalai-Smorodinsky solution satisfies (N-3), but it does not satisfy axiom (N-4). As stated in Theorem 1, the Nash solution is the only one which satisfies all axioms (N-1), (N-2), (N-3), and (N-4).

3   Strategic Approach – Rubinstein's Model

Another approach to bargaining problems, the strategic (also called the dynamic) approach, was introduced by Ariel Rubinstein ([3], [10], [11]). In this section, we recapitulate this approach and Rubinstein's model of alternating offers presented in [3], [10], and [11].

3.1   The Bargaining Game – Stationary Preferences

The bargaining problem can be presented as a kind of game in extensive form, which is referred to as a bargaining game. We model the following situation. Two parties bargain over the division of a certain good. We assume the good to be infinitely divisible and that there is one unit to be divided. An *agreement* is a pair $x = (x_1, x_2)$, in which x_1 is the part of the good received by player 1, and x_2 is the part received by player 2. The *set of all possible agreements* is

$$X = \{(x_1, x_2) \in \mathbb{R}^2 \mid x_1 + x_2 = 1 \text{ and } x_i \geq 0 \text{ for } i = 1, 2\}. \tag{14}$$

First, in period 0, player 1 proposes his way of dividing. Player 2 either accepts this proposal, which means that the game is over, or he rejects the offer, and after a certain unit of time, i.e., in period 1, he proposes his division. Then player 1 either accepts or rejects the proposal. In general, the acceptance of an offer terminates the bargaining, and its rejection leads the rejecting party to submit his proposal in the following period.

The *result* of the game is described by a pair (x, t), where $x = (x_1, x_2)$ is the agreement, and $t \in \mathbb{N}$ is the number of proposals rejected in the bargaining. The players can bargain infinitely, but in this case the value of the good gradually decreases. Disagreement, that is, the situation in which players do not reach agreement, is denoted by the symbol D.

A (pure) *strategy* of a player specifies an action (submitting a proposal or accepting/rejecting an offer) at every node of the extensive form in which it is the turn of the given player to move. We formalize these ideas in the following definition:

Definition 5 *Let X^t denote the set of all t-length sequences of elements from X. A strategy of player 1 is a sequence $f = (f^t)_{t=0}^{\infty}$ of functions such that $f^0 \in X$, $f^t : X^t \to X$ if t is even, and $f^t : X^{t+1} \to \{yes, no\}$ if t is odd.*
By analogy, a strategy of player 2 is a sequence $g = (g^t)_{t=0}^{\infty}$ of functions such that $g^t : X^{t+1} \to \{yes, no\}$ if t is even, and $g^t : X^t \to X$ if t is odd.

Each player i $(i = 1, 2)$ is assumed to have a complete, transitive, and reflexive *preference relation* $\succeq_i$ over the set $(X \times \mathbb{N}) \cup \{D\}$. The *strict preference relation* $\succ_i$, which is irreflexive and transitive, and *the indifference preference relation* $\sim_i$, which is an equivalence relation, are defined in the standard way:

$$\forall x, y \in X \; \forall t_1, t_2 \in \mathbb{N} \; [(x, t_1) \succ_i (y, t_2) \; \Leftrightarrow \; \neg \, (y, t_2) \succeq_i (x, t_1)] \tag{15}$$

$$\forall x, y \in X \; \forall t_1, t_2 \in \mathbb{N} \; [(x, t_1) \sim_i (y, t_2) \; \Leftrightarrow \; ((x, t_1) \succeq_i (y, t_2) \wedge (y, t_2) \succeq_i (x, t_1))] \tag{16}$$

Rubinstein imposed the following conditions on the preference relations, for $i = 1, 2$:

(A-1) *Disagreement is the worst outcome*

$$\forall x \in X \; \forall t \in \mathbb{N} \; [(x, t) \succeq_i D]. \tag{17}$$

(A-2) *The good is attractive*

$$\forall x, y \in X \; \forall t \in \mathbb{N} \; [x_i > y_i \; \Leftrightarrow \; (x, t) \succ_i (y, t)]. \tag{18}$$

(A-3) *Time is valuable*

$$\forall x \in [0, 1] \; \forall t_1, t_2 \in \mathbb{N} \; [(x_i > 0 \wedge t_2 > t_1) \; \Rightarrow \; (x, t_1) \succ_i (x, t_2)]. \tag{19}$$

(A-4) *Continuity*

$$\forall ((x_n, t_1))_{n=1}^{\infty} \; \forall ((y_n, t_2))_{n=1}^{\infty} \; [(x_n \longrightarrow_{n \to \infty} x \; \wedge \; y_n \longrightarrow_{n \to \infty} y \; \wedge$$

$$for \; each \; n \; (x_n, t_1) \succeq_i (y_n, t_2)) \; \Rightarrow \; (x, t_1) \succeq_i (y, t_2)]. \tag{20}$$

(A-5) *Stationarity*

$$\forall x, y \in [0, 1] \; \forall t \in \mathbb{N} \; [(x, 0) \succeq_i (y, 1) \; \Leftrightarrow \; (x, t) \succeq_i (y, t + 1)]. \tag{21}$$

(A-6) *Increasing loss to delay*
  The difference $x_i - \vartheta_i(x_i, 1)$ is an increasing function of x_i, where $\vartheta_i : [0, 1] \times \mathbb{N} \to [0, 1]$ for $i = 1, 2$, is defined as follows:

$$\vartheta_i(x_i, t) = \begin{cases} y_i & if & (y, 0) \sim_i (x, t) \\ 0 & if & (y, 0) \succ_i (x, t) \; for \; all \; y \in X \end{cases} . \tag{22}$$

We refer to $\vartheta_i(x_i, t)$ as the *present value of (x, t) for player i*.

3.2 The Nash Equilibrium

One of the standard solutions used in game theory is the notion of Nash equilibrium. A pair of strategies (f, g) is a *Nash equilibrium* in a two-person game if, given g, no strategy of player 1 results in an outcome that player 1 prefers to the outcome generated by (f, g), and, given f, no strategy of player 2 results in an outcome that player 2 prefers to the outcome generated by (f, g).

Rubinstein showed ([10], ([11]) that each bargaining game of alternating offers, in which the preferences of the players satisfy axioms (A-1) through (A-6), has an infinite number of Nash equilibria. In particular, for every agreement $x \in X$, the outcome $(x, 0)$ is generated by a Nash equilibrium of such a game. This Nash equilibrium appears as follows. Player 1 uses a strategy in which he always proposes x, and accepts an offer y if and only if $y_1 \geq x_1$. The strategy of player 2 is defined analogously.

3.3 Subgame Perfect Equilibrium

Reinhard Selten ([18]) introduced a refinement of the notion of Nash equilibrium, called *subgame perfect equilibrium*. For each node of the bargaining game described above, there is a *subgame*, that is, an extensive game starting at this node. The concept of subgame perfect equilibrium requires that a player's strategy is optimal in every subgame, whether or not the first (beginning) node of this subgame is reached if the players adhere to their strategies. This is expressed precisely in the following definition:

Definition 6 *A strategy pair is a subgame perfect equilibrium of a bargaining game of alternating offers if the strategy pair it induces in every subgame is a Nash equilibrium of that game.*

Rubinstein proved ([11]) the following theorem:

Theorem 2 *Every bargaining game of alternating offers in which the players' preferences satisfy axioms (A-1) through (A-6) has a unique subgame perfect equilibrium (f^*, g^*). In this equilibrium, player 1 always proposes the agreement x^* and player 2 always proposes the agreement y^* such that*

$$y_1^* = \vartheta_1(x_1^*, 1) \quad and \quad x_2^* = \vartheta_2(y_2^*, 1), \tag{23}$$

whenever it is their turn to make an offer; player 1 accepts an offer y submitted by player 2 if and only if $y_1 \geq y_1^*$, and player 2 accepts an offer x submitted by player 1 if and only if $x_2 \geq x_2^*$. The outcome of the game is that player 1 proposes x^* in period 0, and player 2 immediately accepts this offer.

3.4 Preferences with Constant Discount Rate

Rubinstein considered two kinds of models with preferences constant in time. The preferences analyzed by Rubinstein were stationary. In particular, he analyzed a model in which the time preferences of each player i $(i = 1, 2)$ were expressed by a constant discount rate $0 < \delta_i < 1$. We refer to δ_i as the *discount rate (factor)* of player i, $i = 1, 2$. In this case, the utility functions are defined for each $i = 1, 2$ in the following way:

$$u_i(x, t) = x_i \delta_i^t \ for \ every \ (x, t) \in X \times \mathbb{N}, \quad and \quad u_i(D) = 0. \tag{24}$$

Preferences based on these utility functions satisfy all axioms (A-1) through (A-6).

Rubinstein showed ([11]) that, if the players have time preferences with constant discount rates, the game has the unique subgame perfect equilibrium (f^*, g^*) described in Theorem 2, such that

$$x_1^* = \frac{1 - \delta_2}{1 - \delta_1 \delta_2}, \quad x_2^* = \frac{\delta_2(1 - \delta_1)}{1 - \delta_1 \delta_2}, \tag{25}$$

$$y_1^* = \frac{\delta_1(1 - \delta_2)}{1 - \delta_1 \delta_2}, \quad y_2^* = \frac{1 - \delta_1}{1 - \delta_1 \delta_2}. \tag{26}$$

3.5 Preferences with Constant Bargaining Cost

Another model with stationary preferences considered by Rubinstein was the model with constant bargaining costs. In this model, the time preferences of each player i $(i = 1, 2)$ were described by a constant bargaining cost $c_i > 0$. We refer to c_i as the *bargaining cost* or *cost of delay* of player i. The utility functions are defined as follows:

$$u_i(x, t) = x_i - c_i t \ \text{for every} \ (x, t) \in X \times \mathbb{N}, \ \text{and} \ u_i(D) = -\infty. \tag{27}$$

Preferences based on these utility functions satisfy axioms (A-1) through (A-5), but not axiom (A-6). Nevertheless, there is always at least one subgame perfect equilibrium (f^*, g^*) (of the form described in Theorem 2), and moreover, as long as $c_1 \neq c_2$, there is only one. The proposals x^* and y^* of players 1 and 2, respectively, are the following ([10], [11]):

$$if \ c_1 > c_2, \ then \ x_1^* = c_2 \ and \ y_1^* = 0, \tag{28}$$

$$if \ c_1 = c_2, \ then \ c_1 \leq x_1^* \leq 1 \ and \ y_1^* = x_1^* - c_1, \tag{29}$$

$$if \ c_1 < c_2, \ then \ x_1^* = 1 \ and \ y_1^* = 1 - c_1. \tag{30}$$

4 Bargaining Models with Non-stationary Preferences

One of the natural generalizations of Rubinstein's bargaining game is based on removing some of his axioms (A-1)-(A-6). In this section, I present some results of my research into a generalization of Rubinstein's model, in which the players' preferences are non-stationary. I assume the preferences to satisfy only axioms (A-1) through (A-4).

4.1 Subgame Perfect Equilibria for the 'Non-stationary Case'

Let us consider the consequences of removing axioms (A-5) and (A-6) with respect to Nash equilibria and subgame perfect equilibria. This does not change Rubinstein's result with respect to Nash equilibria. The following may be proved ([14]):

Theorem 3 *Suppose that in the bargaining game the players' preferences satisfy axioms (A-1) through (A-3), and $x \in X$ is arbitrary. Then, the pair of strategies (f_1, f_2) defined by*

f_i – *player i ($i = 1, 2$) always proposes x and accepts an offer y submitted by his opponent if and only if $y_i \geq x_i$,*

is a Nash equilibrium of this game.

Removing axioms (A-5) and (A-6) has much more serious consequences with respect to subgame perfect equilibria. First of all, an example may easily be constructed ([14]) showing that Rubinstein's theorem concerning subgame perfect equilibrium (Theorem 2) does NOT hold for the 'non-stationary' case. Let us adopt some assumptions:

(A) *In the bargaining game, the strategies of the players do not depend on the former history of the game, but they depend on the period, and the responses additionally depend on the offers submitted.*

(B) *The pair of strategies (F^*, G^*) is defined in the following way:*
  – *F^* - in each period $2t$ ($t \in \mathbb{N}$), player 1 submits an offer x^{2t}, and in each period $2t + 1$, accepts an offer s by player 2 if and only if $s_1 \geq y_1^{2t+1}$,*
  – *G^* - in each period $2t + 1$, player 2 submits an offer y^{2t+1}, and in each period $2t$, accepts an offer r by player 1 if and only if $r_2 \geq x_2^{2t}$.*

In accordance with the convention adopted, the offer x^{2t} means $x^{2t} = (x_1^{2t}, x_2^{2t})$. By analogy, the offer $y^{2t+1} = (y_1^{2t+1}, y_2^{2t+1})$.

It may be shown ([14]) that, in the class of strategies satisfying (A), the form (F^*, G^*) defined in (B) is a necessary condition for the existence of a subgame perfect equilibrium. This means that, in this class of strategies, there are no subgame perfect equilibria of a form different from (F^*, G^*). In particular, an agreement of any subgame perfect equilibrium independent of the former history of the game can be reached only without delay, i.e., in period $t = 0$.

The following theorem concerning sufficient and necessary conditions for the existence of a subgame perfect equilibrium and its form may be proved (for the proof, see [14], [17]):

Theorem 4 *If, in the bargaining game, the players' preferences satisfy axioms (A-1) through (A-4), and the strategies satisfy assumption (A), then the pair (F^*, G^*) defined in (B) is a subgame perfect equilibrium of this game if and only if, for each $t \in \mathbb{N}$, the offers submitted by the players satisfy the following conditions:*

$$(x^{2t}, 2t) \succeq_2 (y^{2t+1}, 2t+1) \text{ and if } x_2^{2t} > 0, \text{ then } (x^{2t}, 2t) \sim_2 (y^{2t+1}, 2t+1) \quad (31)$$

$$(y^{2t+1}, 2t+1) \succeq_1 (x^{2t+2}, 2t+2) \text{ and if } y_1^{2t+1} > 0, \text{ then}$$

$$(y^{2t+1}, 2t+1) \sim_1 (x^{2t+2}, 2t+2). \quad (32)$$

4.2   Discount Rates Varying in Time

One of the models with non-stationary preferences of the players is the model with discount rates varying in time ([14], [15]). This is a generalization of Rubinstein's model with constant discount rates (see Sect. 3.4). Let us consider a model in which the preferences of player i ($i = 1, 2$) are expressed not by a constant discount rate, but by a *sequence* of discount rates varying in time $(\delta_{i,t})_{t \in \mathbb{N}}$, where $\delta_{i,t}$ denotes the discount rate of player i in period t, $\delta_{i,0} = 1$, $0 < \delta_{i,t} < 1$ for $t \geq 1$. In this case, we have the following utility function:

$$U_i(x, t) = x_i \prod_{k=0}^{t} \delta_{i,k} \text{ for each } (x, t) \in X \times \mathbb{N}, \text{ and } U_i(D) = 0. \tag{33}$$

We then have

$$\forall x, y \in X \ \forall t_1, t_2 \in \mathbb{N} \ [(x, t_1) \succeq_i (y, t_2) \iff x_i \prod_{k=0}^{t_1} \delta_{i,k} \geq y_i \prod_{k=0}^{t_2} \delta_{i,k}]. \tag{34}$$

The preferences described above satisfy all axioms (A-1) through (A-4), but they do not satisfy (A-5). Hence, we cannot apply Rubinstein's theorem (Theorem 2) to this model, but we can obviously apply Theorem 4. We then get the following result (for the proof, see [14], [15]):

Theorem 5 *If, in the bargaining game, the players' preferences are expressed by sequences of discount rates $(\delta_{i,t})_{t \in \mathbb{N}}$, where $\delta_{i,0} = 1$, $0 < \delta_{i,t} < 1$ for $t \geq 1$, $i = 1, 2$, the strategies satisfy (A), and*

$$\prod_{j=1}^{t+1} \delta_{1,2j} \delta_{2,2j-1} \xrightarrow{t \to +\infty} 0, \tag{35}$$

then there is only one subgame perfect equilibrium of the form (F^*, G^*) defined in (B), where the offers submitted by the players are as follows:

$$x_1^0 = 1 - \delta_{2,1} + \sum_{n=1}^{+\infty} (\prod_{k=1}^{n} \delta_{1,2k} \delta_{2,2k-1})(1 - \delta_{2,2n+1}), \tag{36}$$

$$x_1^{2t+2} = \frac{x_1^{2t} + \delta_{2,2t+1} - 1}{\delta_{1,2t+2} \delta_{2,2t+1}} \text{ and } y_1^{2t+1} = x_1^{2t+2} \delta_{1,2t+2} \text{ for each } t \in \mathbb{N}. \tag{37}$$

Example 1 Let us apply Theorem 5 to the model ([14], [15]) in which both players have the same decreasing sequence of discount rates $\delta_{1,t} = \delta_{2,t} = \frac{1}{t+1}$ for $t \in \mathbb{N}$.

$$\prod_{j=1}^{t+1} \delta_{1,2j} \cdot \delta_{2,2j-1} = \frac{1}{(2t+3)!} \xrightarrow{t \to +\infty} 0, \tag{38}$$

$$\sum_{n=1}^{+\infty} (\prod_{k=1}^{n} \delta_{1,2k} \cdot \delta_{2,2k-1}) \cdot (1 - \delta_{2,2n+1}) = \sum_{n=1}^{+\infty} \frac{2n+1}{(2n+2)!} = \frac{1}{3!} - \frac{1}{4!} + \frac{1}{5!} - \frac{1}{6!} + ... \tag{39}$$

Since $1 - \frac{1}{1!} + \frac{1}{2!} - \frac{1}{3!} + \ldots + \frac{(-1)^n}{n!} \pm \ldots = \frac{1}{e}$, we receive

$$a_1^0 = 1 - \frac{1}{e}. \tag{40}$$

The following may also be proved ([14], [15]):

Theorem 6 *If, in the bargaining game, the players' preferences are expressed by sequences of discount rates $(\delta_{i,t})_{t\in\mathbb{N}}$, where $\delta_{i,0} = 1$, $0 < \delta_{i,t} < 1$ for $t \geq 1$, $i = 1,2$, and the strategies satisfy (A), then each pair of strategies (F^*, G^*) defined in (B), such that*

$$x_1^0 \geq 1 - \delta_{2,1} + \sum_{n=1}^{+\infty}(\prod_{k=1}^{n}\delta_{1,2k}\delta_{2,2k-1})(1 - \delta_{2,2n+1}) \quad and \tag{41}$$

$$x_1^0 \leq 1 - \delta_{2,1} + \sum_{n=1}^{+\infty}(\prod_{k=1}^{n}\delta_{1,2k}\delta_{2,2k-1})(1 - \delta_{2,2n+1}) + \prod_{j=1}^{+\infty}\delta_{1,2j}\delta_{2,2j-1}, \tag{42}$$

$$x_1^{2t+2} = \frac{x_1^{2t} + \delta_{2,2t+1} - 1}{\delta_{1,2t+2}\delta_{2,2t+1}} \quad and \quad y_1^{2t+1} = x_1^{2t+2}\delta_{1,2t+2} \quad for \ each \ t \in \mathbb{N}, \tag{43}$$

is a subgame perfect equilibrium of this game.

Example 2 Let $\delta_{i,0} = 1$, $\delta_{i,t} = 1 - \frac{1}{(t+1)^2}$ for $t \geq 1$, $i=1,2$. When applying Theorem 6, we have

$$\prod_{j=1}^{t+1}\delta_{1,2j} \cdot \delta_{2,2j-1} = (1 - \frac{1}{2^2}) \cdot (1 - \frac{1}{3^2}) \cdot \ldots \cdot (1 - \frac{1}{(2t+2)^2}) \cdot (1 - \frac{1}{(2t+3)^2}) =$$

$$\frac{2t+4}{2 \cdot (2t+3)} \xrightarrow{t \to +\infty} \frac{1}{2} > 0, \tag{44}$$

$$\sum_{n=1}^{+\infty}(\prod_{k=1}^{n}\delta_{1,2k} \cdot \delta_{2,2k-1}) \cdot (1 - \delta_{2,2n+1}) = \sum_{n=1}^{+\infty}\frac{1}{2 \cdot (2n+1) \cdot (2n+2)} = \frac{1}{2} \cdot (\ln 2 - \frac{1}{2}), \tag{45}$$

and, therefore,

$$\frac{\ln 2}{2} \leq a_1^0 \leq \frac{1 + \ln 2}{2}. \tag{46}$$

4.3   Bargaining Costs Varying in Time

Another example of a model with non-stationary preferences of the players is the model with bargaining costs varying in time ([14], [16]). This is a generalization of Rubinstein's model with constant bargaining costs (see Sect. 3.5). We analyze the model in which the preferences of player i $(i = 1, 2)$ are described not by a constant bargaining cost, but by a *sequence* of bargaining costs varying in time $(c_{i,t})_{t \in \mathbb{N}}$, where $c_{i,t}$ is the bargaining cost of player i in period t, $c_{i,0} = 0$, $0 < c_{i,t} < 1$ for $t \geq 1$. In this case, we have the following utility function:

$$u_i(x,t) = x_i - \sum_{k=0}^{t} c_{i,k} \ for \ each \ (x,t) \in X \times \mathbb{N}, \ and \ u_i(D) = -\infty. \quad (47)$$

$$\forall x, y \in X \ \forall t_1, t_2 \in \mathbb{N} \ [(x,t_1) \succeq_i (y,t_2) \ \Leftrightarrow \ x_i - \sum_{k=0}^{t_1} c_{i,k} \geq y_i - \sum_{k=0}^{t_2} c_{i,k}]. \quad (48)$$

These preferences satisfy axioms (A-1) through (A-4), but not axiom (A-5). Below, I recapitulate some of the results presented (and proved) in [14], [16]. We consider strategies satisfying (A) and (B).

Theorem 7 *If, in the bargaining game, the preferences of player i ($i = 1,2$) are described by a sequence of bargaining costs varying in time $(c_{i,t})_{t \in \mathbb{N}}$, where $c_{i,0} = 0$, $0 < c_{i,t} < 1$ for $t \geq 1$,*

$$c_{2,2t+1} \geq c_{1,2t+2} \geq c_{2,2t+3} \quad for \ each \ \ t \in \mathbb{N}, \quad (49)$$

and the players use (F^*, G^*) described in (B), then the pair (F^*, G^*), where

$$c_{2,1} \leq x_1^0 \leq 1, \quad (50)$$

$$x_1^{2t+2} = x_1^{2t} + c_{1,2t+2} - c_{2,2t+1} \ and \ y_1^{2t+1} = x_1^{2t+2} - c_{1,2t+2} \ for \ each \ t \in \mathbb{N}, \quad (51)$$

is a subgame perfect equilibrium of this game.

Theorem 8 *If, in the bargaining game, the preferences of player i ($i = 1,2$) are described by a sequence of bargaining costs varying in time $(c_{i,t})_{t \in \mathbb{N}}$, where $c_{i,0} = 0$, $0 < c_{i,t} < 1$ for $t \geq 1$,*

$$c_{2,2t+1} \leq c_{1,2t+2} \leq c_{2,2t+3} \quad for \ each \ \ t \in \mathbb{N}, \quad (52)$$

and the players use (F^*, G^*) described in (B), then the pair (F^*, G^*), where

$$\sum_{i=0}^{+\infty}(c_{2,2i+1} - c_{1,2i}) \leq x_1^0 \leq \sum_{i=0}^{+\infty}(c_{2,2i+1} - c_{1,2i}) + 1 - \lim_{t \to +\infty} c_{1,2t+2}, \quad (53)$$

$$x_1^{2t+2} = x_1^{2t} + c_{1,2t+2} - c_{2,2t+1} \ and \ y_1^{2t+1} = x_1^{2t+2} - c_{1,2t+2} \ for \ each \ t \in \mathbb{N}, \quad (54)$$

is a subgame perfect equilibrium of this game.

Example 3 Let us construct an example ([14], [16]) illustrating one of the theorems presented in this section. Let $c_{2,2t+1} = \frac{1}{3} - \frac{1}{2t+4}$ and $c_{1,2t+2} = \frac{1}{3} - \frac{1}{2t+5}$ for each $t \in \mathbb{N}$. We can apply Theorem 8 here, where

$$\lim_{t \to +\infty} c_{1,2t+2} = \frac{1}{3}, \quad \sum_{i=0}^{+\infty}(c_{2,2i+1} - c_{1,2i}) = \ln 2 - \frac{1}{2}. \quad (55)$$

Hence, we get

$$\ln 2 - \frac{1}{2} \le x_1^0 \le \ln 2 + \frac{1}{6}, \tag{56}$$

and for each $t \in \mathbb{N}$

$$x_1^{2t+2} = x_1^{2t} + \frac{1}{2t+4} - \frac{1}{2t+5}, \quad y_1^{2t+1} = x_1^{2t+2} - \frac{1}{3} + \frac{1}{2t+5}. \tag{57}$$

4.4 'Mixed' Models

Consider 'mixed' models, in which the players' preferences are described *simultaneously* by discount rates and bargaining costs. It appears that, if we mix the two models presented in Sect. 3.4 and Sect. 3.5, we get a model with non-stationary preferences. More precisely, we get a model in which the preferences of each player i $(i = 1, 2)$ are expressed simultaneously by a constant discount rate δ_i, and a constant bargaining cost c_i, where $0 < \delta_i < 1$ and $0 < c_i < 1$. In this case, the utility function is defined as follows:

$$u_i(x, t) = x_i \delta_i^t - c_i t \ \text{ for every } (x, t) \in X \times \mathbb{N}, \ \text{ and } \ u_i(D) = -\infty. \tag{58}$$

Moreover, we have

$$\forall x, y \in X \ \forall t_1, t_2 \in \mathbb{N} \ [(x, t_1) \succeq_i (y, t_2) \ \Leftrightarrow \ x_i \delta_i^{t_1} - c_i t_1 \ge y_i \delta_i^{t_2} - c_i t_2]. \tag{59}$$

Such preferences satisfy axioms (A-1) through (A-4), but not axiom (A-5). In order to find subgame perfect equilibria in this case, Theorem 4 may be applied to this model. A more detailed analysis of this model, in particular, the forms of its subgame perfect equilibria, are given in [14].

Consider also a broader generalization of the models mentioned, that is, a bargaining model in which the preferences of each player i $(i = 1, 2)$ are expressed *simultaneously* by a *sequence* of discount rates varying in time $(\delta_{i,t})_{t \in \mathbb{N}}$, and by a *sequence* of bargaining costs varying in time $(c_{i,t})_{t \in \mathbb{N}}$. In this case, $0 < \delta_{i,t} < 1$ and $0 < c_{i,t} < 1$ refer, respectively, to the discount rate and bargaining cost of player i in period $t \ge 1$, $\delta_{i,0} = 1$, $c_{i,0} = 0$. Such preferences are represented by the following utility function:

$$u_i(x, t) = x_i \prod_{k=0}^{t} \delta_{i,k} - \sum_{k=0}^{t} c_{i,k} \ \text{ for each } (x, t) \in X \times \mathbb{N} \text{ and } u_i(D) = -\infty. \tag{60}$$

Such preferences are not stationary, but they satisfy axioms (A-1) through (A-4). An extensive analysis of this model may be found in [17].

4.5 Delay in Reaching an Agreement

If we relinquish assumption (A) and consider strategies dependent on the former history of the game, a delay in reaching an agreement of subgame perfect equilibrium may appear. Considering, for instance, the model presented in Sect. 4.3, if the bargaining costs are sufficiently small, then a subgame perfect equilibrium in which agreement is reached in period $t \geq 1$ may exist. Let us consider the bargaining game in which the preferences of player i $(i = 1, 2)$ are expressed by a sequence of bargaining costs varying in time $(c_{i,1})_{t \in \mathbb{N}}$, where $c_{1,1} \geq c_{2,1}$, $c_{1,1} + c_{2,1} \leq \frac{1}{2}$ and $c_{2,2t+1} \geq c_{1,2t+2} \geq c_{2,2t+3}$ for each $t \in \mathbb{N}$.

This example was presented in [16]. We know from Theorem 7 that the pair of strategies (F^*, G^*), in which the offers of the players satisfy (50) and (51), is a subgame perfect equilibrium of this game. This means that, for this game, there is an infinite number of subgame perfect equilibria in which agreement is reached in period $t = 0$. On the other hand, for this model, there is an additional subgame perfect equilibrium with a delayed agreement. The following theorem may be proved ([16]):

Theorem 9 *In the bargaining game, in which the preferences of player i $(i = 1, 2)$ are expressed by a sequence of bargaining costs varying in time $(c_{i,t})_{t \in \mathbb{N}}$, where $c_{i,0} = 0$, $0 < c_{i,t} < 1$ for $t \geq 1$,*

$$c_{1,1} \geq c_{2,1}, \; c_{1,1} + c_{2,1} \leq \frac{1}{2}, \; c_{2,2t+1} \geq c_{1,2t+2} \geq c_{2,2t+3} \text{ for each } t \in \mathbb{N}, \quad (61)$$

the following pair of strategies (F', G') is a subgame perfect equilibrium:

F' *– strategy of player 1:*

- *in period 0, he submits an offer x^0 such that $x_1^0 = 1$,*
- *in period $2t$ $(t \geq 1)$, he submits an offer x^{2t},*
- *in period $2t + 1$ $(t \in \mathbb{N})$, he accepts an offer s if and only if $s_1 \geq y_1^{2t+1}$.*

G' *– strategy of player 2:*

- *if an offer in period 0 equals 1, then in period 1 he submits an offer y^1 such that $y_1^1 = c_{1,1} + 2c_{2,1}$,*
- *if an offer in period 0 does not equal 1, then in period 1 he submits y^1 such that $y_1^1 = c_{2,1}$,*
- *in period $2t + 1$ $(t \geq 1)$, he submits an offer y^{2t+1},*
- *in period 0, he accepts an offer r by player 1 if and only if $r_1 \leq 2c_{2,1}$,*
- *in period $2t$ $(t \geq 1)$, he accepts an offer r by player 1 if and only if $r_1 \leq x_1^{2t}$,*

where

$$y_1^{2t+1} = x_1^{2t+2} - c_{1,2t+2} \text{ for each } t \in \mathbb{N}, \text{ and} \quad (62)$$

$$x_1^{2t+2} = x_1^{2t} + c_{1,2t+2} - c_{2,2t+1} \text{ for each } t \geq 1. \quad (63)$$

The agreement is reached in period 1, and it assigns $c_{1,1} + 2c_{2,1}$ to player 1.

5   Other Generalizations of Rubinstein's Model

In the literature, many other generalizations of Rubinstein's bargaining model have been analyzed. In this section, I consider a few of them. Some of the generalizations presented in [10] are recapitulated.

5.1 Models with Outside Options

I present two modifications of the structure of the bargaining game in which player 2 has the option of leaving player 1 and terminating the game. The value of such an outside option is equal to some b, $b < 1$, for player 2, and 0 for player 1. We assume that the players have time preferences with the same constant discount rate δ, $0 < \delta < 1$. The two options differ in the times at which player 2 may quit. If player 2 may quit only after he has rejected an offer, then the game has a unique subgame perfect equilibrium. If player 2 may quit only after player 1 rejects his offer, then, for some values of the outside option b, the game has multiple subgame perfect equilibria.

Opting Out Only When Responding to an Offer

In this game, player 2 can opt out only when he responds to an offer. The bargaining structure of this game is as follows. First, player 1 proposes a certain division $x = (x_1, x_2)$, meaning x_1 for player 1, and x_2 for player 2. Next, player 2 responds, and he has three possibilities:

- to accept the offer and, by this, terminate the game (in this case, player 1 receives x_1, and player 2 gets x_2)
- to reject the offer, and opt out, which also means the end of the game (the players' payoffs are then 0 for player 1, and b for player 2)
- to reject the offer and continue bargaining - this means that the play passes into the following period, in which player 2 submits his proposal. Next, player 1 either accepts or rejects this offer. In the case of rejection, another period passes, in which player 1 makes an offer, etc.

In general, if player 2 opts out in period t, then players 1 and 2 get 0 and $b\delta^t$, respectively, where $b < 1$. Hence, if $b > 0$, then player 2 seems to have an advantage over player 1. In [10], the following theorem was proved:

Theorem 10 *Consider the bargaining game described above, in which player 2 may opt out only when responding to an offer. Assume that the players have time preferences with the same constant discount rate $0 < \delta < 1$, and that their payoffs in the event that player 2 opts out in period t are 0 for player 1 and $b\delta^t$ for player 2, where $b < 1$.*

1. *If $b < \frac{\delta}{1+\delta}$, then the game has a unique subgame perfect equilibrium, which coincides with the subgame perfect equilibrium of the game in which player 2 has no outside option. In this equilibrium,*
   - *player 1 always proposes $\frac{1}{1+\delta}$ for himself, and $\frac{\delta}{1+\delta}$ for player 2, and he accepts a proposal y if and only if $y_1 \geq \frac{\delta}{1+\delta}$,*
   - *player 2 always proposes $\frac{\delta}{1+\delta}$ for player 1, and $\frac{1}{1+\delta}$ for himself, and he accepts a proposal x if and only if $x_2 \geq \frac{\delta}{1+\delta}$, and never opts out.*

   The outcome is then $((\frac{1}{1+\delta}, \frac{\delta}{1+\delta}), 0)$.

2. If $b > \frac{\delta}{1+\delta}$, then the game has a unique subgame perfect equilibrium, in which
   - player 1 always proposes $1 - b$ for himself, and b for player 2, and he accepts a proposal y if and only if $y_1 \geq (1 - b)\delta$,
   - player 2 always proposes $(1-b)\delta$ for player 1, and $1-(1-b)\delta$ for himself, and he accepts a proposal x if and only if $x_2 \geq b$, and opts out if $x_2 < b$.

   The outcome is then $((1 - b, b), 0)$.
3. If $b = \frac{\delta}{1+\delta}$, then in every subgame perfect equilibrium, the outcome is $((1 - b, b), 0)$.

Opting Out Only after an Offer Is Rejected by Player 1

In this game, we assume that player 2 may opt out only after player 1 has rejected an offer. The bargaining procedure is as follows. First, in period 0, player 1 submits his proposal, and player 2 responds. If player 2 accepts the offer, then the game is over. If he rejects it, then the game passes into the following period, in which player 2 makes his offer, and player 1 responds. Acceptance of the offer terminates the game. However, if player 1 rejects the offer, then player 2 can either opt out or continue bargaining. In [10], the following theorem was proved:

Theorem 11 *Consider the bargaining game described above, in which player 2 can opt out only after player 1 has rejected an offer. Assume that the players have time preferences with the same constant discount rate δ, $0 < \delta < 1$, and that their payoffs in the event that player 2 opts out in period t are 0 for player 1 and $b\delta^t$ for player 2, where $b < 1$.*

1. If $b < \frac{\delta^2}{1+\delta}$, then the game has a unique subgame perfect equilibrium, which coincides with the subgame perfect equilibrium of the game in which player 2 has no outside option. In this equilibrium,
   - player 1 always proposes $\frac{1}{1+\delta}$ for himself, and $\frac{\delta}{1+\delta}$ for player 2, and he accepts a proposal y if and only if $y_1 \geq \frac{\delta}{1+\delta}$,
   - player 2 always proposes $\frac{\delta}{1+\delta}$ for player 1, and $\frac{1}{1+\delta}$ for himself, and he accepts a proposal x if and only if $x_2 \geq \frac{\delta}{1+\delta}$, and never opts out.

   The outcome is then $((\frac{1}{1+\delta}, \frac{\delta}{1+\delta}), 0)$.
2. If $\frac{\delta^2}{1+\delta} \leq b \leq \delta^2$, then there are many subgame perfect equilibria. In particular, for every $a \in [1 - \delta, 1 - \frac{b}{\delta}]$, there is a subgame perfect equilibrium with the outcome $((a, 1 - a), 0)$. In every subgame perfect equilibrium, player 2's payoff is at least $\frac{\delta}{1+\delta}$.
3. If $\delta^2 < b < 1$, then there is a unique subgame perfect equilibrium, in which
   - player 1 always proposes $1-\delta$ for himself, and δ for player 2, and accepts any proposal,
   - player 2 always proposes 0 for player 1, and 1 for himself, and he accepts any proposal x in which $x_2 \geq \delta$, and always opts out.

5.2   Bargaining Game with Three Players

One of the natural generalizations of Rubinstein's model is to increase the number of bargainers. Let us consider a game of alternating offers, in which not two, but three players bargain over the division of one unit of a certain good ([10]). In the first period, player 1 proposes a partition $x = (x_1, x_2, x_3)$, where x_i $(i = 1, 2, 3)$ means the part of the good falling to player i, and $x_1 + x_2 + x_3 = 1$. Players 2 and 3 either accept or reject this proposal. If either of them rejects the offer, then the play passes to the following period, in which player 2 proposes his partition, and players 3 and 1 respond. If at least one of them rejects the proposal, then again the play passes to the following period, in which player 3 submits an offer, and players 1 and 2 respond. In general, agreement requires the approval of all three players, that is, a bargain terminates after two players accept the offer of the third one.

Theorem 12 *Suppose that the preferences of the players satisfy all six axioms (A-1)-(A-6), and $\vartheta_i(1,1) \geq \frac{1}{2}$ for $i = 1, 2, 3$, where $\vartheta_i(x_i, t)$ is the present value to player i of the agreement x in period t, as defined by equation (22). Then, for any partition x^*, there is a subgame perfect equilibrium of the three-player bargaining game defined above, in which the outcome is immediate agreement on the partition x^*.*

5.3   Bargaining Game with Short Periods

Consider a bargaining game of alternating offers, in which the delay between offers is Δ, where $\Delta \to 0$ (see [10]). This means that an offer can be made only at a time belonging to $\{0, \Delta, 2\Delta, ...\}$. Let us denote this game by $\Gamma(\Delta)$. We require that each player $i = 1, 2$ has a complete, transitive, and reflexive preference ordering $\succeq_i$ over $(X \times T_\infty) \cup \{D\}$, where $T_\infty = [0, \infty)$, X is the set of all agreements, and D is disagreement. For each $\Delta > 0$, such an ordering induces an ordering over the set $(X \times \{0, \Delta, 2\Delta, ...\}) \cup \{D\}$. We may impose on this preference ordering conditions similar to the axioms (A-1) through (A-6) adopted in Sect. 3.1, but with the difference that now $t \in T_\infty$. Let us denote these new conditions (similar to (A-1) through (A-6), but with $t \in T_\infty$) by (C-1) through (C-6).

Since axiom (C-3) is stronger than (A-3), we deduce that, for each $(x, t) \in X \times T_\infty$, there exists an agreement $y \in X$ such that $(y, 0) \sim_i (x, t)$. If the preference ordering $\succeq_i$ of player i over $(X \times T_\infty) \cup \{D\}$ satisfies (C-1) through (C-6), then for any value Δ, the ordering induced over $(X \times \{0, \Delta, 2\Delta, ...\}) \cup \{D\}$ satisfies (A-1) through (A-6), and hence, we can apply Theorem 2 to the game $\Gamma(\Delta)$. For any value of $\Delta > 0$, let $(x^*(\Delta), y^*(\Delta)) \in X \times X$ be the unique pair of agreements satisfying

$$(y^*(\Delta), 0) \sim_1 (x^*(\Delta), \Delta) \quad and \quad (x^*(\Delta), 0) \sim_2 (y^*(\Delta), \Delta). \tag{64}$$

We then have

Theorem 13 *Suppose that each player's preference ordering satisfies axioms (C-1) through (C-6). Then, for each $\Delta > 0$, the game $\Gamma(\Delta)$ has a unique subgame perfect equilibrium. In this equilibrium, in period 0, player 1 proposes the agreement $x^*(\Delta)$ defined in (64), which player 2 accepts.*

There is also a relation between the agreement $x^*(\Delta)$ and the Nash solution:

Theorem 14 *If the preference ordering of each player satisfies (C-1) through (C-6), then the limit, as $\Delta \to 0$, of the agreement $x^*(\Delta)$ reached in the unique subgame perfect equilibrium of $\Gamma(\Delta)$ is the agreement given by the Nash solution to the bargaining problem (S, d), where $d = (0, 0)$, and*

$$S = \{(s_1, s_2) \in \mathbb{R}^2 \mid (s_1, s_2) = (u_1(x_1), u_2(x_2)) \ for \ some \ (x_1, x_2) \in X\}. \quad (65)$$

5.4 Bargaining Model with a Risk of Breakdown

In the literature, many other generalizations of Rubinstein's bargaining model may be found. Consider, for instance, a model of alternating offers with a risk of breakdown. Such a game differs in two respects from the original game presented in Sect. 3. First, at the end of each period, after an offer has been rejected, there is a chance that the negotiation will end with a breakdown event B. More precisely, this event occurs independently with (exogenous) probability $0 < p < 1$ at the end of each period. The breakdown may be interpreted as the result of the intervention of a third party, who exploits the mutual gains. Second, each player is indifferent to the period in which an agreement is reached. Let us denote the resulting extensive game by $\Gamma(p)$. In this game, the risk of breakdown, rather than the players' impatience (as assumed in Sect. 3), motivates the players to reach an agreement as soon as possible. A strategy for each player in the game $\Gamma(p)$ is defined in exactly the same way as in the 'traditional' game of alternating offers (see Sect. 3.1). The result of the game $\Gamma(p)$ is a *lottery* denoted by $[[x, t]]$, in which the agreement $x \in X$ occurs with probability $(1 - p)^t$, and the breakdown event B occurs with probability $1 - (1 - p)^t$. We assume that each player $i = 1, 2$ has a complete, transitive, and reflexive preference ordering $\succeq_i$ over lotteries on $X \cup \{B\}$, which can be represented by the expected value of a continuous utility function $u_i : X \cup \{B\} \to \mathbb{R}$. We assume the following conditions:

(B-1) *The good is desirable:*
 For any $x, y \in X$, we have $x \succ_i y$ iff $x_i > y_i$ for $i = 1, 2$.
(B-2) *Breakdown is the worst outcome:*
 $(0, 1) \sim_1 B$ and $(1, 0) \sim_2 B$.
(B-3) *Risk aversion:*
 For any $x, y \in X$, and $\alpha \in [0, 1]$, each player $i = 1, 2$ either prefers the outcome $\alpha x + (1 - \alpha)y \in X$ to the lottery in which the outcome is x with probability α, and y with probability $1 - \alpha$, or is indifferent to the two.

It appears that assumptions (B-1), (B-2), and (B-3) are sufficient to allow application of both the Nash solution and Theorem 2 to the game $\Gamma(p)$. The preferences over lotteries of the form $[[x,t]]$ induced by the orderings $\succeq_i$ over lotteries on $X \cup \{B\}$ satisfy (A-1) through (A-6), when we replace the symbol (x,t) by $[[x,t]]$, and the symbol D by B. Each preference ordering over outcomes $[[x,t]]$ is complete and transitive, and for each $i = 1,2$

$$[[x,t_1]] \succ_i [[y,t_2]] \quad iff \quad (1-p)^{t_1} u_i(x) > (1-p)^{t_2} u_i(y). \tag{66}$$

For every lottery $[[x,t]]$ there is an agreement $y \in X$ such that $[[y,0]] \sim_i [[x,t]]$. Let $(x^*(p), y^*(p))$ be the unique pair of agreements satisfying

$$[[y^*(p),0]] \sim_1 [[x^*(p),1]] \quad and \quad [[x^*(p),0]] \sim_2 [[y^*(p),1]]. \tag{67}$$

We have the following theorems:

Theorem 15 *For each $p \in [0,1]$, the game $\Gamma(p)$ has a unique subgame perfect equilibrium. In this equilibrium, in period 0, player 1 proposes the agreement $x^*(p)$ defined by (67), which player 2 accepts.*

Theorem 16 *The limit, as $p \to 0$, of the agreement $x^*(p)$ reached in the unique subgame perfect equilibrium of $\Gamma(p)$ is the agreement given by the Nash solution to the bargaining problem (S,d), where $d = (0,0)$, and*

$$S = \{(s_1, s_2) \in \mathbb{R}^2 \mid (s_1, s_2) = (u_1(x_1), u_2(x_2)) \text{ for some } (x_1, x_2) \in X\}. \tag{68}$$

6   Applications of the Bargaining Models

There are many applications of the dynamic bargaining problem to the modelling of economic and social situations (see, for instance, [12], [13], [14], [15]).

Bargaining models can be applied to the problem of the utilization of common resources. Common exploitation is frequently the cause of arguments between countries, which, in order to solve the problem, negotiate the quantities of their shares. A good example concerning negotiations between two countries on the assignment of fishing quotas is presented in [4].

There are also many applications of a multi-party bargaining game. An example of such an application is described in [7], where a multi-party bargain is applied to a three-member legislative body with two rounds of offers. The members representing different districts bargain over the division of a certain 'legislative good', and the purpose of the bill is to propose the division of the 'good' among three players. Two cases are considered concerning the structure of the legislative game: one with the possibility of introducing an amendment and the other without such a possibility. Baron and Ferejohn ([1]) presented some applications of the n-party bargain to the legislative procedure, where $n > 3$.

In this paper, in particular, the bargaining models with preferences varying in time were presented. Making a decision using a bargaining game with preferences varying in time is important in situations with changing institutional conditions. The bargaining model with preferences varying in time can be applied, in particular, to modelling the negotiations on commission between a broker and a customer ([12], [14], [15]). The commission for the broker's services can be agreed on in negotiations with the institutional customer. Most broker agencies allow the possibility of negotiating the commission if the value of the transaction exceeds a certain amount of money. Similarly, the model analyzed may be applied to the negotiations between a bank and a customer with respect to the amount of credit interest and commission ([12], [14], [15]).

As another application of the bargaining model with preferences varying in time, we consider the example of a strike in a factory ([17]). The management and trade unions negotiate the division of the profits of a company. Both parties would like to reach an agreement, because they bear the costs of such a bargain. Because of the strike, the company does not fulfil contracts, the employer has to pay some fines, and workers may get no wages. The discount rates of both players are also perceptible in this example. A protracted strike is disadvantageous for the management and for the trade unions. On the one hand, the company's goodwill decreases in the eyes of potential clients, and, on the other hand, a lack of expected effects of negotiations decreases the credibility of the trade unions.

References

1. Baron, D.P., Ferejohn, J.A.: Legislatures. American Political Science Review (1989)
2. Fershtman, Ch.: A Note on Multi-Issue Two Sided Bargaining: Bilateral Procedures. Tel Aviv University, Working Paper No. 6-98 (1989)
3. Fishburn, P.C., Rubinstein, A.: Time Preferences. International Economic Review **23** (1982) 667–694
4. Houba, H., Sneek, K., Vardy, F.: Can Negotiations Prevent Fish Wars. Discussion Paper, Tinbergen Institute (1996)
5. Kalai, E., Smorodinsky, M.: Other Solutions to Nash's Bargaining Problem. Econometrica **43** (1975) 513–518
6. Klemisch-Ahlert, M.: Bargaining in Economic and Ethical Environments. An Experimental Study and Normative Solution Concepts. Springer - Verlag, Berlin (1996)
7. Morrow, J.D.: Game Theory for Political Scientists. Princeton University Press (1994)
8. Muthoo, A.: Bargaining Theory with Aplications. Cambridge University Press (1999)
9. Nash, J.: The Bargaining Problem. Econometrica **18** (1950) 155–162
10. Osborne, M.J., Rubinstein, A.: Bargaining and Markets. San Diego, Academic Press (1990)
11. Rubinstein, A.: Perfect Equilibrium in a Bargaining Model. Econometrica **50** (1982) 97-109
12. Rusinowska, A.: Bargaining Problems with Preferences Varying in Time. Journal of the College of Economic Analyses, No. 6 - Problems of Mathematical Economics and Financial Mathematics (in Polish), Warsaw (1998)
13. Rusinowska, A.: Bargain. In: Collective Decisions. Theory and Applications (in Polish). Publishing house - Scholar, Warsaw (1999) 139–156
14. Rusinowska, A.: Bargaining Problem with Non-stationary Preferences of the Players. Ph.D. Thesis (in Polish), Warsaw School of Economics, Warsaw (2000)

15. Rusinowska, A.: On Certain Generalization of Rubinstein's Bargaining Model. In: Petrosjan, L.A., Mazalov, V.V. (eds.): Game Theory and Applications, Vol. 8. Nova Science Publishers, Inc. New York (2002) 159–169

16. Rusinowska, A.: Subgame Perfect Equilibria in Model with Bargaining Costs Varying in Time. Mathematical Methods of Operations Research **56** (2002) 303–313

17. Rusinowska, A.: Bargaining Model with Sequences of Discount Rates and Bargaining Costs. International Game Theory Review. To appear (2002)

18. Selten, R.: Reexamination of the Perfection Concept for Equilibrium Points in Extensive Games. International Journal of Game Theory **4** (1975) 25–55

Categoric and Ordinal Voting:
An Overview

Harrie de Swart[1], Ad van Deemen[2], Eliora van der Hout[1], and Peter Kop[3]*

[1] Tilburg University, Faculty of Philosophy, P.O. Box 90153
5000 LE Tilburg, The Netherlands
H.C.M.deSwart@uvt.nl
http://www.uvt.nl/faculteiten/fww/medewerkers/swart
[2] Nijmegen School of Management, University of Nijmegen, The Netherlands
A.vanDeemen@nsm.kun.nl
[3] koppmgm@iclon.leidenuniv.nl

Abstract. There are many ways to aggregate individual preferences
to a collective preference or outcome. The outcome is strongly dependent on the aggregation procedure (election mechanism), rather than
on the individual preferences. The Dutch election procedure is based on
proportional representation, one nation-wide district, categoric voting
and the Plurality ranking rule, while the British procedure is based on
non-proportional representation, many districts, categoric voting and the
Plurality choice rule to elect one candidate for every district. For both
election mechanisms we indicate a number of paradoxes. The German
hybrid system is a combination of the Dutch and British system and
hence inherits the paradoxes of both systems. The STV system, used in
Ireland and Malta, is based on proportional representation (per district)
and on ordinal voting. Although designed with the best intentions - no
vote should be wasted - , it is prone to all kinds of paradoxes. May be the
worst one is that more votes for a candidate may cause him to lose his
seat. The AV system, used in Australia, is based on non-proportional representation (per district) and on ordinal voting. It has all the unpleasant
properties of the STV system. The same holds for the French majority-
plurality rule. Arrow's impossibility theorem is presented, roughly saying
that no 'perfect' election procedure exists. More precisely, it gives a characterization of the dictatorial rule: it is the only preference rule that is
IIA and satisfies the Pareto condition. Finally we mention characterizations of the Borda rule, the Plurality ranking rule, the British FPTP
system and of k-vote rules.

1 Introduction

In this overview, we give an analysis of election procedures and their properties.
An election mechanism can serve, given individual preference orderings of the

---

* We thank Marc Roubens for some useful suggestions and Sven Storms for his help
in translating the original Dutch manuscript. This paper is an extended version of
the original Dutch booklet 'Verkiezingen', published in 2000 by Epsilon Uitgaven,
Utrecht, The Netherlands. We thank Epsilon Uitgaven for permission to do so.

H. de Swart et al. (Eds.): TARSKI, LNCS 2929, pp. 147–195, 2003.

alternatives, to select one alternative, for instance, a travel goal or a chairman. In these cases, we speak of a (collective) *choice rule*. An election procedure can also be used to select a set of alternatives, for instance, a parliament or a set of potential bus stops. In this case, we speak of a (collective) *choice correspondence*. Finally, an election mechanism can be used to determine an order of collective preference regarding the alternatives, for instance, of candidates for the Eurovision Song Contest. In this case, we speak of a (collective) *preference rule*.

In section 2, it becomes clear that the outcome of elections is strongly dependent on the election procedure used. We consecutively consider: Most votes count (Plurality Rule), Pairwise comparison (Majority Rule), the Borda rule, and Approval voting. There are numerous other election procedures, too many to name here.

In sections 3 and 4, we distinguish four different kinds of election procedures that are used in most Western European countries to elect parliament and government. Subsequently, we show that each of the four globally distinguished election procedures is subject to paradoxes. By 'paradox' we mean an outcome that is contrary to what one would prima facie expect or contrary to our sense of justice and honesty. For instance, it is a paradox that more votes for a candidate or party under a specific election procedure can mean that the candidate or party is worse off. (This is the Negative Responsiveness paradox for the election procedure designated by STV.)

In Section 3, we compare the Dutch election procedure to the British one. In section 3.2, paradoxes are discussed that occur, or can occur, within the Dutch system. Section 3.3 considers the paradoxical properties of the British system. The hybrid election procedure that is used in Germany is treated in section 3.4. This system, which is a combination of the Dutch and the English systems, also has its own paradoxes.

The Single Transferable Vote (STV) and the Alternative Vote (AV) systems are discussed in Section 4. Section 4.2 elaborates on the properties of the STV election procedure that is used in Ireland and Malta. Section 4.4 considers some paradoxes that may occur in the election procedure that is used in Australia. Finally, in section 4.5, we deal with the French election system, which is very similar to the AV system that is used in Australia and is similarly the cause of several paradoxes.

Naturally, the question then arises if there are any 'good' election procedures, that is, election procedures that, at any rate, do not have the unwanted properties that we noted in the chapters mentioned above. Kenneth Arrow addressed this question over fifty years ago. In Section 5, we examine Arrow's result, which is essentially a characterization of the dictatorial rule. Although no 'perfect' election procedure exists, some procedures are 'better' than others. One way to decide on this, is by studying the characteristic properties of these procedures. We mention characterizations of the Plurality ranking rule, of the Borda rule and of k-vote rules.

2   Other Procedure, Other Outcome

In this section, we consider a number of election procedures. These are procedures by which the outcome of an election is determined. At first glance, you might think that this is simple: most votes count. Doesn't that seem fairest? However, we will see that there are objections to the 'Most votes count' (Plurality Rule) election procedure. Hence, we also look at other procedures: the Majority Rule, the Borda rule, and Approval Voting.

For all our examples, we assume that we know the individual preferences of the voters. A survey (the whole) of all individual preferences is called a *(voter)profile*, denoted by the symbol p or q. An election procedure is a procedure that assigns to each (voter)profile an outcome (of the election).

In Example 1 (see below), we will show that different election procedures may produce different outcomes. This means that one can doubt whether the outcome generated by any single procedure is the 'best' or 'correct' outcome. In other words, one can doubt the appropriateness and quality of the used procedure.

Example 1: A group of secondary school students is given the choice between Venice, Florence, and Siena as the destination of their school trip. Each student is allowed to give his or her order of preference, for instance,

Venice Siena Florence.

This means that Venice is the first preference of this student, Siena the second, and Florence the third. Now suppose that there are 31 students with the following individual preferences.

Florence	Venice	Siena	: 5 students
Florence	Siena	Venice	: 7 students
Venice	Florence	Siena	: 3 students
Venice	Siena	Florence	: 7 students
Siena	Florence	Venice	: 3 students
Siena	Venice	Florence	: 6 students

Such a survey of individual preferences is called a *profile*, usually denoted by the letter p. Election procedures aggregate profiles of individual preferences to an outcome.

In this example, if each student is allowed to give his or her first preference and the procedure 'Most votes count' is applied, then Florence, with $5 + 7 = 12$ votes, will be selected. Later, we will see that other election procedures might assign different outcomes to this same (voter)profile.

We also discuss a number of important properties of election procedures in this chapter, such as the Pareto condition, the condition of Independence of Irrelevant Alternatives (IIA), and the monotonicity-condition. We will explain these conditions using examples.

2.1   Plurality Rule

The election procedure 'Most votes count' only considers the first preferences of the voters; second, third, etc., preferences are not considered. For 'Most votes count' (Plurality Rule), alternative x is *collectively* (by the community) *preferred* to alternative y if the number of persons that prefer x is greater than the number of persons that prefer y. In particular, if one choice is needed, the alternative that is put first by most people will be elected. We call x and y (collectively) *indifferent* if the number of individuals that prefer x is equal to the number of individuals that prefer y.

If there are just two alternatives, or candidates x and y, 'x is collectively preferred over y' means that x gets more than half of the (first) votes.

In the (voter)profile of Example 1, Florence is mentioned 12 times as first preference, Venice 10 times, and Siena 9 times. Therefore, on application of the 'Most votes count' election procedure, Florence will become - as we already saw - the destination of our class. In other words, Florence is the (collective) choice of our class under application of the 'Most votes count' election procedure.

Not only can 'Most votes count' be used to determine a collective choice, but also to determine a collective order of preferences. In that case one speaks of the *Plurality ranking rule*. Given the profile of Example 1, the collective order of preference on application of 'Most votes count' will be

Florence Venice Siena.

This corresponds to the fact that Florence gets more first votes than Venice, which in turn gets more first votes than Siena.

Suppose that later on it turns out that Venice is so expensive that it was not a realistic alternative. One could then argue that a new vote is not needed, as Venice was not the chosen destination anyway. However, if Venice is no longer an alternative and the preferences of the students remain unchanged as far as the other alternatives are concerned, the preferences of the 31 students will be as follows:

Florence	Siena	: 5 students
Florence	Siena	: 7 students
Florence	Siena	: 3 students
Siena	Florence	: 7 students
Siena	Florence	: 3 students
Siena	Florence	: 6 students

Now there are 15 students with Florence as first preference and 16 with Siena as first preference. So, on application of 'Most votes count', Siena would be elected as the destination instead of Florence.

We say that 'Most votes count' is not *Independent of Irrelevant Alternatives* (not IIA): although Venice is an irrelevant alternative, because of the cost, the outcome is not independent of this alternative. The property 'Independent of Irrelevant Alternatives (IIA)' can also be described as follows: adding irrelevant (non eligible) alternatives does not influence the outcome.

'Most votes count' is frequently used in real life: it is the foundation of many election systems that are in current use, such as the Dutch and British systems (see section 3.1). Nonetheless, this system has some serious drawbacks, as we will explain below and in subsections 3.2 and 3.3. (Here, we will follow the exposition of Van Deemen, 1997).

In the first place, a choice made using the procedure 'Most votes count' is not necessarily a majority choice. This remarkable fact was discovered as early as 1781 by the Frenchman J.-C. de Borda (1781), one of the founders of Social Choice Theory. To clarify this, we consider the voter profile of Example 1.

- Check that there are $0 + 0 + 3 + 7 + 0 + 6 = 16$ students that prefer Venice to Florence, and 15 that prefer Florence to Venice. In other words, if the students have to choose between Florence and Venice, they will choose Venice.
- Check that there are $0 + 0 + 0 + 7 + 3 + 6 = 16$ students that prefer Siena to Florence, and 15 that prefer Florence to Siena.
- Also check that there are $0 + 7 + 0 + 0 + 3 + 6 = 16$ students that prefer Siena to Venice, and 15 that prefer Venice to Siena.

Hence, we can conclude that

1) On pairwise comparison, Florence has a minority of the votes with respect to both Venice and Siena: for this reason, Florence is called a *Condorcet loser*, after the French Marquis de Condorcet (1743 - 1794).

2) On pairwise comparison, Siena has a majority of the votes with respect to both Florence and Venice, and, hence, Siena is the majority choice of our class; for that reason, Siena is called the *Condorcet winner*.

From the above, it follows that the winner on application of 'Most votes count' (Florence) need not be the majority choice (Siena). In other words, the *majority principle* is violated by 'Most votes count'.

To clarify the second drawback of 'Most votes count', we consider the following voter profile.

Florence	Paris	London	Venice	Siena	: 10 voters
Siena	Paris	Venice	Florence	London	: 8 voters
Venice	Siena	Paris	London	Florence	: 7 voters

As neither Paris nor London is the first preference of any voter, they are collectively indifferent on application of 'Most votes count': for each city, the number of individuals for whom it is first choice is 0. However, everyone prefers Paris to London. How odd! Everyone prefers Paris to London, but this is not shown in the outcome: Paris and London are equally preferred in the outcome.

The aforementioned comes down to the fact that the election procedure 'Most votes count' violates the so-called Pareto condition. This *Pareto condition* goes as follows: if every individual prefers alternative x to alternative y, then, in the outcome, x must also be (collectively) preferred to y.

The third drawback of 'Most votes count' is that it does not have the monotony property. This *monotony property* (positive responsiveness) says that if an alternative x is raised vis-a-vis an alternative y in someone's preference ordering, and x goes down in no one's preference ordering vis-a-vis y, then x must also be raised vis-a-vis y in the collective preference ordering. To see that 'Most votes count' does not have this monotony property, we consider the following (voter)profile p. With (xy) we mean that x and y are indifferent in the preference ordering.

Profile p: Florence (Paris London)   Venice   Siena   : 10 students
            Siena   (Paris London)   Venice   Florence : 8 students
            Venice   (Paris London)   Siena   Florence : 7 students

Because neither Paris nor London occurs as first preference in the preference ordering of the students, on application of 'Most votes count' both are indifferent.

But now consider the following profile q, identical to profile p except for the fact that everybody now prefers Paris to London in his or her preference ordering.

Profile q: Florence Paris   London   Venice   Siena   : 10 students
            Siena   Paris   London   Venice   Florence : 8 students
            Venice   Paris   London   Siena   Florence : 7 students

Comparing the profiles p and q, in profile q everybody has ranked Paris higher in his or her preference ordering than London. So, according to the monotony property, Paris should now be (collectively) preferred by the community to London. However, on application of 'Most votes count', this is not the case, since neither Paris nor London is the first preference of an individual and, hence, they are indifferent in the collective preference (if this is determined by 'Most votes count'). Consequently, the election procedure 'Most votes count' may not react to changes in the individual preferences, which seems at odds with the idea of democracy.

Given these results, it is no wonder that Borda and Condorcet had little faith in 'Most votes count'!

2.2   Profiles, Choice, and Preference Rules

In this subsection, we will formulate in a mathematically precise way a number of properties that were introduced informally in the previous subsection, as well as add some new mathematical notions. Amongst others, the following concepts will be defined: relation, weak and linear ordering, profile, choice rule, choice correspondence, preference rule, and Independence of Irrelevant Alternatives.

The individual order of preference 'Florence Venice Siena' can be rendered by the following *(preference-)relation* R:

$$R = \{<\text{Florence, Venice}>, <\text{Venice, Siena}>, <\text{Florence, Siena}>\}.$$

Here $< x, y >$ is an *ordered pair*, and $< x, y > \in R$ is read as 'x is *at least as good* as y'. Instead of $< x, y > \in R$, we usually write xRy.
'x is (strictly) *preferred to* y' now corresponds with 'xRy and not yRx', while 'xRy en yRx' states that 'x and y are *indifferent*', which is often denoted by (xy).

Suppose that A is a set of alternatives, for instance, $A = \{$Florence, Venice, Siena$\}$ and that N is a set of individuals, for instance, $N = \{$student 1, ..., student 31$\}$. Then we can identify for every individual i in N his or her individual preference ordering with respect to the alternatives in A by means of a *relation* R_i on A, also called a *preference-relation* on A.

Definition 1 R is a *(preference-)relation* on A if R is a set of ordered pairs $< x, y >$ with $x, y \in A$. Instead of writing $< x, y > \in R$, one can also write xRy.

Definition 2 Let R be a (preference-)relation on A.

R is *complete* if, for all $x, y \in A$, xRy or yRx. That is, a relation on A is complete if every alternative in A is comparable to every alternative in A, including itself. Recall that xRy is read as 'x is at least as good as y'.

R is *transitive* if, for every $x, y, z \in A$, if xRy and yRz, then xRz. That is, if x is at least as good as y by R and y is at least as good as z by R, then x is at least as good as z by R. Thus, in transitive preference relations, the preferences are consequent.

R is *antisymmetric* if, for every $x, y \in A$ with $x \neq y$, if xRy, then not yRx. That is, a relation is antisymmetric if indifference between two distinct alternatives does not occur. 'xRy and not yRx' is read as: x is (strictly) preferred to y by R.

Definition 3 A preference relation R is a *weak ordering* on A if R is complete and transitive. R is a *linear ordering* on A if R is complete, transitive, and antisymmetric. Hence, there can be indifference in weak orderings, but not in linear orderings.

Definition 4 $C(A)$ is, by definition, the set of all complete relations on A. $W(A)$ is, by definition, the set of all weak orderings on A. $L(A)$ is, by definition, the set of all linear orderings on A. Because every linear ordering is, by definition, also a weak ordering, it follows that $L(A)$ is a subset of $W(A)$, while $W(A)$, in its turn, is a subset of $C(A)$: $L(A) \subseteq W(A)$ and $W(A) \subseteq C(A)$.

For the sake of simplicity, we will limit ourselves to individual preferences R_i that are linear orderings. With a *profile* p, we mean a combination of individual linear orderings.

Definition 5 A *profile* p associates with every individual i in N a linear ordering R_i on A, in other words,

$$\text{a } profile \text{ is a function } p : N \to L(A).$$

$p(i)$ or R_i is the individual linear ordering of individual i in profile p. $L(A)^N$ is the set of all profiles.

So, in Example 1, a profile is given for which

$N = \{$student 1,..., student 31$\}$ and $A = \{$Florence, Venice, Siena$\}$.

A group of individuals can make three kinds of collective decision on the basis of a given voter profile (a combination of individual preferences).

1. It can choose one alternative, for instance, a travel destination, a chairman, a president, or a location for a sporting facility.
2. It can choose a collection of alternatives, for instance, a parliament, a food package, or a set of potential locations for a waste dump.
3. It can determine an order of preference of the alternatives, for instance, of applicants or of candidates for the Eurovision Song Festival.

In Case (1), we call the election procedure a *(collective) choice rule*, in Case (2), we call it a *(collective) choice correspondence*, and, in Case (3), we call it a *(collective) preference rule*.

Definition 6 Let N be a set of individuals and A a set of (at least 3) alternatives.

1. A *(collective) choice rule* is a function $K : L(A)^N \to A$. Thus, a choice rule K assigns to each profile $p \in L(A)^N$ a collective choice $K(p)$ in A.
2. A *(collective) choice correspondence* is a function $C : L(A)^N \to P(A)$, where $P(A)$ is the powerset of A. This is the collection of all subsets of A. Therefore, a choice correspondence C assigns to each profile $p \in L(A)^N$ a set $C(p)$ of collective choices in A.
3. A *(collective) preference rule* is a function $F : L(A)^N \to C(A)$. Thus, a preference rule F assigns to each profile $p \in L(A)^N$ a complete preference relation $F(p)$ on A.

The election procedure 'Most votes count' can be seen as a (collective) choice rule or choice correspondence and as a (collective) preference rule.

Definition 7 Suppose N is a set of individuals and A is a set of alternatives. Given a profile p and an alternative x in A, we define $t(x, p)$ as the number of individuals i in N that have x as the *first* preference in $p(i)$ (i.e., for which there is no alternative y in A that is more preferred by i than x in $p(i)$).

'Most votes count' as a preference rule is now rendered by the function Pl (Plurality) from $L(A)^N$ to $W(A)$, defined as

$$x Pl(p) y \text{ if and only if } t(x, p) \geq t(y, p).$$

In other words, $x Pl(p) y$ if and only if the number of individuals that prefer x most in p is greater than or equal to the number of individuals that prefer y most in p. Note that $Pl(p)$ is a weak ordering on A and, in general, not a linear ordering, because there can be two or more alternatives that occur equally often as first preference in p.

Definition 8 The collective preference rule Pl gives rise to the collective choice correspondence P (Plurality), $P : L(A)^N \to P(A)$, with $P(p)$, by definition, the set of all x in A such that, for all y in A, $x Pl(p) y$. Therefore, $P(p)$ is the set of all alternatives x in A for which there is no alternative y in profile p which is more frequently preferred most in p.

Definition 9 Let $F : L(A)^N \to C(A)$ be a (collective) preference rule. F is *Independent of Irrelevant Alternatives* (IIA) if, for all $x, y \in A$ and for all profiles $p, q \in L(A)^N$, if p limited to x and y is equal to q limited to x and y, then $F(p)$ limited to x and y is equal to $F(q)$ limited to x and y.

So, if p is the profile from Example 1 and q is the same profile but without Venice or with Venice as last choice, then p limited to Florence and Siena is equal to q limited to Florence and Siena. Now, let Pl (Plurality) be the (collective) preference rule that corresponds to 'Most votes count'. $Pl(p) = \{$<Florence, Venice>, <Venice, Siena>, <Florence, Siena>$\}$ and $Pl(q) = \{$<Siena, Florence>, <Siena, Venice>, <Florence, Venice>$\}$. Then, $Pl(p)$ limited to Florence and Siena would be $\{$<Florence, Siena>$\}$, but $Pl(q)$ limited to Florence and Siena would be $\{$<Siena, Florence>$\}$. So, Pl, which is 'Most votes count', is not Independent of Irrelevant Alternatives (not IIA).

2.3   Majority Rule (Pairwise Comparison)

The *majority principle* states that if the number of voters that prefer alternative x to alternative y is larger than the number of voters that prefer y to x (in other words, if x *defeats* y), then x must also be preferred to y in the outcome. It follows from this that, if there is an alternative x that defeats every other alternative in pairwise comparison, this alternative x must win. Such an alternative is called a *Condorcet winner*.

In the previous section, we saw how, given a voter profile, the Condorcet winner is determined and that this Condorcet winner need not be the winner under application of 'Most votes count'. In fact, with the voter profile of Example 1, the winner under application of 'Most votes count' (Florence) is the *Condorcet loser*: on pairwise comparison, Florence loses from both Venice and Siena.

It is difficult to justify the fact that a candidate or party preferred by a minority, may get elected or receive more seats than a candidate or party that is preferred by a majority. Therefore, Borda (1781) and Condorcet (1788) concluded that the procedure 'Most votes count' is seriously defective, because it does not satisfy the majority principle.

As the majority principle seems so plausible, one could wonder why we still use other procedures. The answer is simple: there are profiles that have no Condorcet winner. The most famous example is the following so called *Condorcet profile* p (in which k is a random natural number, $k \geq 1$):

Florence	Venice	Siena	: k students
Venice	Siena	Florence	: k students
Siena	Florence	Venice	: k students

The Majority Rule (pairwise comparison) applied to the above Condorcet profile leads to a collective order of preference that is *not transitive*, meaning that alternatives x, y, and z exist, in our example, respectively, Florence, Venice, and Siena, such that x defeats y and y defeats z, but x does not defeat z. Hence, for the above Condorcet profile, no Condorcet winner can be found.

The absence of a Condorcet winner for a profile is also called the *Condorcet paradox* or *voting paradox*.

One might wonder if the probability of an occurrence of the Condorcet paradox in actual elections is significantly large. Bill Gehrlein (1981) showed that, under certain assumptions, the probability, in the case of three alternatives, is $\frac{1}{16}$ if the number of individuals is large. For more than three alternatives, the probability of the Condorcet paradox occurring increases; see [16].

Despite the Condorcet paradox, the Majority Rule (pairwise comparison) has a number of properties that come close to the ideal of a democracy. In [10], 3.2.1, Van Deemen notes that the Majority Rule (pairwise comparison) has the following properties.

- *Anonymity:* Individuals are treated equally. It does not matter from whom the preferences originated, the only thing that counts are the preferences themselves. Personal qualifications of the individuals are irrelevant to the determination of the collective choice. Anonymity prevents unequal treatment of individuals: it erects a barrier to any form of discrimination. Note that 'Most votes count' is also an anonymous election procedure.
- *Neutrality:* The alternatives are treated equally. Every opinion counts, independent of its content. Note that 'Most votes count' also has this property.
- *Independence of Irrelevant Alternatives (IIA):* The determination of the collective preference with respect to two alternatives x and y is not influenced by a third (irrelevant) alternative. In Section 2.1, we have seen that 'Most votes count' is not IIA.
- *Pareto condition:* If everybody prefers alternative x to alternative y, then x will also be collectively preferred to y. In Section 2.1, we have seen that 'Most votes count' does not satisfy the Pareto condition.
- *Monotony:* If an alternative x is raised vis-a-vis an alternative y in someone's preference ordering and x goes down in no one's preference vis-a-vis y, then, on pairwise comparison, x will also be raised vis-a-vis y in the collective order of preference. A voting procedure that does not have this property can be regarded as having a certain inertia: it cannot register changes in the profiles and adapt its outcome in accordance with these changes. In Section 2.1, we showed that 'Most votes count' is not monotonic.

We can speak of an election procedure even in the case of dictatorship. An individual is called a *dictator* if, for every voter profile p, the collective preference is exactly the preference of that individual. The *dictatorial preference rule* with dictator i assigns to each voter profile p the preference of i. See [34], pp. 70-72.

For instance, consider a class with individual preferences as in Example 1 and a teacher with preference ordering Venice Florence Siena. If the teacher plays the role of dictator, the class will go to Venice.

Check that a dictatorial preference rule is not anonymous, but neutral, IIA, and satisfies the Pareto Condition. (In Section 5, we will see that the dictatorial preference rule is the only preference rule that is IIA and satisfies the Pareto condition.)

In the next section, we will formulate the above mentioned properties in a mathematically precise way.

2.4   Properties of the Majority Rule

Definition 10 Given a profile p, an alternative x *defeats* an alternative y *on pairwise comparison* if the number of individuals that prefer x to y in profile p is greater than the number of individuals that prefer y to x in p. Given a profile p, we write 'x defeats y on pairwise comparison' as $xM(p)y$ (the M stands for *Majority Rule*). This defines the collective preference rule $M : L(A)^N \rightarrow C(A)$.

A *Condorcet winner* is an alternative that defeats any other alternative on pairwise comparison.

Note that the relation $M(p)$ need not be transitive, for instance, if p is a Condorcet profile. Also note that there can be several Condorcet winners. For instance, in the following profile p, where (xz) means that x and z are indifferent.

$$
\begin{array}{llll}
z & x & y & : 3 \\
y & x & z & : 3 \\
& (xz)y & : 1
\end{array}
$$

Definition 11 A permutation σ of N is a bijective function from N to N. We can see a permutation σ of N as a name change for all individuals in N. After application of σ, individual i is named $\sigma(i)$.

Let p be a profile in $L(A)^N$. Then $p \circ \sigma$ is, by definition, the profile in which each individual i plays the role of $\sigma(i)$ in p. So, for all i in N, $(p \circ \sigma)(i) = p(\sigma(i))$.

Example: Suppose that $N = \{a(d), b(ob), c(ees)\}$ and that $\sigma(a) = b$, $\sigma(b) = c$ and $\sigma(c) = a$. Suppose also that $A = \{\text{Florence, Venetië, Siena}\}$ and that profile p is given by

$$
\begin{array}{llll}
p(a) : & \text{Florence} & \text{Venice} & \text{Siena} \\
p(b) : & \text{Florence} & \text{Siena} & \text{Venice} \\
p(c) : & \text{Venice} & \text{Florence} & \text{Siena}
\end{array}
$$

Then, $p \circ \sigma$ is the following profile:

$$
\begin{array}{llll}
p \circ \sigma(a) = p(b) : & \text{Florence} & \text{Siena} & \text{Venice} \\
p \circ \sigma(b) = p(c) : & \text{Venice} & \text{Florence} & \text{Siena} \\
p \circ \sigma(c) = p(a) : & \text{Florence} & \text{Venice} & \text{Siena}
\end{array}
$$

It can be easily seen that $M(p \circ \sigma) = M(p) = \{<\text{Florence, Venice}>, <\text{Venice, Siena}>, <\text{Florence, Siena}>\}$.

Definition 12 A collective preference rule $F : L(A)^N \rightarrow C(A)$ is *anonymous* if, for all profiles p in $L(A)^N$ and for every permutation σ of N, $F(p \circ \sigma) = F(p)$.

Definition 13 Suppose τ is a permutation of A and R is a complete relation on A. Then τR is, by definition, the set of all pairs $< \tau(x), \tau(y) >$ with $< x, y >$ in R. So, in τR, $\tau(z)$ plays the role of z in R.

Let p be a profile in $L(A)^N$. Then, τp is, by definition, the profile with $(\tau p)(i) = \tau(p(i))$ for all i in N. τp originates from p by applying the permutation τ on the alternatives.

Example: Suppose that $N = \{$a, b, c$\}$ and $A = \{$Florence, Venice, Siena$\}$. Suppose τ is the permutation of A given by $\tau($Florence$) = $ Venice, $\tau($Venice$) = $ Florence en $\tau($Siena$) = $ Siena. And suppose that p is the following profile:

$p($a$)$:   Florence   Venice   Siena
$p($b$)$:   Siena   Venice   Florence
$p($c$)$:   Venice   Siena   Florence

Then, $M(p) = \{<$Venice, Siena$>$, $<$Siena, Florence$>$, $<$Venice, Florence$>\}$.

The profile τp now originates from profile p by interchanging the alternatives Florence and Venice:

$\tau p($a$) = \tau(p($a$))$: Venice   Florence   Siena
$\tau p($b$) = \tau(p($b$))$: Siena   Florence   Venice
$\tau p($c$) = \tau(p($c$))$: Florence   Siena   Venice

It can now be easily seen that $M(\tau p) = \tau(M(p)) = \{<$Florence, Siena$>$, $<$Siena, Venice$>$, $<$Florence, Venice$>\}$.

Definition 14 A collective preference rule $F : L(A)^N \rightarrow C(A)$ is *neutral* if, for every permutation τ of A and for every profile p, $F(\tau p) = \tau(F(p))$.

Definition 15 A collective preference rule $F : L(A)^N \rightarrow C(A)$ satisfies the *Pareto condition* if, for every profile p in $L(A)^N$ and for all alternatives x, y in A, if for every $i \in N$ $xp(i)y$ (and hence not $yp(i)x$), then $xF(p)y$ and not $yF(p)x$.

Definition 16 A collective preference rule $F : L(A)^N \rightarrow C(A)$ is *monotonic* if, for all profiles p, q in $L(A)^N$ and for all alternatives x, y in A, if

1. for all $i \in N$, if $xp(i)y$ (and hence not $yp(i)x$), then $xq(i)y$ (and hence not $yq(i)x$), and

2. there is an individual $k \in N$ such that $yp(k)x$ and $xq(k)y$, then $xF(p)y$ implies that $xF(q)y$ and not $yF(q)x$.

As was mentioned earlier, the following theorem is easy to see.

Theorem 1 The collective preference rule M (Majority Rule) is anonymous, neutral, IIA, monotonic, and satisfies the Pareto condition, but it is not transitive.

In order to avoid the voting paradox or non-transitivity, Copeland modified the Majority Rule in the following way. The Copeland score of an alternative x

given profile p is by definition the number of alternatives y such that x defeats y on pairwise comparison given p. The Copeland preference rule $F_{Copeland}$ is now defined by $xF_{Copeland}y$ if and only if the Copeland score of x given p is greater than or equal to the Copeland score of y given p. So, x is more preferred than y by $F_{Copeland}(p)$ if and only if x defeats more alternatives than y given p. Evidently, the Copeland preference rule is transitive, as well as anonymous and neutral, it satisfies the Pareto Condition, but it is not IIA.

2.5 Borda (Preference) Rule

In 1781, the Frenchman J.C. de Borda noted that, with 'Most votes count', the second, third, etc., preferences of the individuals have no weight in determining the outcome. Borda proposed giving weight to all the positions of the alternatives in the individual preferences. Hence, not only the first preference of the individuals is taken into account, but also their second, third, etc. If an individual i has 'Florence Venice Siena' as individual preference ordering, Florence gets 3 points, Venice 2, and Siena 1. Subsequently, a decision is made based on the total score of every alternative in a given profile p. For n alternatives, every individual gives n points to his or her most preferred alternative, $n - 1$ points to his or her second choice, etc., and 1 point to his or her least preferred alternative.

If we apply the Borda preference rule to Example 1 (see page 149), Florence, Venice, and Siena will get the following numbers of points:

Florence: $5 \times 3 + 7 \times 3 + 3 \times 2 + 7 \times 1 + 3 \times 2 + 6 \times 1 = 61$
Venice: $5 \times 2 + 7 \times 1 + 3 \times 3 + 7 \times 3 + 3 \times 1 + 6 \times 2 = 62$
Siena: $5 \times 1 + 7 \times 2 + 3 \times 1 + 7 \times 2 + 3 \times 3 + 6 \times 3 = 63$

The *Borda score* of an alternative x for a given profile p is now, by definition, the total number of points that the individuals have given to x. In Example 1, the Borda score of Florence is 61, the Borda score of Venice is 62, and the Borda score of Siena is 63.

According to the *Borda (preference) rule*, the collective ordering of the alternatives will then be

Siena Venice Florence.

Note that, for 'Most votes count', the outcome for the profile of Example 1 is exactly the opposite,

Florence Venice Siena,

because, in Example 1, Florence is preferred 12 times, Venice 10 times, and Siena 9 times. Also note that Siena, with the highest Borda score, happens to be the Condorcet winner in Example 1.

The obvious question now is if the Condorcet winner, if one exists, will always have the highest Borda score. Unfortunately, this is not the case, as is shown by the following example.

A group of seven people go out for dinner. The restaurant offers three menus: a, b, and c. As there is a reduction if they all take the same menu, they decide to choose collectively. But which menu should be chosen? The individual preferences are given in the profile below.

c a b : 3 persons
a b c : 2 persons
a c b : 1 person
b c a : 1 person

1) Check that c is the Condorcet winner for this profile!
2) Now check that c, when the Borda procedure is applied to this profile, only receives 15 points, while a gets 16 points under these circumstances. Thus, an alternative with the highest Borda score need not be the Condorcet winner.

The profile just mentioned also illustrates that, like 'Most votes count', the Borda procedure is not Independent of Irrelevant Alternatives (not IIA).

1) On application of the Borda procedure on the profile just given, the collective order of preference is a c b.
2) When they want to order menu a, the waiter tells them this is very convenient, as menu b cannot be served today. You might think this information is unimportant. However, if the Borda procedure is applied in this new situation (only a and c), the collective order of preference will become c a.

So, for the Borda procedure, the presence of the (irrelevant) alternative b influences the preference between a and c. Hence, the collective choice between a and c, on application of the Borda procedure, is dependent on all alternatives, in particular on the irrelevant alternative b.

Note that, when there are two alternatives, the Borda procedure corresponds to 'Most votes count' as well as to the Majority Rule. Suppose that there are two alternatives, x and y, and $m+n$ individuals, and that the individual preferences are given in the following profile:

x y : m voters
y x : n voters

Then, the Borda score of x equals $2m+n$ and the Borda score of y equals $2n+m$. Now $2m + n > 2n + m$ if, and only if, $m > n$. Thus, the Borda score of x is greater than that of y precisely when the number of voters (m) that prefer x to y is greater than the number of voters (n) that prefer y to x.

The reader may easily verify the following theorem.

Theorem 2 The Borda (preference) rule is anonymous, neutral, not IIA, monotonic, transitive and satisfies the Pareto condition.

It is worth mentioning that the Majority Rule and the scoring procedure (generally ascribed to Borda) were in fact first proposed respectively by Ramon Lull ($\pm$ 1235 - 1315) and Nicolas Cusanus (1401 - 1464), as reported in [26] and [28].

2.6   Strategic Behavior

We have already seen that the Borda preference rule does not necessarily pick out the Condorcet winner, if there is one. Another drawback of the Borda rule is that it is sensitive to strategic behavior. This means that individuals can profit from giving an insincere preference instead of their true preference. To illustrate this, we consider the following profile (17 voters):

Florence	Venice	Siena	: 7 students
Venice	Florence	Siena	: 6 students
Siena	Venice	Florence	: 4 students

The Borda score for Siena is $7 \times 1 + 6 \times 1 + 4 \times 3 = 25$. The Borda score for Florence is $7 \times 3 + 6 \times 2 + 4 \times 1 = 37$. The Borda score for Venice is $7 \times 2 + 6 \times 3 + 4 \times 2 = 40$. So the outcome is

$$\text{Venice Florence Siena.}$$

Venice ends above Florence. The first group of 7 students, preferring Florence to Venice, can now act strategically: instead of giving their true preferences, they can vote as follows:

$$\text{Florence Siena Venice}$$

Venice now gets 7 points less: $40 - 7 = 33$, while Siena gets an extra 7 points: $25 + 7 = 32$. As the score of Florence remains unaltered, 37 points, the resulting collective ordering is now

$$\text{Florence Venice Siena.}$$

This is exactly the outcome desired by the first group of seven students.

In this example, a coalition of seven voters acts strategically and benefits from this. One could remark that the strategic behavior of a coalition presupposes internal attunement and, hence, would be difficult to realize in practice. The next example shows that one person can also benefit from strategic behavior, assuming that the other voters give their true preferences.

Suppose that there are five alternatives, x, y, z, u, and v, and seven voters. Also suppose that the (sincere) individual preferences are given in the following profile:

x	y	z	u	v	: 3 persons
z	x	y	u	v	: 2 persons
y	z	x	u	v	: 2 persons

Now, the Borda score of x is 29, that of y 28, of z 27, of u 14, and of v 7. So, on application of the Borda preference rule, the outcome for the above profile will be

$$x \ y \ z \ u \ v.$$

Now, suppose that one of the last two voters foresees this outcome. Now this voter can accomplish a new outcome, which is more attractive to this voter than the original outcome, by means of strategic behavior, by giving the insincere preference $y\ z\ u\ v\ x$, where the Borda winner x is put at the lowest position.

Thus, the Borda procedure gives a voter the possibility to get his or her preferred outcome by giving an insincere order of preference. Hence, on application of the Borda procedure, cheating can be advantageous. The Borda procedure is *not immune to strategic behavior*, or the Borda procedure is *manipulable*.

When Borda was informed of the fact that his procedure was sensitive to strategic behavior, he apparently answered that his procedure was only intended for honest men (Black, 1958, p. 238).

Despite the fact that there are obvious objections to the Borda procedure, it scores relatively well in a comparison of many election procedures (Brams and Fishburn). We would like to quote the following passage from the conclusions of [6]:
'Among ranked positional scoring procedures to elect one candidate, Borda's method is superior in many respects, including susceptibility to strategic manipulation, propensity to elect Condorcet candidates, and ability to minimize paradoxical possibilities. ... Despite Borda's superiority in many respects, it is easier to manipulate than many other procedures. For example, the strategy of ranking the most serious rival of one's favorite candidate last is a transparent way of diminishing the rival's chances.'

'Most votes count' is also sensitive to strategic behavior. This can be seen as follows. For the profile p in Example 1 (page 149), Florence is chosen on application of 'Most votes count'. However, the seven students with individual preference orderings 'Venice Siena Florence' would rather go to Siena than to Florence. This coalition of seven students can accomplish that, on application of 'Most votes count', Siena becomes the collective destination, by giving the insincere individual preference ordering 'Siena Venice Florence'.

The obvious question now is whether the Majority Rule is sensitive to strategic behavior. It can be shown that the possible strategic behavior of a *coalition S*, a group of voters, in determining a Condorcet winner would be disadvantageous for at least one of the members of that coalition. So, on application of the Majority Rule (pairwise comparison) for any coalition, there will be at least one member that is disadvantaged due to the strategic behavior of his or her coalition.

Theorem 3 Suppose S is a coalition. Suppose that profile p renders the true preferences of the voters and that q is the profile in which the individuals in S give insincere preference orderings instead of true preference orderings. Let alternative x be the Condorcet winner for the true profile p and alternative y the Condorcet winner for the insincere profile q. Also suppose that $x \neq y$. Then there is an individual i in coalition S that prefers alternative x to y. Hence, for that individual, the strategic behavior of the coalition S is disadvantageous, as y will be the Condorcet winner for q, while individual i prefers x.

Proof: Suppose S is a coalition, that is, a (sub)set of individuals. Also suppose that x is the Condorcet winner for the true profile p and that y is the Condorcet winner for the insincere profile q, in which only the individuals in S do not give their true preference orderings. Furthermore, suppose that $x \neq y$. Because x is the Condorcet winner for profile p, for profile p it holds that x defeats y on pairwise comparison. And because y is the Condorcet winner for q, it holds for profile q that y defeats x on pairwise comparison. Hence, there is an individual i such that

1. i prefers x to y for p, and
2. i prefers y to x for q. (Somebody must have switched preferences.)

As only voters from coalition S give different preference orderings, individual i must be in coalition S. Since i prefers x to y, i is punished for the strategic behavior of the coalition S to which he or she belongs. □

2.7   Approval Voting

Approval Voting assumes that the voter can divide the alternatives into two classes: the candidates that he or she approves of and the ones that he or she disapproves of. The number of candidates that is found to be acceptable can vary, depending on the voter. In the ultimate case, someone can find all alternatives acceptable. The candidate who gets the most votes this way, is the winner. Because the voter mentions all candidates that he approves of, he enlarges the chance that a candidate he finds acceptable will win.

We will divide the acceptable and not acceptable alternatives by means of $\gg$. With, for example,

Florence Siena $\gg$ Venice

we indicate that an individual orders the alternatives from left to right in descending order of acceptability, and that the individual in question only finds Florence and Siena to be acceptable alternatives.

Approval voting is elaborately discussed and propagated by Brams and Fishburn [4].

Now consider the following profile $\hat{p}$, which differs from profile p in Example 1 only in the appearance of the division mark $\gg$.

Florence		Venice	$\gg$	Siena	: 5 students
Florence		Siena	$\gg$	Venice	: 7 students
Venice	$\gg$	Florence		Siena	: 3 students
Venice	$\gg$	Siena		Florence	: 7 students
Siena		Florence	$\gg$	Venice	: 3 students
Siena		Venice	$\gg$	Florence	: 6 students

Florence then gathers $5 + 7 + 0 + 0 + 3 + 0 = 15$ votes. Siena is good for $0 + 7 + 0 + 0 + 3 + 6 = 16$ votes. And Venice now gets $5 + 0 + 3 + 7 + 0 + 6 = 21$ votes. So, for this profile $\hat{p}$, Venice is the collective choice under Approval Voting. The collective preference ordering is

Venice Siena Florence.

In order to see that the winner under Approval Voting need not necessarily be the Condorcet winner, consider the following profile with three alternatives a (Ann), b (Bob), and c (Coby) and nine voters.

a b c : 5 voters
b a c : 2 voters
c b a : 2 voters

Then a is the Condorcet winner. Now suppose that, under Approval Voting, all voters give their approval only to the first two alternatives in their respective preference orderings. Then, under Approval Voting, b is the winner, while a is the Condorcet winner.

Approval Voting is also sensitive to strategic behavior. In the example at the beginning of this section, the last group of six students prefers Siena to Venice. Now, by not giving their true preference

$$\text{Siena Venice} \gg \text{Florence}$$

but their insincere preference

$$\text{Siena} \gg \text{Venice Florence,}$$

they ensure that Venice gets 6 votes less, $21 - 6 = 15$, and, hence, Siena, with 16 votes, is the collective choice, which is the preferred alternative for these six students.

However, the strategic behavior of one individual or a group of individuals may have the consequence that alternatives which are acceptable to this individual or group get less votes or that unacceptable alternatives get more votes.

We quote from the conclusions of [6]:

'Among non-ranked voting procedures to elect one candidate, approval voting distinguishes itself as more sincere, strategy proof, and likely to elect Condorcet candidates than other procedures ... . Its use in earlier centuries in Europe [...], and its recent adoption by a number of professional societies - including the Institute of Management Sciences [...], the Mathematical Association of America [...], the American Statistical Association [...], the Institute of Electrical and Electronics Engineers [...], and the American Mathematical Society - augurs well for its more widespread use, including possible adoption in public elections [...]. Bills have been introduced in several U.S. state legislatures for its enactment for state primaries, and its consideration has been urged in such countries as Finland [...] and New Zealand [...].'

The reader may easily verify the following theorem.

Theorem 4 Approval Voting is anonymous, neutral, IIA, transitive, not monotonic and does not satisfy the Pareto condition.

2.8 Summary

There are many ways to aggregate individual preferences to a collective preference or outcome. Some of the more frequently occurring ones have been discussed here. For the same individual preferences of the voters, in general, the outcome strongly depends on the election mechanism used.

'*Most votes count*' (Plurality Rule) is very frequently used and is the foundation of the Dutch and British election systems. We have shown that this election mechanism has many disadvantages: it does not satisfy the majority principle (if the number of voters that prefer x to y is greater than the number of voters that prefer y to x, then alternative x must also end above y in the outcome), it does not satisfy the Pareto condition (if everybody prefers x to y, then x must also be collectively preferred to y), and it does not have the monotonicity property (if alternative x is raised vis-a-vis an alternative y in someone's preference ordering and x goes down in no one's preference vis-a-vis y, then x must also be raised vis-a-vis y in the outcome). 'Most votes count' can even give an alternative as winner that is defeated by all other alternatives.

The *Majority Rule* (or pairwise comparison) is based on the majority principle. In comparison to 'Most votes count', the Majority Rule has many advantages: not only is it anonymous and neutral, but it is also monotonic and IIA, and it satisfies the Pareto condition. The Majority Rule has just one serious disadvantage: in some situations, it may happen that no winner can be selected, for instance, in the case of three alternatives x, y, and z, where x defeats y, y defeats z, but also z defeats x. In other words, the Majority Rule (pairwise comparison) is not transitive.

The *Borda preference rule* also takes into account the second, third, etc., preferences of individuals in the determination of the collective preference. Frequently, but not always, the Borda preference rule generates the Condorcet winner (if there is one), which is the alternative that defeats all other alternatives on pairwise comparison. The Borda preference rule is not independent of irrelevant alternatives, but perhaps the greatest objection that can be raised against the Borda procedure is its sensitivity to strategic behavior. Nonetheless, the Borda preference rule is, with respect to choosing one single candidate, in many ways superior to other procedures that also weigh second, third, etc., preferences.

Approval Voting gives the voter the opportunity to distinguish between the candidates he or she approves of and the ones he or she does not approve of. It is sensitive to strategic behavior of the voter(s). However, among non-ranked voting procedures to elect one candidate, Approval Voting distinguishes itself as more sincere, more strategy proof, and more likely to elect Condorcet candidates than other procedures. However, it is known (see [16]) that the chance of selecting a Condorcet winner, if there is one, under Approval Voting is significantly smaller than under the Borda procedure.

3 Categoric Voting

Generally, in Western Europe four different election procedures can be distinguished. This is the result of a division in the way of representation (proportional or non-proportional) and in the way of voting (categoric or ordinal). As to the way of representation, we distinguish (see [13], p. 4):

- *Proportional Representation (PR):* the distribution of seats is proportional to the number of votes.
- *Non-Proportional Representation:* the distribution of seats is not proportional to the number of votes.

Concerning the way of voting we distinguish (see [32], pp. 17, 126):

- *categoric voting:* the voters cast one vote, meaning that they select one candidate or party.
- *ordinal voting:* the voters give a preferential order of candidates or parties. For instance, in Australia, Ireland and Malta the voters are allowed, instead of casting just one vote, to give their first, second, third, etc. preference.

On the basis of the aforementioned distinctions in the way of representation and the way of voting, we can, in general, distinguish four different categories of election procedures, as given in the scheme below.

	Way of voting	
Representation	Categoric	Ordinal
Proportional	NL and most European countries	Ireland (STV), Malta (STV)
Non- Proportional	UK, US, Canada, New Zealand	Australia (AV), France (two voting rounds)

Here STV stands for *Single Transferable Vote*, to be considered in section 4.1 and AV for *Alternative Vote*, to be considered in section 4.3. Besides the above-mentioned election procedures, there are also the so-called hybrid systems, such as the 'two-vote' system in Germany, which we will discuss in section 3.4.

For each category we will discuss a particular election procedure, and we will show the paradoxes this procedure can give rise to. With a paradox, we mean an outcome that is completely contrary to what we would expect or contrary to our sense of righteousness.

In section 3.1 we discuss the main ideas of the Dutch and British election procedures. In section 3.2 a number of paradoxes that may occur in the Dutch system are considered, while in section 3.3 the paradoxes in the British system are discussed.

In sections 4.1 and 4.2 the Single Transferable Vote system is considered. Finally, in sections 4.3 and 4.4 the Alternative Vote system, as applied in, for instance, Australia and to a certain degree also in France, is discussed.

3.1   The Netherlands vs. the United Kingdom

The Dutch election procedure is characterized, among others, by:

- proportional representation, where the parties receive a number of seats more or less proportional to the number of votes (according to the d'Hondt method; see the end of this section).
- One district, containing the entire nation.

Because of this, there exist more parties and the government usually consists of a coalition of a number of parties.

The British election procedure, on the contrary, is characterized by:

- a division in (approximately) 659 districts for (approximately) 659 seats.
- in each district precisely one representative is elected, by means of 'Most votes count' (Plurality Rule): in each district the party with the most votes wins the seat.

Because of this, the United Kingdom (England, Wales, Scotland and Northern-Ireland) has a two (recently three) party system and the government is usually formed by the party with a majority of the seats.

In general, the Dutch and British election procedures produce different outcomes given the same individual preferences of the voters. To illustrate the difference in outcome between the Dutch and British election procedures, we consider the following example in [29], table 12.1:

Party	vote percentage
A	30
B	25
C	20
D	15
E	10

In the Dutch system, every party will get a number of seats more or less proportional to the number of votes. Party A will then get approximately 30% of the seats, part B 25%, etc. Hence, it is to be expected that a multiparty system evolves (five in this example). Because, in general, none of the parties will get a majority of the seats in parliament, government will usually consist of a coalition of several parties.

Now suppose that the same distribution as given in the above-mentioned table occurs in every district in the United Kingdom. Then, in the British system, the seat for each district is given to party A, because this party has most votes in every district. So, in the British system, the other parties would get no seat at all! Because of the nature of the British election procedure, where only large parties have a realistic chance of a seat, and because of the strategic behavior of voters who do not want to waste their vote on a party that has no chance at all, it is to be expected that the British system will give rise to a two- (or three-) party system. This phenomenon is called *Duverger's law*. Since one party usually gets a majority of the seats in parliament, British government usually consists of one party.

Notice that in the Netherlands 'Most votes count' (Plurality Rule) is used to generate a collective (order of) *preference*: A is collectively preferred to B, B to C, etc. Contrary to this, in the United Kingdom 'Most votes count' is used to establish for each district a collective *choice*: the candidate for party A.

As far as appreciation of the Dutch and British election procedures is concerned, Lijphart remarks in [24], page 144:

1. If much weight is given to the representation of minorities, then proportional representation and more than two parties seem to be the best choice.
2. If, on the contrary, much weight is given to government responsibility, then 'Most votes count' (Plurality Rule) and a two-party system seem to be the best choice. The voter then knows that the ruling party is responsible for the achievements of the past and can hold this party responsible for them.

According to [27], pp. 173-175, different considerations concerning representation (democracy) are at the basis of the Dutch and British systems:

1. The Dutch system corresponds to the *reflection model* of representation. The underlying consideration is that the composition of parliament must be a reflection of the composition of the constituency. Various groups and interests are to be proportionally represented, as in a representative test sample. Ideally, proportionally, there will be as many liberals, socialists, etc. in parliament as there are in society.
2. The British system corresponds to the *principal agent model* of representation. According to this model, representatives are agents that act in the interest of others. Parliament does not have to be a reflection of society, but has to honestly defend the interests of the constituency. Not the composition of parliament but its decisions are important.

The *Ostrogorski paradox* shows that the formation of parties and voting for them may give results that deviate from voting for issues, as is done in referenda. Suppose that there are two parties: X and Y. Also, suppose that these two parties have different points of view concerning three issues, numbered 1, 2 and 3. Finally, suppose that there are four groups of voters, named A (20%), B (20%), C (20%) and D (40%), whose positions concerning the three issues are given in the table below, taken from [8], p. 205. For instance, the voters in group A share the position of party X concerning issues 1 and 2, and the position of Y on issue 3.

We now distinguish between two forms of voting:

1. *issue-by-issue voting:* a voting round is held for each separate issue.
2. *voting by platform:* a party or candidate is chosen on the grounds of its policy.

Voters	Issues			Elected party
	1	2	3	
A (20%)	X	X	Y	X
B (20%)	X	Y	X	X
C (20%)	Y	X	X	X
D (40%)	Y	Y	Y	Y
	Y: 60%	Y: 60%	Y: 60 %	

For the situation given in the table, these two forms of voting result in completely different outcomes. If we take issue-by-issue voting, party Y will get 60% of the votes for all issues and will, hence, be able to impose its position on society.

With voting by platform, the voter chooses the party that approximates his or her own position best. The voters in group A will then vote for party X because this party holds their position on two of the three issues. Given this form of voting, party X will get 60% of the votes and party Y only 40%. So now party X has a majority and is able to impose its position concerning the issues on society.

The conclusion is that the outcome of issue-by-issue voting can be completely different from the outcome of voting by platform.

Notice that for three issues that have to be decided on by yes or no, there are $2^3 = 8$ different possibilities to answer these questions. So there would have to be at least 8 different parties to give the voter the possibility to vote for a party that holds his position on all issues. Because in the Netherlands there are more parties than in Britain, the probability of the occurrence of the Ostrogorski paradox will be slightly smaller in the Netherlands.

As has been noticed before, the Netherlands has a system of proportional representation. However, it seldom occurs that the seats can be distributed precisely proportionally to the number of votes. To appoint seats to parties in a more or less proportional manner, the *d'Hondt formula* is used in the Netherlands.

This formula uses the numbers 1, 2, 3, 4, ... to divide the total number of votes a party has received, every time the party gets a seat. The first seat goes to the largest party, whose number of votes is then divided by two. The second seat is allocated to the party that now has the most votes, given that the number of votes the largest party had has now been divided by two. When the largest party receives a second seat, its total number of votes is then divided by three, and so on. The effect of the d'Hondt formula is illustrated by means of the following example from [24], p. 154, in the case of six seats.

Party	v (= votes)	$v/2$	$v/3$	number of seats
A	41,000 (1)	20,500 (3)	13,667 (6)	3
B	29,000 (2)	14,500 (5)	9,667	2
C	17,000 (4)	08,500		1
D	13,000			0

3.2 Paradoxes in the Dutch System

This section is based on [9] and on [11]. In the first article, Van Deemen shows that a number of paradoxes may occur in the Dutch system. Next, the authors of the second article show, by means of empirical research, that most of these paradoxes do, in fact, occur more than once.

To start with, let us look at the distribution of votes and seats in the elections of September 6, 1989 for the House of Commons, given in the following table. Here SR stands for Small Right, a coalition of some smaller parties.

Party	Percentage of votes	Number of seats
CDA	35.3	54
PvdA	31.9	49
VVD	14.6	22
D66	7.9	12
GL	4.1	6
SR	5.0	7

Now consider the following profile, where the distribution of first votes for the parties corresponds precisely to the election results of September 6, 1989. Notice that the profile, though fictitious, is not unrealistic.

CDA	D66	VVD	SR	PvdA	GL	: 35.3%
PvdA	GL	D66	CDA	VVD	SR	: 31.9%
VVD	PvdA	D66	SR	CDA	GL	: 14.6%
D66	PvdA	CDA	VVD	GL	SR	: 07.9%
GL	PvdA	D66	CDA	VVD	SR	: 04.1%
SR	VVD	CDA	D66	PvdA	GL	: 05.0%

Paradox 1: The reader can check for himself that application of pairwise comparison to the above-mentioned profile, yields the following result. Here $\#(X)$ stands for the number of seats given to party X.

- PvdA defeats CDA with 58.5% (31.9 + 14.6 + 7.9 + 4.1) to 40.3% (35.3 + 5.0), while $\#(CDA) = 54 > \#(PvdA) = 49$.
- VVD defeats PvdA with 54.9 % (35.3 + 14.6 + 5.0) to 43.9% (31.9 + 7.9 + 4.1), while $\#(PvdA) = 49 > \#(VVD) = 22$.
- D66 defeats VVD with 79.2% to 19.6%, while $\#(VVD) = 22 > \#(D66) = 12$.
- D66 defeats CDA with 59.5% to 40.3%, while $\#(CDA) = 54 > \#(D66) = 12$.

Van Deemen calls this the *More-Preferred, Less-Seats paradox*: a party that (in a pairwise comparison) is more preferred than another party may still get fewer seats!

The Dutch election procedure is also sensitive to strategic behavior. If for the above profile it is expected, on the grounds of predictions of voting outcomes, that CDA and PvdA will form a coalition, then voters with SR as first preference (the last group of 5%) could choose strategically and mention VVD (their actual second choice) as their (insincere) first choice, hoping to make a coalition of CDA and VVD possible. Such a coalition would then represent 35.3 + 14.6 + 5 = 54.4% of all voters.

Also, if predictions concerning the voting outcome show that one's most preferred party will not make the election threshold (the minimal percentage of votes needed to get a seat, 0.67% in the Netherlands), this voter could decide not to vote for his or her true first preference in order to avoid wasting the vote.

Paradox 2: In the elections of September 6, 1989 there was a party, called the Groenen (Greens), that did not get sufficient votes to win a seat. We denote this party with the letter G. Now consider the following profile:

CDA	G	D66	VVD	SR	PvdA	GL	: 35.3%
PvdA	G	GL	D66	CDA	VVD	SR	: 31.9%
VVD	G	PvdA	D66	SR	CDA	GL	: 14.6%
D66	G	PvdA	CDA	VVD	GL	SR	: 07.9%
GL	G	PvdA	D66	CDA	VVD	SR	: 04.1%
SR	G	VVD	CDA	D66	PvdA	GL	: 05.0%

This profile originates from the previous profile by placing party G second in every row. So, it is supposed that every Dutchman has party G as his second preference. The reader can easily check that, given this profile, party G will defeat every other party in a pairwise comparison and, hence, is a Condorcet winner. However, under the Dutch election procedure, party G will get no seat at all!

Van Deemen calls this the *Condorcet-Party-Turns-Loser paradox*. A Condorcet winner does not necessarily get the largest number of seats; it may even happen that the Condorcet winner gets no seat at all.

Paradox 3: The following result is even more amazing. Consider the following profile:

CDA	GL	SR	D66	VVD	PvdA	: 35.3%
PvdA	GL	SR	D66	VVD	CDA	: 31.9%
VVD	GL	SR	D66	PvdA	CDA	: 14.6%
D66	GL	SR	VVD	CDA	PvdA	: 07.9%
GL	SR	D66	VVD	PvdA	CDA	: 04.1%
SR	GL	D66	VVD	PvdA	CDA	: 05.0%

For this profile it holds that in a pairwise comparison

GL defeats SR, SR defeats D66, D66 defeats VVD,
VVD defeats PvdA, and PvdA defeats CDA

in other words, GL has a majority over SR, which has in its turn a majority over D66, etcetera. But this is precisely the inverse of the collective preference as given by the distribution of seats:

$$\#(CDA) > \#(PvdA) > \#(VVD) > \#(D66) > \#(SR) > \#(GL)$$

in other words, on the above-mentioned fictitious profile CDA gets more seats than PvdA on application of the Dutch system, PvdA will get more seats than VVD, et cetera.

Van Deemen calls this the *Reversal-of-Majority paradox*: the order given by application of the Majority Rule (pairwise comparison) is exactly the inverse of the order given by the actual distribution of seats in the Dutch system.

Notice that the last of the three paradoxes given is the strongest: an occurrence of the Reversal-of-Majority paradox entails the occurrence of the Condorcet-Party-Turns-Loser paradox; and the latter entails the occurrence of the More-Preferred, Less-Seats paradox.

We cite here A. van Deemen, [9], page 240:

'It is hard to find reasons that justify the possibility that a candidate or party which is preferred by a minority is elected or has more seats than a candidate or party which is preferred by a majority. Borda (1781) and Condorcet (1788) rightly concluded that for this reason the plurality systems are 'seriously defective' ([2], p. 44). The paradoxes presented in this paper lead to the same conclusion for list systems of proportional representation.'

Above, we have constructed, behind our desks, three situations or profiles in which the Dutch system, based on 'Most votes count', leads to paradoxical results. The obvious question now is if such situations also occur in real life. Well then, in [11] the authors describe the results of their empirical research concerning the occurrence of situations (profiles) in Dutch elections that could lead to one of the above-mentioned paradoxes. In short, their findings are that they could not find an occurrence of the strongest paradox, the Reversal-of-Majority paradox, but that the other paradoxes occur frequently.

The reader may wonder how such empirical research is possible, since the voters are only asked to give their first preference. How can we know what their second, third, etcetera preferences are? During the run-up to every election so-called voter research is done, in which a number of voters is asked to give their individual preference ordering with respect to *all* parties. By making the number of participants sufficiently large, reliable information concerning the individual preferences of the voters over all parties can be gathered.

- [11], page 484: For the elections of 1982, 1986 and 1994, the More-Preferred, Less-Seats paradox frequently occurred: *a party that has a majority over an other party can still get fewer seats*. This paradox also occurred in 1989, be it to a lesser extent.
- [11], page 485: The Condorcet-Party-Turns-Loser paradox occurred in the elections of 1982 and 1994: *it can happen that a Condorcet winner does not get the largest number of seats, or perhaps even no seats at all*. In 1994, D66 was Condorcet winner, but PvdA got most of the seats. Also, CDA and VVD got more seats than D66. A second case occurred in 1982 when CDA got more seats than the Condorcet winner PvdA. The possibility of a Condorcet winner getting no seats at all did not occur.
- [11], page 485: the Reversal-of-Majority paradox did not occur in the elections of 1982, 1986, 1989 or 1994.

3.3   Paradoxes in the British System

In what follows we will analyze three paradoxes of the British system and pay some attention to the May 1948 election in South-Africa.

Condorcet-Loser-Wins paradox. In the separate districts, where 'Most votes count' is used, the paradoxes we have seen in the Dutch system will, of course, also occur. Suppose, for instance, that the preferences of the voters in a district with respect to three candidates Ad (*a*), Bob (*b*) and Carol (*c*) are as follows:

a b c : 30%
b a c : 30%
c b a : 40%

For this profile, on application of 'Most votes count' (Plurality Rule), c is elected, while b is the Condorcet winner. Worse, c is the *Condorcet loser*, meaning that all other candidates have a majority over c. There is a majority (60%) that prefers a to c and a majority (60%) that prefers b to c.

A second paradox in the British system is caused by the division in districts and is, therefore, called the *districts paradox*. Suppose that there are three districts, two parties A and B, twenty voters in each district and that the votes are divided over the candidates for the two parties as follows.

	candidate for A	candidate for B	elected
district 1	11 votes	9 votes	A
district 2	11 votes	9 votes	A
district 3	5 votes	15 votes	B

When 'Most votes count' (Plurality Rule) is applied, the candidate for party A will win in districts 1 and 2, and in district 3 the candidate for party B will win. According to the British system, party A will then have a majority in the House of Commons and, hence, form a government. But B has 33 votes, which is more than the 27 votes for A. So, on direct elections, B would have won and formed the government.

The majority that party A acquires is called, in [32], pp. 74-75, a *manufactured majority*: a majority in the legislative power, won by a party that has got fewer votes than the other party. According to empirical research of [24], page 74, British parliamentary elections over the period 1945-1990 produced manufactured majorities in 92.3 percent of all cases.

A similar situation occurred in the *elections in South-Africa in May 1948*. There were two parties, the United Party, against Apartheid, and the National Party, in favor of Apartheid. The United Party got 50.9%, so more than half of all votes, but due to the district system only 71 seats. The National Party only got 41.2% of all votes, but 79 seats in parliament. This party won the elections due to the single-member district plurality system: a district system where in each district one candidate is chosen by means of 'Most votes count'. This shows again that the choice of the election procedure can have far-reaching consequences.

The *districts paradox* that may occur in the British system is illustrated again by means of the following example from [29], pp. 221-222. Suppose there are five parties A, B, C, D and E, ten districts and in each district 10% of the voters, and the following distribution of votes (in percentages).

PARTY	DISTRICT									
	1	2	3	4	5	6	7	8	9	10
A	3	3	3	3	3	3	3	3	3	3
B	3	3	2	2	3	0	3	3	3	3
C	0	0	0	0	0	4	4	4	4	4
D	0	0	4	4	4	3	0	0	0	0
E	4	4	1	1	0	0	0	0	0	0
Total	10	10	10	10	10	10	10	10	10	10

Because party C gets most votes in districts 6 to 10, this party will get 50% of the seats in the British system, even though the party only gets 20% of the votes nationally. Party D wins in the three districts 3, 4 and 5 and so gets 30% of the seats, even though the party only gets 15% of the votes nationally. Party E wins in the districts 1 and 2 and gets 20% of the seats in the British system, while the party only gets 10% of the votes nationally. Party A, on the contrary, that by far gets most votes nationally, 30%, gets no seat at all in the British system.

In the above-mentioned example, parties C, D and E together get all seats in parliament, while only getting 20 + 15 + 10 = 45% of the votes. Parties A and B get no seat in parliament at all, while getting 30 + 25 = 55% of all votes. It is now obvious that parties A and B will dissolve themselves, leaving only three parties. In reality we also see that only 2 to 3 parties are active in Great Britain. Hence, the above paradox illustrates why one can expect that 'Most votes count' (Plurality Rule) combined with a district system will generate a system with few parties (*Duverger's law*). However, as this example shows, it will not necessarily be the small parties that disappear.

In the separate districts 'Most votes count' is used, which - as we already saw - is sensitive to strategic behavior. Therefore also *the British election system is sensitive to strategic behavior*, as is illustrated in the following example. Suppose that in a certain district there are three candidates A (Ann), B (Bob) and C (Cod) and the preferences of the eighteen voters are as follows.

A B C : 6 voters
C A B : 5 voters
B C A : 4 voters
B A C : 3 voters

For 'Most votes count' this means that candidate B will be elected in this district. An opinion poll preceding the elections could cause the five voters with candidate C as most preferred and candidate B as least preferred candidate to change their vote and put candidate A first. The result would then be that A defeats B with 11 to 7 votes and this outcome is preferred to the original outcome by these five voters.

Note that in the British system a party that wins the district's seat after joining two districts, does not necessarily win in both original districts. In order to see this, consider the following two profiles for district 1 and district 2.

District 1:   *a*   *b*   *c*   : 9 voters   District 2:   *a*   *b*   *c*   : 6 voters
               b   *c*   *a*   : 5 voters   *b*   *c*   *a*   : 9 voters
               c   *b*   *a*   : 3 voters   *c*   *b*   *a*   : 2 voters

The Dutch and British system are not Independent of Irrelevant Alternatives. In order to see this, consider the following profile.

a   *b*   *c*   : 4 voters
c   *b*   *a*   : 3 voters
b   *c*   *a*   : 2 voters

1) Check that in the Dutch system the number of seats for *c* will be greater than the number of seats for *b*, but that on absence of *a* the inverse will be the case. The collective preference concerning *b* and *c*, as expressed in the distribution of seats, is, hence, influenced by the presence of the (irrelevant) alternative *a*.

2) Check that in the British system the district's seat is given to neither *b* nor *c* and that these two parties are hence collectively indifferent, but that on absence of *a* the district's seat is given to *b* and not to *c*. So also in the British system the collective preference, as expressed in the distribution of seats, between *b* and *c* is influenced by the presence of the (irrelevant) alternative *a*.

3.4   Hybrid Systems

A country like *Germany* has a hybrid election procedure, the so-called *two-vote system*. It combines two ideals, namely district representation, as in England, and proportional representation, as in The Netherlands. Every voter has two votes: a first vote (Erststimme) for one candidate from his or her district and a second vote (Zweitstimme) for one of the national parties. Half of the Bundestag consists of district candidates and half of representatives of the national parties. Furthermore, there is an election threshold of 5%. The voter may vote with his first vote for a candidate that does not necessarily belong to the party he or she gives the second vote to. (For reasons of simplicity we do not consider Überhangmandate.)

Similar systems are used in Mexico, South-Korea, Taiwan and Venezuela. Versions of this system have recently been applied in Hungary, Italy, Japan, New Zealand and Russia (see [13], p. 87). In the Netherlands such a two-vote system was proposed in 1995. The proposal was to give each voter two votes. Parliament will then consist of 75 national seats and 75 district seats. The country would be divided into five districts, each of which distributes 15 seats among the district candidates. Every voter's first vote would be for a candidate from his district, the second vote would be for a national party (as in the current system).

A. van Deemen showed that the proposed election procedure could lead to what has been named the *two-vote paradox*: a party B could get a majority over party A in each district, while party A gets a greater number of seats in parliament than party B. To see this, we suppose four parties A, B, C and D and a distribution of the national seats as follows:

Party	Number of seats	Percentage
A	30	40,0%
B	20	26,7%
C	15	20,0%
D	10	13,3%

We also suppose that the district elections lead to the following result.

Districts:	D1	D2	D3	D4	D5	Total
Party A	5	5	5	4	4	23
Party B	6	8	6	6	5	31
Party C	2	1	2	2	2	09
Party D	2	1	2	3	4	12
Total:	15	15	15	15	15	75

For the above-mentioned outcome, party A will get 30 national and 23 district seats, so the party will have 53 seats in parliament. Party B will get 20 national and 31 district seats, so a total of 51. According to the second table, however, party B has a majority over party A in each district.

Because the proposed election procedure is a combination of the British and the current Dutch system, it is also prone to the paradoxes that may occur in either of these systems (see section 3.2 and 3.3).

The proposed system is an amelioration of the existing one in the sense that more information is asked of the voter and processed. The function of an election mechanism is to aggregate information received from the voter to distribute seats. Such an election procedure, however, cannot serve to, for instance, close the gap between voters and politicians.

3.5   Summary

The Dutch election mechanism is based on proportional representation and categoric voting. In section 3.2 we have discussed a number of paradoxes that may occur in this system: the More-Preferred, Less-Seats paradox, the Condorcet-Party-Turns-Loser paradox and the Reversal-of-Majority paradox.

The British system is based on non-proportional representation, as well as on categoric voting. In section 3.3 we have found some paradoxes that may occur in this system: the Condorcet-Loser-Wins paradox and the districts paradox.

In the Netherlands 'Most votes count' (Plurality Rule) is used to generate a collective (order of) *preference*. Contrary to this, in the United Kingdom 'Most votes count' is used to establish for each district a collective *choice*: the candidate for a certain party.

The German hybrid election procedure is a combination of the Dutch and British system and, hence, inherits the paradoxes of both systems. The German two-vote system also creates its own paradox, which we have called the two-vote paradox.

4 Ordinal Voting

In ordinal voting, voters are asked to list the candidates or parties in order of preference. In section 4.1, the Single Transferable Vote (STV) system, used in Ireland and Malta, is considered. That this system gives rise to a number of very akward paradoxes is made clear in section 4.2. In section 4.3 we consider the Alternative Vote (AV) system that is used in Australia. That this system is also subject to paradoxes is the subject of section 4.4. The French election system and the paradoxes inherent in it are considered in section 4.5.

4.1 The Single Transferable Vote (STV) System

In Ireland and Malta, the Single Transferable Vote (STV) system is used. This system was first proposed by Thomas Hare (1861) in England and Carl George Andrae in Denmark around 1850. Hare presented his system as a way to secure the proportional representation of important minorities. His idea was that no vote should be wasted: even if someone's vote does not aid in electing his or her first choice, the vote can still count for his or her lower choices. John Stuart Mill (1862) classified STV 'among the greatest improvements yet made in the theory and practice of government' ([6], 11.1).

The members of parliament are elected per district. In Ireland, there are about 40 districts that have to elect approximately 150 members of parliament. This means that each district must appoint several representatives. Even though STV is a system of proportional representation, it differs from proportional systems with lists (parties) because the voters choose individual candidates instead of parties. Furthermore, they are asked to give an order of preference regarding the candidates, independent of the parties they belong to, instead of casting just one vote for one candidate (or party), so that the voting is ordinal instead of categoric. To gain a seat, the candidate must pass a certain election threshold.

Let us begin by considering an example that will make clear what moved Hare to develop his STV system. The example is taken from [19], page 211. Consider an imaginary district in which two of the four candidates must be elected. Two candidates, Ann (a) and Bob (b), are conservative, the other two, Coby (c) and Donald (d), are progressive. Suppose that the preferences of the 23 voters are as follows:

a	b	c	d	: 7 voters
b	a	c	d	: 6 voters
c	d	b	a	: 6 voters
d	c	b	a	: 4 voters

In an election where each voter may vote for two candidates, a and b will win with 13 votes each. Hence, the 10 progressive voters remain unrepresented, even though they constitute 43% of the electorate. The 13 conservative voters get a 100% representation.

Before giving a general description of Hare's STV system, let us use the above example to see how it works.

The *election threshold* in this example is 8, because

1. for 23 voters, there can be no more than 2 candidates that get eight votes of first choice: three candidates with eight votes of first choice would require 3 × 8 = 24 voters;
2. with an election threshold of 7, three instead of two candidates could get seven votes of first choice: 3 × 7 = 21.

Thus, the *election threshold* is the smallest number of votes of first choice so that the maximum amount of candidates that can reach the election threshold corresponds to the available number of seats.

Because the election threshold is 8, in the above example every candidate is short of votes. The least popular candidate, Donald (d), is then (under STV) eliminated and his four supporters then transfer their votes to Coby (c), their second choice. In the second round, the list of preferences then looks as follows:

a b c : 7 voters
b a c : 6 voters
c b a : 10 voters

Now c exceeds the election threshold by two votes and is hence elected. Her remaining votes are transferred to b, the second choice of this group. The situation in the third round then becomes:

a b : 7
b a : 6 + 2 = 8

Now b reaches the election threshold and is elected. Note that on application of STV both the progressive candidate c and the conservative candidate b are elected, while without STV the two conservative candidates a and b would have been elected. So, on application of STV, both groups of voters are more or less proportionally represented.

In the above example, in which 2 (of the 4) candidates had to be elected according to the STV procedure by 23 voters, the election threshold was 8.

Now suppose there are n voters and k available seats. The election threshold q is then, by definition, the smallest natural number such that $kq \leq n$ and $(k+1)q > n$. Thus, $q = [\frac{n}{k+1}] + 1$, where $[\frac{n}{k+1}]$ is the integer obtained by rounding down $\frac{n}{k+1}$.

So, in the above example, where 23 voters have to elect 2 candidates, the election threshold $q = [\frac{23}{2+1}] + 1 = 7 + 1 = 8$.

We now give a general *description of STV*, the system of Single Transferable Vote, as found in [19], page 212. If we assume that at least one candidate reaches the election threshold and at least one seat remains, the winning votes that pass the election threshold are proportionally transferred to the second choice of these voters. If as a result of this transfer another candidate reaches the election threshold, this candidate is elected; and if seats remain, the remaining votes (the ones that passed the election threshold) are once again transferred proportionally. This process continues until all seats are occupied. If at any point there is a seat unoccupied without there being votes to be transferred, the candidate with the least number of votes is eliminated and the supporters of this candidate transfer their votes to their most preferred candidate amongst those that are still in the running (that is, not eliminated and not yet occupying a seat).

The idea is that no vote is wasted: each vote beyond the number a candidate needs to be elected must be counted elsewhere; a vote that is wasted on the least popular candidate must be counted elsewhere.

Let us illustrate how STV works by means of a more complex example, taken from [24], page 158. Suppose that one district has to elect three representatives. It has a hundred voters and seven candidates: P, Q, R, S, T, U, and V. Suppose that the individual preference orderings are given in the following profile.

P Q R : 15 voters
P R Q : 15 voters
Q R P : 8 voters
R P Q : 3 voters
S T   : 20 voters
T S   : 9 voters
U   : 17 voters
V   : 13 voters

Hence, the election threshold is $[\frac{100}{3+1}] + 1 = 26$. On application of STV, the following occurs. In the first round, P has $15 + 15 = 30$ votes and is the only candidate to reach the election threshold. In the second round the four $(30 - 26)$ surplus votes P had are proportionally transferred to the second choice of these voters. In this case, two to Q and two to R, because half of the original thirty preferences with P as first choice had Q as second choice and half had R as second choice. In the second round, Q then has $8 + 2 = 10$ (first) votes, R has $3 + 2 = 5$, S 20, T 9, U 17, and V 13 votes. So, none of these candidates reaches the election threshold in the second round. Hence, in the third round the weakest candidate, R, is eliminated and the five votes R had are transferred to the first candidate that is still in the running, which is Q. In the third round, Q then has $10 + 5 = 15$ votes, S 20, T 9, U 17, and V 13 votes. This means that also in the third round, no candidate reaches the election threshold. In the fourth round, according to the STV procedure, again the weakest candidate is eliminated, in this case T, and T's nine votes are transferred to S, the second choice of these voters. In the fourth round, Q has 15 votes, S $20 + 9 = 29$, U 17, and V 13 votes. S now passes the election threshold and is elected. At this stage, P and S are elected while R and T have been eliminated. Because three candidates must be elected, a fifth round is necessary: the three $(29 - 26)$ surplus votes S had must be transferred to the next preference of the voters, in this case T. Since T has already been eliminated, the three votes are not transferable. So, nobody reaches the election threshold in the fifth round, and, in the sixth round, the weakest candidate is removed. This is V, whose 13 votes are not transferable. Only Q with 15 and U with 17 votes remain. This means that in the seventh round, the weakest candidate, Q, is eliminated and that U becomes the third candidate elected.

4.2   Paradoxes in the STV System

Even though Hare developed his Single Transferable Vote system with the best of intentions, this system frequently leads to results that are contrary to what one would expect. In this section we will expound on a number of these paradoxes.

STV is majority-inconsistent: A Condorcet winner may exist without being elected by STV. We have already seen in sections 3.2 and 3.3 that the Dutch and British systems, both based on 'Most votes count', are also majority-inconsistent: the Condorcet-Party-Turns-Loser paradox in the Dutch system and the Condorcet-Loser-Wins paradox in the British system. Probably, the first to recognize the majority-inconsistency of STV were Hoag and Hallett (see [18]).

Suppose that STV is to be used to elect one of three candidates, Ann (a), Bob (b), and Coby (c), by eight voters. The election threshold is then $[\frac{8}{1+1}] + 1 = 5$. Suppose that the individual preferences are given in the following profile:

a	b	c	: 3 voter
b	a	c	: 2 voters
c	b	a	: 3 voters

Because nobody reaches the election threshold in the first round, STV requires that b, being the candidate with the fewest (first) votes, be eliminated and his two votes transferred to the second choice of these voters, a. Consequently, a reaches the election threshold with $3 + 2 = 5$ votes and is elected.

However, on application of the Majority Rule (pairwise comparison), we see that b defeats a with $2 + 3 = 5$ to 3 votes and b defeats c with $3 + 2 = 5$ to 3 votes. Thus, b is the Condorcet winner, while b is the first to be eliminated in the STV procedure.

STV is sensitive to strategic behavior. One of the instructions for the elections of the American Mathematical Society (AMS) was 'there is no tactical advantage to be gained by marking few candidates.' However, in 1982, Steven Brams showed, by means of an example, that, on application of STV, it may well be beneficial to some voters to mention fewer candidates in their preference ordering. In [6], Example 11.1, he gives the following example.

Suppose that two of the four candidates, x, a, b, and c, have to be elected by 17 voters who have the following preferences (divided over three (preference) classes I, II, and III).

I	x	a	b	c	: 6 voters
II	x	b	c	a	: 6 voters
III	x	c	a	b	: 5 voters

The election threshold for STV is now $[\frac{17}{2+1}] + 1 = 6$. On application of the STV procedure, x and a are elected. This goes as follows. In the first round, x, with all 17 votes, is the only one to reach the election threshold and is elected. The $17 - 6 = 11$ surplus votes for x are then transferred in the ratio $6 : 6 : 5$ to the second preferences of the voters in group I, II, and III. Groups I and II each receive $\frac{6}{17} \times 11 = \frac{66}{17}$ votes and III receives $\frac{5}{17} \times 11 = \frac{55}{17}$ votes.

I	a	b	c	: $\frac{66}{17}$ votes
II	b	c	a	: $\frac{66}{17}$ votes
III	c	a	b	: $\frac{55}{17}$ votes

Since after the transfer of votes no candidate reaches the election threshold, in the third round, the candidate with the fewest votes, c, is eliminated, resulting in this preference profile:

I	a	b	: $\frac{66}{17}$
II	b	a	: $\frac{66}{17}$
III	a	b	: $\frac{55}{17}$

a now has $\frac{66}{17} + \frac{55}{17}$ votes, reaches the election threshold, and is hence elected with x.

The voters in class II see their least preferred candidate a elected alongside their most preferred candidate x. Now suppose that two of the six class II voters had only given their first choice x. The profile would then look like this.

I	x	a	b	c	: 6
II.1	x				: 2
II.2	x	b	c	a	: 4
III	x	c	a	b	: 5

The reader can easily verify that on application of the STV procedure candidate c is now elected alongside x (see also [19], pp. 214-215).

Note that the outcome in which c and x are elected is more attractive to class II voters, as they prefer c to a. We have already seen that if all voters in class II had given their complete preference ordering, a would have been elected. So, the two class II voters who gave only their first choice have thereby gained a tactical advantage.

Negative Responsiveness paradox: On application of STV, a candidate's position can deteriorate when he gets extra votes; more votes can even turn a winner into a loser! In other words, *STV is not monotonic*. Doron and Kronick have clarified this by means of the following example (see [12]). Suppose that 26 voters must elect two of four candidates: Ann (a), Bob (b), Coby (c), and Donald (d). Further, suppose that their individual preferences are given in the following profile.

I	a	b	c	d	: 9 voters
II	c	d	b	a	: 6 voters
III	d	c	b	a	: 2 voters
IV	d	b	c	a	: 4 voters
V	b	c	d	a	: 5 voters

As the election threshold is 9, a is elected. In the second round, b is dropped (he has the least votes), after which c is elected with $6 + 5 = 11$ votes.

Now suppose that the two voters in class III come to prefer c over d and leave the rest of their preferences unchanged. Then c gains two votes. The new situation is then:

I	a	b	c	d	: 9
II	c	d	b	a	: 6
III*	c	d	b	a	: 2
IV	d	b	c	a	: 4
V	b	c	d	a	: 5

Once again, a is elected in the first round. Consequently, d is removed (having the least votes), after which b is elected with $5 + 4 = 9$ votes.

In the above example, we see that candidate c turns into a loser by getting two more votes. If the two voters in class III* had ranked c second instead of first, c would have been elected.

Two-districts paradox ([19], pp. 220-221): A candidate can win in each separate district and still lose, on application of STV, an election in a combination of those districts.

To see this, consider two districts with 21 voters and the same four candidates Ann (a), Bob (b), Coby (c), and Donald (d). Suppose that one candidate must be elected in each district, so that the election threshold in each district is 11. The individual preferences for districts 1 and 2 are given by the following profiles.

I	a	b	c	d	: 8 voters
II	b	c	d	a	: 4 voters
III	c	a	d	b	: 3 voters
IV	d	c	b	a	: 6 voters

I	a	b	c	d	: 8 voters
II	b	c	d	a	: 4 voters
III	c	a	d	b	: 6 voters
IV*	d	a	b	c	: 3 voters

In both districts, STV will elect a: originally nobody reaches the threshold, but in the second round c and d will be eliminated, after which a reaches the election threshold in both districts with $8 + 3 = 11$ votes.

If the two districts are combined into one district and we assume that the individual preferences of the 42 voters remain the same and that once again one candidate must be elected, the STV procedure has a surprise in store for a: STV now elects c! The election threshold is now 22. Initially, nobody reaches the election threshold and b is eliminated. Hence, c gets $3 + 6 + 4 + 4 = 17$ votes. In the next round, d is eliminated. Six of his votes go to c, who gets $17 + 6 = 23$ votes and reaches the election threshold.

In their article 'Paradoxes of Preferential Voting; What can go wrong with so-phisticated voting systems designed to remedy problems of simpler systems' ([14]), Fishburn and Brams give a beautiful example of what they call the no-show paradox. Because a couple cannot participate in a mayoral election due to the breakdown of their car, they unwittingly prevent their least preferred candidate from winning. Had they been able to vote, their least preferred candidate would have won on application of STV.

No-show paradox: By adding individual preferences with x as the least preferred alternative, x can turn into a winner (while originally being a loser).

Fishburn and Brams describe a situation in which a village has to elect a mayor according to the STV procedure, whereby a car breakdown prevents Mr. and Mrs. Smith from participating in the election. Both have the preference ordering: Ann Bob Cod, where they strongly dislike Cod. Owing to their absence, there are only 1608 voters that have to elect one mayor according to the STV procedure. The individual preferences are given in the following profile.

Amount	Preference
417	Ann Bob Cod
82	Ann Cod Bob
143	Bob Ann Cod
357	Bob Cod Ann
285	Cod Ann Bob
324	Cod Bob Ann
1608	

On application of STV, the election threshold is $[\frac{1608}{1+1}] + 1 = 805$. Ann has $417 + 82 = 499$ first votes, Bob $143 + 357 = 500$, and Cod has $285 + 324 = 609$ first votes. So, in the first round, none of the candidates reaches the election threshold and the candidate with the fewest first votes, Ann, is eliminated. The votes for Ann are then transferred to Bob and Cod. This means that, in the second round, Bob gets $500 + 417 = 917$ votes, reaches the election threshold, and is elected mayor. Note that Bob is the second preference of the Smith couple, who are very happy that Cod, whom they despise, did not win.

Now let us see what would have happened had the Smith's car functioned properly and they had been able to participate in the election. The election result would then look like the one above, only with 419 voters with preference 'Ann Bob Cod' and a total of 1610 voters.

Because there are two more voters with the preference 'Ann Bob Cod', Ann has two more first votes: 501 instead of 499. This means that, in the first round, Bob, who received only 500 first votes, is eliminated instead of Ann and that his votes are proportionally transferred to Ann and Cod. Hence, Cod gets $609 + 357 = 966$ votes, reaches the election threshold $[\frac{1610}{1+1}] + 1 = 806$, and is elected mayor. Mr. and Mrs. Smith are perturbed: by not showing up (no-show), they unwittingly prevented their least preferred candidate from winning the election.

It is easy to verify that all other paradoxes given in this section can be illustrated by means of the above example:

a) The winner in the above example according to STV is also the Condorcet winner in the case of a car breakdown, but not in the case where the Smith couple was able to participate in the election. (*STV is majority-inconsistent*).

b) Consider the case in which the Smith couple is able to vote. If two or more of the 82 voters with preference 'Ann Cod Bob' put Cod in first place (Cod Ann Bob), then Ann is eliminated first instead of Bob, and Bob wins instead of Cod.

An increase in support for Cod turns him into a loser (*Negative Responsiveness paradox*).

c) The Smith couple's village has two districts: East and West. The votes were distributed over the districts as follows.

Amount	Preference	East	West
417	Ann Bob Cod	160	257
82	Ann Cod Bob	0	82
143	Bob Ann Cod	143	0
357	Bob Cod Ann	0	357
285	Cod Ann Bob	0	285
324	Cod Bob Ann	285	39
1608		588	1020

Ann would have won in both separate districts (on application of STV), while she loses in a combined vote for the two districts (*Two-districts paradox*). Note that both Bob and Cod have a majority over Ann on a merging of the districts.

4.3   The Alternative Vote (AV) System

In Australia, the members of parliament are chosen per district (as in England), where each district elects one member. As the Australian parliament has 148 seats, the number of districts is 148 (see [24], page 17). The voters are asked to give their preference ordering of the candidates. To aggregate the individual preferences per district to an election of one candidate, the Alternative Vote procedure is used.

With the Alternative Vote (AV) procedure, the candidate who gets more than 50% of the first votes is elected. If no such candidate exists, the candidate with the fewest first votes is eliminated. The votes for that candidate are then transferred to the other candidates in accordance with the second preferences. This procedure is repeated until one of the candidates gets more than half of all votes. In other words, as long as no candidate gets more than 50% of the votes, the candidate with the fewest first votes is eliminated and his votes are transferred in accordance with the second, third, etc. preferences.

The following example illustrates this procedure. Suppose that there are 24 voters and five candidates: Ann (a), Bob (b), Coby (c), Donald (d), and Edward (e). The individual preferences are given in the following profile.

a	e	c	b	d	: 4 voters
b	a	d	c	e	: 5 voters
c	d	b	e	a	: 8 voters
d	a	e	b	c	: 2 voters
d	c	b	e	a	: 1 voters
e	a	b	d	c	: 2 voters
e	d	b	c	a	: 2 voters

In the first round, nobody has more than half (that is, 12) of the first votes, and candidate d is eliminated with the fewest first votes. Two votes are then transferred to a and one to c, the second preferences of the voters with d as first preference. In the second round, a has $4 + 2 = 6$ votes and c has $8 + 1 = 9$. Since still none of the candidates has more than half (12) of all votes (24), candidate e is eliminated with the fewest first votes. Two votes for e are then transferred to a, the second preference of these voters, and, because candidate d has already been eliminated, two votes for e are transferred to candidate b. In the third round, a then has $6 + 2 = 8$ votes and b has $5 + 2 = 7$ votes.

	a	b	c	d	e
number of votes in first round	4	5	8	3	4
second round	6	5	9	-	4
third round	8	7	9	-	-
fourth round	13	-	11	-	-

In the third round, still none of the candidates has more that half of the votes and candidate b is eliminated with the fewest (7) votes. Of the seven votes for b, five are transferred to a, the second choice of the voters with b as first preference. The other two votes for b are transferred to candidate c. These are the two votes that candidate b acquired after candidate e was eliminated. So, even the fourth preference of these voters is taken into account. With two remaining candidates a and c, a is elected in the fourth round because, with 13 votes, he now has more that half of the votes.

4.4   Paradoxes in the AV System

The Alternative Vote system also leads to unexpected outcomes in some situations, as is illustrated by the following paradoxes.

AV is majority-inconsistent: A Condorcet winner may exist without being elected by the Alternative Vote (AV) procedure. This was already shown by Hoag and Hallett in [18]. Van Deemen in [9], p. 237, gives the following example concerning one district with nine voters and three candidates Ann (a), Bob (b), and Coby (c).

a b c : 4 voters
c b a : 3 voters
b c a : 2 voters

It is easily seen that, in the above example, b is the Condorcet winner, while the Alternative Vote procedure eliminates b first and then elects c.

AV is sensitive to strategic behavior: In the above profile, the four voters with preference ordering $a\,b\,c$ prefer b over c. If these four voters vote strategically by giving the untrue preference ordering $b\,a\,c$ instead of their true preference ordering, on application of AV, b will have more than half of the votes and will hence be elected. This outcome is preferred by our four voters over the outcome c that results when they give their true preference.

The above example can also be used to show that *AV is not Independent of Irrelevant Alternatives* (not IIA). To see this, consider the profile that is generated from the above profile by dropping a. We then have two profiles that are equal as far as b and c are concerned. But AV applied to the original profile gives c as the outcome, while AV applied to the modified profile gives b as the outcome. The presence of the (irrelevant) alternative a therefore influences the preference between b and c on application of AV. Similarly, this example also shows that *STV is not IIA* .

In the case that only one candidate has to be elected, the election threshold on application of STV equals $[\frac{n}{2}] + 1$, where n is the number of voters. So, when electing one candidate by means of STV, he or she will be elected when he or she gets more than half of the votes. Hence, when only one candidate has to be elected, the Alternative Vote (AV) procedure corresponds to the STV procedure. Therefore, the examples in section 4.2 show that the *no-show paradox* and the *Negative Responsiveness paradox* can also occur on application of the *AV procedure*.

The reader can easily check that the *two-districts paradox*, considered in section 4.2, can also occur in the Alternative Vote (AV) system. To see this, consider again the example used to illustrate the two-districts paradox for the STV system.

The *districts paradox* that can occur in the British system can also occur on application of STV or AV. Each can be illustrated by the same example; see section 3.3. In this example, party A will gain a *manufactured majority* in parliament also on application of STV and AV respectively: a majority of the number of seats gained with a minority of (first) votes.

4.5   The French Election System

In Australia, seats are distributed on the basis of absolute majority. This means that the seats in parliament are not distributed proportionally to the number of votes a party managed to get. The structure of voting is ordinal: apart from their first preference, voters can also give their second, third, etc. preferences.

In **France**, a similar election procedure is used, one that combines non-proportional representation with a form of ordinal voting. In France, 555 members of parliament are elected in 555 districts by means of the so-called *majority-plurality rule*. According to [24], p. 18, an (absolute) majority (more than half of all votes) is needed in the first round to gain the seat. When none of the candidates gets an absolute majority of votes in the first round, a second and final election round is organized, where the criterion 'Most votes count' (plurality) is used. In general, only two candidates compete in this second round, because the weakest candidates (those with less than 17% of all votes) are forced to withdraw and other candidates are allowed to withdraw in favor of another candidate from an allied party. The French presidential elections are decided by the *majority-runoff* formula. This formula is comparable to the majority-plurality formula, except that only two candidates may participate in the second round, i.e., the ones that got the most votes in the first round. In addition to France, this formula is also used for direct presidential elections in Portugal and Austria.

The French election procedure with two rounds is also majority-inconsistent. We illustrate this paradox by means of an example taken from [19], p. 222. Suppose there are three candidates, *Pro*(gressive), *Cen*(ter), and *Con*(servative), and the preference ordering of the voters is given by the following profile.

Pro	Cen	Con	: 49%
Cen	?	?	: 10%
Con	Cen	Pro	: 41%

Cen defeats *Con* with 49 + 10 = 59% of the votes and *Cen* defeats *Pro* with 10 + 41 = 51% of the votes. So, *Cen* is the Condorcet winner. However, in the first round, nobody gets more than half of the (first) votes, and *Cen*, with the fewest votes, will have to withdraw. The second round will then be between *Pro* and *Con*, despite the fact that *Cen* is the Condorcet winner.

Consider (again) the following example:

a	b	c	: 4 voters
c	b	a	: 3 voters
b	c	a	: 2 voters

It is easy to see that the French election mechanism is *not IIA and sensitive to strategic behavior.*

In their book *Approval Voting* [4], Brams and Fishburn show that the *Negative Responsiveness paradox* may also occur in the French election procedure. Consider three candidates, Ann (*a*), Bob (*b*), and Coby (*c*), and seventeen voters that have the following preferences.

I	a	b	c	: 6 voters
II	c	a	b	: 5 voters
III	b	c	a	: 4 voters
IV	b	a	c	: 2 voters

Then *a* and *b* with six votes each will reach the second round, where *a* will win from *b* with 6 + 5 = 11 against 4 + 2 = 6 votes. Now suppose that the voters in class IV promote *a* from their second to their first preference. This gives rise to the following profile.

I	a	b	c	: 6 voters
II	c	a	b	: 5 voters
III	b	c	a	: 4 voters
IV*	a	b	c	: 2 voters

Then *a* and *c* will reach the second round with 6 + 2 = 8 and 5 votes respectively, where *c* will win from *a* with 5 + 4 = 9 against 6 + 2 = 8 votes. Greater support for *a* costs her the victory!

The reader can easily check that the example Fishburn and Brams use to illustrate the no-show paradox on application of STV (see section 4.2) can also be used to illustrate that the *no-show paradox* may occur for presidential elections in France.

4.6   Summary

The STV (Single Transferable Vote) system is used in Ireland and Malta. This system is based on proportional representation (per district) and on ordinal voting (meaning that the voters give a preference ordering of candidates or parties). It is a fairly complex procedure that, in spite of all the good intentions (proportional representation), gives rise to a number of paradoxes, as seen in section 4.2: STV is majority-inconsistent, sensitive to strategic behavior, and subject to the Negative Responsiveness paradox. Especially the latter seems damning: more votes for a candidate can cause him to lose his seat. Besides, on application of STV, the no-show paradox can occur: adding individual preferences with x as the least preferred candidate can turn x from a loser into a winner. STV is also prone to the districts paradox and the two-districts paradox.

The Alternative Vote (AV) system that is used in Australia (and implicitly in France) is based on non-proportional representation (per district) and on ordinal voting. This system is likewise not free of paradoxes: all the above mentioned paradoxes occur in this system, as we have seen in section 4.4. Like STV, AV is also not Independent of Irrelevant Alternatives.

In section 4.5, we have shown that the French election procedure is also subject to the paradoxes discussed above.

5   Arrow's Theorem

In the previous chapters, we have considered several election procedures: 'Most votes count' (Plurality Rule), pairwise comparison (Majority Rule), Borda procedure, Approval Voting, the Dutch and British systems both based on 'Most votes count', the Single Transferable Vote (STV) system, and the Alternative Vote (AV) system. For each of these election procedures, we have ascertained a number of properties that we consider to be unwanted or to have negative consequences. Of course, the question arises as to whether any 'good' election procedures exist. However, what are 'good' election procedures?

We have already seen a number of requirements which we could say an election procedure needs to satisfy: anonymity, neutrality, IIA, Pareto condition, monotonicity, not sensitive to strategic behaviour, non dictatorial (which means that there is no individual - called the *dictator* - such that the election procedure selects the preference of that individual in all cases). Furthermore, an election procedure should be *transitive*, meaning that, if the procedure prefers x to y and y to z, then the procedure will also prefer x to z. We have seen previously that the Majority Rule does not have this property: for the Condorcet profile in section 2.3, we have seen that, on pairwise comparison, Florence defeats Venice, Venice defeats Siena, but Florence does not defeat Siena (on the contrary, Siena defeats Florence). 'Most votes count' is a transitive election procedure: if x has more votes than y and y has more votes than z, then x has more votes than z.

The question remains if all these properties are unquestionably positive. For instance, anonimity (of the voters) can be considered a great good, but it is sometimes irritating that the vote of a specialist is counted only as heavily (or lightly) as that of the ignorant. On the other hand, if an election procedure is manipulable (meaning that it is sensitive to strategic behaviour), that can be considered a negative property. However, if the possibility of manipulation is small and the procedure has many other positive properties, we might still consider it a 'good' election procedure.

In order to choose from the multitude of available election procedures, it seems wise to determine the properties of the procedures. Based on these properties, positive and negative, it will then hopefully be possible - keeping in mind the purpose of the procedure to be selected - to make a well-considered choice. Even better, but also more difficult, would be to characterise several procedures by means of a number of properties. This means that you show that a procedure with such and such properties must of necessity be one particular procedure. In [25], K. May gave the following characterisation of pairwise comparison (Majority Rule): Majority Rule (pairwise comparison) is the only election procedure that is anonymous, neutral, and monotonic.

There are deep mathematical-logical results at the base of the perpetual wonderment when studying election procedures for more than three candidates ([6], Introduction).

The first observation, by Kenneth Arrow around 1950, is that, for three or more alternatives, there can not be an election procedure that satisfies a number of conditions which could be considered desirable properties for such a procedure. Arrow's theorem states that, for at least three alternatives, there can not be a transitive election procedure - more precisely, there is no transitive (collective) preference rule - that at the same time satisfies the Pareto condition, is Independent of Irrelevant Alternatives (IIA), and non-dictatorial. In other words, K. J. Arrow proved that every transitive election procedure (more precisely, every transitive preference rule) that satisfies the Pareto condition and is Independent of Irrelevant Alternatives (IIA) must necessarily be dictatorial. With this, he gives a characterisation of dictatorial election procedures (preference rules). As a dictatorial election procedure is generally considered to be unwanted, Arrow's result is also called an *impossibility theorem*: a transitive election procedure (preference rule) can never at the same time satisfy the Pareto condition, be Independent of Irrelevant Alternatives, and be non-dictatorial. In other words, the Pareto condition, IIA, and non-dictatoriality are incompatible; it is impossible to construct a transitive election procedure (preference rule) that has all three of these properties. This is quite surprising, as each of these three properties appears to be reasonable at first sight. Arrow's theorem shows us that reaching a collective decision is much more complicated than it would seem.

The second observation, by Gibbard (1973) and Satterthwaite (1975), says that all reasonable election procedures - more precisely, choice rules - for three or more alternatives are sensitive to strategic behaviour. The Gibbard-Satterthwaite theorem states that, for at least three alternatives, there can not be an election procedure (choice rule) that is at the same time Pareto optimal, not sensitive to strategic behaviour, and non-dictatorial. In other words, for at least three alternatives, every Pareto optimal and non-manipulable choice rule is dictatorial.

Notice that Arrow's impossibility theorem is about preference rules, whereas that of Gibbard and Satterthwaite is about choice rules. As we have seen in the Introduction, a *choice rule* selects one alternative for every voter profile, while a *preference rule* assigns to each voter profile a collective order of preference of the alternatives.

A choice rule K is called Pareto optimal if it selects for every profile p an alternative y $(= K(p))$ that is not collectively improvable (meaning that there does not exist an alternative x that everyone prefers to y).

A choice rule K is *non-manipulable* or *not sensitive to strategic behaviour* if cheating does not pay, or more precisely, if strategic behaviour of one of the individuals is not beneficiary to that individual independent of what the other individuals do.

The reader interested in the proofs of these theorems, is referred to [10]. Notice that, in both Arrow's theorem and that of Gibbard and Satterthwaite, the condition is 'at least three alternatives'. For two alternatives, 'Most votes count' and the Borda rule coincide with the Majority Rule, and hence are, like the Majority Rule, neutral, anonymous, monotonic, IIA, non-dictatorial and satisfy the Pareto condition.

The challenge posed by the two above mentioned impossibility theorems is not to devise a perfect election procedure. The theorems put forward by Arrow and Gibbard and Satterthwaite show that this is impossible. The challenge is to identify those procedures that aggregate the wishes of the voters to a collective choice or preference (outcome) as loyally as possible to the voter preferences. We would want an election procedure that ([6], Introduction)

- encourages sincere voting (based on true preferences)
- is relatively immune to strategic manipulation
- avoids obvious paradoxes, like the Negative responsiveness paradox, that occurs when increased support for a candidate turns it into a loser where it previously was a winner.

Furthermore, both impossibility theorems assume that all possible individual preference orderings (all possible profiles) are allowed, while, in practice, the voter profiles will often be subject to several restrictions. For instance, someone who has a person from the right-wing as favorite politician is unlikely to have someone from the extreme left-wing as second choice. Under some restrictions, for instance, single peakedness of the profiles, some nice (that is, satisfying a number of desired properties) election procedures (more precisely, choice rules) are possible; see [1] or [34], Chapter 4. Single peakedness of a profile roughly means that the individual preferences in the profile can be ordered along a line in such a way that, if we follow the line from left to right, the preference of each individual grows to a peak and then diminishes. For instance, the individual preferences regarding the temperature in a room are of the following form: every individual has one optimal temperature and, as one deviates from this temperature, the preference will diminish.

The next example is taken from [34], pp. 97-99. Suppose that, at an office, five persons work in the same space, for which the central heating can be set between 15 and 30 degrees Celcius (no decimals). The employees have to decide collectively on the temperature. It is very likely that every individual will have one optimal temperature and that the preference will diminish as one deviates from this optimum. Now consider the profile below, in which the preference concerning the temperature for each of the employees is given. Call this profile p. It is an example of a single peaked profile.

The optimum of individual 1 is 15 degrees, that of individual 2 is 20 degrees, 3 and 4 have 23 degrees as optimum, and individual 5 has 30 degrees as optimum. Notice that three of the five individuals, namely 3, 4 and 5, (strictly) prefer 23 degrees to all temperatures x lower than 23 degrees. In other words, for all $x \in \{15, 16, \ldots, 22\}$, 23 defeats x on pairwise comparison, given profile p.

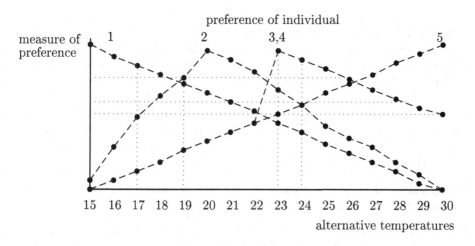

Notice also that four of the five individuals, namely 1, 2, 3 and 4, (strictly) prefer 23 degrees to all temperatures x higher than 23 degrees. In other words, for each $x \in \{24, 25, \ldots, 30\}$, 23 defeats x on pairwise comparison, given profile p.

Hence, given the above (single peaked) profile p, 23 defeats every other alternative on pairwise comparison. In other words, 23 is the Condorcet winner given profile p.

In general, it can be shown that, for single peaked profiles, there is always a Condorcet winner if there is an uneven number of individuals. The choice rule that assigns to those profiles the Condorcet winner is not sensitive to strategic behaviour (non-manipulable), Pareto optimal, and anonymous. This is also the only choice rule for these profiles that has these properties. See [1], [34].

5.1   The Impossibility Theorems

In this section, we will give a mathematically precise formulation of the impossibility theorems of Arrow and of Gibbard and Satterthwaite. For proofs of these theorems, we refer to [10] or to [34], Chapter 3. We start with Arrow's theorem.

Definition 17 Given a set N of individuals, a set A of alternatives, and a (collective) preference rule $F : L(A)^N \to C(A)$, we call i a *dictator* for F if, for every profile $p \in L(A)^N$, $F(p) = p(i)$.
A (collective) preference rule F is called *dictatorial* if there is an individual i such that i is a dictator for F.

Hence, a collective preference rule F is non dictatorial if, for every individual i, there exists a profile p such that the collective preference $F(p)$ is not equal to the individual preference $p(i)$ of i ($F(p) \neq p(i)$).

It seems reasonable to suppose that the collective preference must be transitive: for all alternatives x, y, and z, if the community prefers x to y and y to z, then the community will also prefer x to z. A collective preference rule F that generates, for every profile p, a transitive collective preference $F(p)$, is called transitive.

Definition 18 A preference rule $F : L(A)^N \to C(A)$ is called *transitive* if, for every profile p in $L(A)^N$, $F(p)$ is a transitive relation. So, $F : L(A)^N \to C(A)$ is transitive if $F : L(A)^N \to W(A)$, where $W(A)$ is the set of all weak orderings (meaning, complete and transitive) on A.

Theorem 5 *Arrow's impossibility theorem:* Suppose that there are at least three alternatives in A and $F : L(A)^N \to W(A)$ is a transitive (collective) preference rule. If F satisfies the Pareto condition and is Independent of Irrelevant Alternatives (IIA), then F is dictatorial.

In conclusion, we present a mathematically precise formulation of the Gibbard and Satterthwaite theorem.

Definition 19 A choice rule $K : L(A)^N \to A$ is called *Pareto optimal* if, for every profile $p \in L(A)^N$, there exists no x in A with $x \neq K(p)$ such that for all $i \in N$, $x\ p(i)\ K(p)$. In other words, a choice rule K is called Pareto optimal if it assigns to each profile p an alternative $K(p)$ that can not be collectively ameliorated.

Definition 20 A choice rule $K : L(A)^N \to A$ is called *dictatorial* if there exists an individual $i \in N$ such that, for each profile $p \in L(A)^N$, the collective choice $K(p)$ is the best alternative in $p(i)$.

A choice rule $K : L(A)^N \to A$ is called *non-manipulable* or *not sensitive to strategic behaviour* if cheating does not pay, or more precisely, if strategic behaviour of an individual is not beneficiary to that individual, independent of what other individuals do.

Definition 21 $K : L(A)^N \to A$ is *non-manipulable* if, for all $i \in N$ and for all $p, q \in L(A)^N$, if $p(j) = q(j)$ for all $j \neq i$, then $K(p)\ p(i)\ K(q)$. Therefore, K is non-manipulable if, for every profile p, unilateral deviation of an individual i from p to q is not beneficiary to i ($K(p)$ is at least as good as $K(q)$ for i).

Theorem 6 *Gibbard/Satterthwaite's impossibility theorem:* Suppose there are at least three alternatives in A. Let $K : L(A)^N \to A$ be a Pareto optimal and non-manipulable choice rule. Then K is dictatorial.

5.2 Other Characterization Theorems

The theorem of Arrow gives a characterization of the dictatorial preference rule: it is the only transitive preference rule that is IIA and satisfies the Pareto condition (if there are at least three alternatives).

Similarly, the Gibbard/Satterthwaite theorem gives a characterization of the dictatorial choice rule: it is the only choice rule that is Pareto optimal and non-manipulable (if there are at least three alternatives).

In order to decide which are the better election procedures, it may be useful to have characterizations of the different procedures. We mention a few below.

In [35] H.P. Young gave a characterization of the Borda choice correspondence: it is the only choice correspondence that is neutral, consistent, faithful and has the cancellation property. The *consistency* condition relates choices made by disjoint subsets of voters to choices made by their union. It says that if two disjoint subsets of voters choose the same alternatives, using a choice correspondence, then their union should choose exactly the same alternatives, using this same choice correspondence. *Faithfulness* demands of a choice correspondence that, if society consists of a single individual, it must choose the most preferred alternative of this individual. A third desirable property introduced by Young is the cancellation property. A choice correspondence has the *cancellation property* if and only if it declares a tie between all alternatives if, for all pairs (p, q) of alternatives, the number of voters who prefer p to q equals the number of voters preferring q to p.

Recently, Eliora van der Hout et al. ([20]) gave a characterization of the Plurality ranking rule, on which the Dutch elections are based: it is the only preference rule that is consistent, faithful and has the FS-cancellation property. In the context of a social preference rule F, *consistency* demands that if two disjoint sets of voters I and J both socially prefer party p to party q, using F, then their union should also socially prefer party p to party q, using F. Similarly, it requires that if party p is socially preferred to party q by voter set I, using F, and voter set J is socially indifferent between party p and party q, using F, then party p should also be socially preferred by the union of I and J, using F. A social preference rule is *faithful* if, in case society consists of a single individual whose most preferred party is party p, it orders this party p first. A social preference rule is said to have the first score (FS) *cancellation* property if and only if it declares a tie between party p and party q if the number of individuals who prefer party p most (order p first) equals the number of individuals who prefer party q most (order q first).

In [21] Eliora van der Hout and Harrie de Swart gave a characterization of the British (so-called, First Past The Post) election system.

Finally, in [3], Rob Bosch gave a characterization of k-vote rules. These rules are like Approval Voting, but now everyone has to approve of a fixed number k of candidates.

References

1. D. Black, *The Theory of Committees and Elections*. Cambridge University Press, Cambridge, 1958.
2. J.-C. de Borda, Mémoire sur les Elections au Scrutin, 1781. English translation by A. De Grazia, Mathematical Derivation of an Election System. *Isis*, 44: 42-51, 1953.
3. R. Bosch, Characterization of k-vote rules. KMA and Tilburg University, Faculty of Philosophy. Submitted for Publication.
4. S. J. Brams and P. C. Fishburn, *Approval Voting*. Birkhäuser, Boston, 1983.
5. S. J. Brams, *Rational Politics: Decisions, Games, and Strategy*. CQ Press, 1985; reprinted by Academic Press, 1989.
6. Steven J. Brams and Peter C. Fishburn, Voting Procedures. In: Kenneth Arrow, Amartya Sen and Kotaro Suzumura (eds.), *Handbook of Social Choice and Welfare*. Elsevier Science, Amsterdam, 2002.
7. Condorcet, *Sur les Elections et autres Textes*. Corpus des Oeuvres de Philosophie en Langue Francaise. Librarie Artheme Fayard, Paris, 1986.
8. H. Daudt, De politieke toekomst van de verzorgingsstaat. In: J.J.A. van Doorn and C.J.M. Schuyt (red.), *De stagnerende verzorgingsstaat*. Meppel, 1978.
9. A. van Deemen, Paradoxes of Voting in List Systems of Proportional Representation. *Electoral Studies*, 12: 234-241, 1993.
10. A. van Deemen, *Coalition Formation and Social Choice*. Kluwer, Dordrecht, 1997.
11. A. van Deemen and N. Vergunst, Empirical evidence of paradoxes of voting in Dutch elections. *Public Choice*, 97: 475-490, 1998.
12. G. Doron and R. Kronick, Single transferable vote: An example of a perverse social choice function. *American Journal of Political Science*, 21: 303-311, 1977.
13. D.M. Farrell, *Comparing Electoral Systems*. Prentice Hall/Harvester Wheatsheaf, 1997.
14. P. C. Fishburn and S. J. Brams, Paradoxes of preferential voting. *Mathematics Magazine* 56: 207-214, 1983.
15. W. V. Gehrlein, The expected probability of Condorcet's paradox. *Economics Letters* 7: 33-37, 1981.
16. W.V. Gehrlein, Condorcet's paradox and the Condorcet efficiency of voting rules. *Mathematica Japonica*, 45: 173-199, 1997.
17. Melvin J. Hinich and Michael C. Munger, *Analytical Politics*. Cambridge University Press, 1997.
18. C. G. Hoag and G. H. Hallett, *Proportional Representation*, Macmillan, New York, 1926.
19. P. Hoffman, *Archimedes' Revenge: the joys and perils of mathematics*. Norton, New York, 1988.
20. E. van der Hout, H. de Swart and A. ter Veer, Axioms Characterizing the Plurality Ranking Rule. Tilburg University, Faculty of Philosophy. Submitted for publication.
21. E. van der Hout and H. de Swart, Axioms for FPTP systems. Tilburg University, Faculty of Philosophy. Submitted for Publication.
22. Paul E. Johnson, *Social Choice; Theory and Research*. SAGE Publications, London, 1998.
23. A. Lijphart, *Democracies*. Yale University Press, New Heaven and London, 1984.
24. A. Lijphart, *Electoral Systems and Party Systems: A study of 27 Democracies 1945-1990*. Oxford University Press, Oxford, 1994.

25. K. O. May, A set of independent, necessary and sufficient conditions for simple majority decision. *Econometrica* 20, 680-684, 1952.
26. I. McLean, The Borda and Condorcet Principles: Three Medieval applications. *Social Choice and Welfare* 7, 99-108, 1990.
27. I. McLean, Forms of Representation and Systems of Voting. In: D. Held (ed.), *Political Theory Today*, Polity Press, Cambridge, 1991.
28. I. McLean and A. Urken (Eds.), *Classics of Social Choice*. The University of Michigan Press, 1995.
29. D. C. Mueller, *Public Choice II: A revised edition of Public Choice*. Cambridge University Press, Cambridge, 1989.
30. Hannu Nurmi, *Rational Behaviour and the Design of Institutions: Concepts, Theories and Models*. Edward Elgar, 1998.
31. Hannu Nurmi, *Voting Paradoxes and How to Deal with Them*. Springer-Verlag, 1999.
32. D. W. Rae, *The Political Consequences of Electoral Laws*, Revised Edition. Yale University Press, New Haven, Conn., 1971.
33. Kenneth A. Shepsle and Mark S. Bonchek, *Analyzing Politics: Rationality, Behavior, and Institutions*. W.W. Norton, 1997.
34. A.J.A. Storcken and H.C.M. de Swart, *Verkiezingen, Agenda's en Manipulatie*. Epsilon Uitgaven, Utrecht, 1992.
35. Young, H. P., An Axiomatization of Borda's Rule, *Journal of Economic Theory* 9: 43-52, 1974.

Relational Models of Lambek Logics

Wojciech Buszkowski

Faculty of Mathematics and Computer Science
Adam Mickiewicz University, Poznań Poland
buszko@amu.edu.pl

Abstract. Lambek logics are substructural logics related to the Syntactic Calculus of Lambek [17]. In this paper we prove several representation theorems for algebras of Lambek logics (residuated semigroups, residuated monoids and others) with respect to certain algebras of binary relations. First results of this kind were obtained by Andréka and Mikulás [1], using a method of labeled graphs. Other results were proved in [9, 28], using a method of labeled formulas. In the present paper, we prove these and other results, using a construction of chains of partial representations; this idea was announced earlier in the abstract [5]. We also provide a simpler construction which works for right and left pregroups [8].

1 Lambek Logics and Their Models

Substructural logics are logics whose Gentzen-style axiomatization abandons some (all) structural rules: Contraction, Weakening, Exchange ,etc.; a good introduction into this area is Restall [27]. Historically, one of the earliest substructural logics is the Syntactic Calculus of Lambek [17]: a type processing system for categorial grammars. Other well known representatives are systems of relevant logic. Models and proof techniques for substructural logics have been studied by numerous authors; see e.g. [11, 13, 10, 22, 24]. Models of Lambek logics have been studied in e.g. [3, 2, 12, 25, 1, 29, 16, 9, 6]; see also survey papers [4, 21].

A standard application of relational methods in modelling substructural logics are Kripke frames: an $n-$ary logical connective is interpreted by means of an $n + 1-$ary relation on the universe. This kind of models is very popular and will not be considered here. We shall be interested in models whose elements are binary relations on a fixed universe, and logical connectives are interpreted by means of basic operations on binary relations. Models of that kind are significant for at least two reasons. First, they show a link between substructural logics and dynamic algebras. An early application of Lambek logics (essentially, but not explicitly) to modelling program specification is [15]. Another example is Pratt [26], where residuation operations, characteristic to substructural logics, affixed to regular operations result in an elegant equational axiomatization of action logic. Second, they allow to interpret Lambek logics within Tarski's relational calculus, which makes it possible to employ the latter as a proof system for the former; this line has been developed by Orłowska [23, 24, 20]. (Orłowska's relational proof systems can also be designed on the basis of Kripke frames.)

H. de Swart et al. (Eds.): TARSKI, LNCS 2929, pp. 196–213, 2003.

The aim of this paper is to provide a new, uniform method of proof of representation theorems of the following form: every algebraic structure from some class is isomorphically embeddable into an algebra of binary relations. Algebraic structures, considered here, are abstract models of different Lambek logics (associative, nonassociative, with meet, unit, top, etc.). In [1, 9, 28], several representation theorems of this kind were proved, using other methods: labeled graphs in [1] and labeled formulas in [9, 28] (the latter papers employed some ideas of Kurtonina [16]). The method, presented here, seems to be conceptually simpler: we avoid logical techniques, as e.g. consistent and complete sets of formulas, appearing in [16, 9, 28], and a cumbersome labeling of graphs from [1]. A representation morphism is constructed as a limit of a chain of partial representation morphisms such that each morphism in the chain is a conservative extension of the preceding ones. Some technical details are similar to those in [16, 9, 28], but the contents of lemmas are different. This method was announced in [5]. We also show that the axiom of choice can be eliminated from proofs of some representation theorems. Finally, a much simpler construction is given for right and left pregroups [8].

We describe the Syntactic Calculus, further called *Associative Lambek Calculus* (AL). Its formulas (sometimes called: types) are formed out of denumerably many variables (atoms) $p, q, r, \ldots$ and binary logical connectives $\otimes$ (product), $\to$ (right arrow) and $\leftarrow$ (left arrow). A, B, C denote formulas, and Γ, Δ denote finite strings of formulas. Sequents are expressions $\Gamma \vdash A$. Axioms are the sequents:

$$(\text{Id}) \ A \vdash A,$$

and the inference rules are:

$$(\otimes \text{L}) \ \frac{\Gamma, A, B, \Delta \vdash C}{\Gamma, A \otimes B, \Delta \vdash C}, \quad (\otimes \text{R}) \ \frac{\Gamma \vdash A; \ \Delta \vdash B}{\Gamma, \Delta \vdash A \otimes B},$$

$$(\leftarrow \text{L}) \ \frac{\Gamma, B, \Gamma' \vdash C; \ \Delta \vdash A}{\Gamma, B \leftarrow A, \Delta, \Gamma' \vdash C}, \quad (\leftarrow \text{R}) \ \frac{\Gamma, A \vdash B}{\Gamma \vdash B \leftarrow A},$$

$$(\to \text{L}) \ \frac{\Gamma, B, \Gamma' \vdash C; \ \Delta \vdash A}{\Gamma, \Delta, A \to B, \Gamma' \vdash C}, \quad (\to \text{R}) \ \frac{A, \Gamma \vdash B}{\Gamma \vdash A \to B},$$

$$(\text{CUT}) \ \frac{\Gamma, A, \Gamma' \vdash B; \ \Delta \vdash A}{\Gamma, \Delta, \Gamma' \vdash B}.$$

In rules ($\leftarrow$R) and ($\to$R), one assumes $\Gamma \neq \epsilon$ (by ϵ we denote the empty string). Lambek [17] proves the cut elimination theorem for AL: every provable sequent has a proof without (CUT). Notice that $\Gamma \neq \epsilon$, for every provable sequent $\Gamma \vdash A$. We write $\Gamma \vdash_{AL} A$ if $\Gamma \vdash A$ is provable in AL, and similarly for other systems.

We also consider *Nonassociative Lambek Calculus* (NL), introduced in [18]. The formulas are the same as for AL. Formula structures (structures, for short) are defined by recursion: (i) all formulas are (atomic) structures, (ii) if X, Y are structures, then $(X \circ Y)$ is a structure. $X[Y]$ denotes a structure X with a designated occurrence of a substructure Y, and in this context $X[Z]$ denotes the result of replacing Y with Z in X. X, Y, Z range over structures. Sequents are expressions $X \vdash A$. The axioms are (Id), and the inference rules are:

$$(\otimes L) \ \frac{X[A \circ B] \vdash C}{X[A \otimes B] \vdash C}, \quad (\otimes R) \ \frac{X \vdash A; \ Y \vdash B}{X \circ Y \vdash A \otimes B},$$

$$(\leftarrow L) \ \frac{X[B] \vdash C; \ Y \vdash A}{X[(B \leftarrow A) \circ Y] \vdash C}, \quad (\leftarrow R) \ \frac{X \circ A \vdash B}{X \vdash B \leftarrow A},$$

$$(\rightarrow L) \ \frac{X[B] \vdash C; \ Y \vdash A}{X[Y \circ (A \rightarrow B)] \vdash C}, \quad (\rightarrow R) \ \frac{A \circ X \vdash B}{X \vdash A \rightarrow B},$$

$$(\text{CUT}) \ \frac{X[A] \vdash B; \ Y \vdash A}{X[Y] \vdash B}.$$

Again, the cut elimination theorem holds for NL [18]. NL admits no structural rules; it is regarded as the basic substructural logic in [21]. AL is equivalent to NL enriched with the structural rule of associativity: from $X[(Y_1 \circ Y_2) \circ Y_3] \vdash A$ infer $X[Y_1 \circ (Y_2 \circ Y_3)] \vdash A$, and conversely.

A natural extension of AL is AL0 which admits $\Gamma = \epsilon$ in rules $(\leftarrow R)$ and $(\rightarrow R)$. We have $\vdash_{AL0} p \rightarrow p$, by (Id) and $(\rightarrow R)$, hence $(p \rightarrow p) \rightarrow p \vdash_{AL0} p$, by (Id) and $(\rightarrow L)$, but the latter sequent is not provable in AL. So, AL0 is a nonconservative extension of AL. NL0 admits the empty structure 0 which satisfies $0 \circ X = X \circ 0 = X$, for every X. By the same example, NL0 is a nonconservative extension of NL. Strictly related systems are AL1 and NL1 which admit a logical constant 1. AL1 is AL0 with 1, the new axiom $\vdash 1$, and the new inference rule:

$$(1L) \ \frac{\Gamma, \Delta \vdash A}{\Gamma, 1, \Delta \vdash A}.$$

NL1 is NL0 with 1, the new axiom $0 \vdash 1$, and the new inference rule:

$$(1L) \ \frac{X[0] \vdash A}{X[1] \vdash A}.$$

The cut elimination theorem holds for each of these systems, and consequently, AL1 and NL1 are conservative extensions of AL0 and NL0, respectively.

The structural rule of exchange for associative systems can be written:

$$(\text{EXC}) \ \frac{\Gamma, A, B, \Delta \vdash C}{\Gamma, B, A, \Delta \vdash C},$$

and for nonassociative systems:

$$(\text{NEXC}) \ \frac{X[Y \circ Z] \vdash A}{X[Z \circ Y] \vdash A}.$$

AL with (EXC) is denoted by ALE, and AL0E, AL1E are used in a similar sense. NLE is NL with (NEXC), and similarly for NL0E, NL1E. We say that formulas A, B are *deductively equivalent* in system S, if both $A \vdash_S B$ and $B \vdash_S A$.

One easily shows that $A \to B$ and $B \leftarrow A$ are deductively equivalent in all systems with exchange, mentioned above, hence $\leftarrow$ or $\to$ can be removed from the language. The product-free fragment of AL0E is equivalent to the logic BCI whose Hilbert-style axiomatization consists of the axioms:

(B) $(B \to C) \to ((A \to B) \to (A \to C))$,
(C) $(A \to (B \to C)) \to (B \to (A \to C))$,
(I) $A \to A$,

and the rule Modus Ponens: from $A \to B$ and A infer B. AL0E is a conservative fragment of Intuitionistic Linear Logic [14]. The product-free fragment of ALE was studied by van Benthem [2] as a logic of semantic types. As shown in [6], Multiplicative-Additive Linear Logic (MALL) is faithfully interpretable in BCI with constants $\wedge$ and $\top$, and Cyclic MALL [30] is faithfully interpretable in AL0 with $\wedge$, $\top$ and an additional rule for a distinguished constant. For associative systems without exchange, the rules for $\wedge$ are:

$$(\wedge \text{L1}) \ \frac{\Gamma, A, \Delta \vdash C}{\Gamma, B \wedge A, \Delta \vdash C}, \ (\wedge \text{L2}) \ \frac{\Gamma, A, \Delta \vdash C}{\Gamma, A \wedge B, \Delta \vdash C},$$

$$(\wedge \text{R}) \ \frac{\Gamma \vdash A; \ \Gamma \vdash B}{\Gamma \vdash A \wedge B}.$$

For associative systems with exchange, Δ can be dropped in rules ($\wedge$L1) and ($\wedge$L2). The axiom for $\top$ is: $\Gamma \vdash \top$. The formulation of analogous rules for nonassociative systems is left to the reader.

We survey basic algebraic frames for Lambek logics. A *residuated groupoid* is a structure $(M, \leq, \cdot, \backslash, /)$ such that $(M, \leq)$ is a poset, and $\cdot, \backslash, /$ are binary operations on M, satisfying the equivalences:

$$(\text{RES}) \ ab \leq c \text{ iff } b \leq a \to c \text{ iff } a \leq c \leftarrow b,$$

for all $a, b, c \in M$. As a consequence, $(M, \leq, \cdot)$ is a partially ordered groupoid (p.o. groupoid), that is, a groupoid with a partial ordering $\leq$, satisfying the following condition:

$$(\text{MON}) \text{ if } a \leq b \text{ then } ca \leq cb \text{ and } ac \leq bc.$$

A *unit* in M is an element 1 such that $1a = a = a1$, for all $a \in M$. A residuated groupoid such that $\cdot$ is an associative operation is called *a residuated semigroup*, and a residuated semigroup with unit is called *a residuated monoid*.

Let M be a residuated groupoid. A model over M is a pair (M, μ) such that μ is a mapping from the set of formulas into M which satisfies the following conditions:

$$\mu(A \otimes B) = \mu(A)\mu(B),$$

$$\mu(A \to B) = \mu(A)\backslash\mu(B), \ \mu(B \leftarrow A) = \mu(B)/\mu(A),$$

for any formulas A, B. Then, μ is extended to formula structures, by setting $\mu(X \circ Y) = \mu(X)\mu(Y)$. If 1 is the unit in M, then we set $\mu(0) = 1$ and $\mu(1) = 1$.

For associative systems, models are based on residuated semigroups M; then, μ can be defined on nonempty strings of formulas, by setting:

$$\mu(A_1 \ldots A_n) = \mu(A_1) \cdots \mu(A_n).$$

If 1 is the unit in M, we also set $\mu(\epsilon) = 1$. A sequent $\Gamma \vdash A$ (resp. $X \vdash A$) is said to be *true* in model (M, μ), if $\mu(\Gamma) \leq \mu(A)$ (resp. $\mu(X) \leq \mu(A)$). A sequent is said to be *valid* in a frame M if it is true in all models over M.

It can easily be shown that the sequents provable in NL are precisely those which are valid in all residuated groupoids. Thus, NL is complete with respect to residuated groupoids. The soundness is easy, and the completeness can be proved by the Lindenbaum model. Analogously, NL0 and NL1 are complete with respect to residuated groupoids with unit, AL is complete with respect to residuated semigroups, and AL0, AL1 are complete with respect to residuated monoids [3, 4].

An element $\top \in M$ is called *a top* of M if $a \leq \top$, for all $a \in M$. A binary operation $\wedge$ on M is called *a meet* if the following equivalence:

(MEET) $a \leq b \wedge c$ iff $a \leq b$ and $a \leq c$,

holds, for all $a, b, c \in M$. If M admits meet and/or $\top$, then one assumes:

$$\mu(A \wedge B) = \mu(A) \wedge \mu(B), \ \mu(\top) = \top,$$

provided that the corresponding operation symbols appear in the language. Now, each of the afore mentioned systems, enriched with $\wedge$ and/or $\top$, is complete with respect to the corresponding frames supplied with meet and/or top.

In this paper, we focus on special frames which consist of binary relations on a set. For $R, S \subseteq U^2$, one sets:

$$R \circ S = \{(x, y) \in U^2 : \exists z((x, z) \in R \& (z, y) \in S)\},$$

$$R \backslash S = \{(x, y) \in U^2 : \forall z((z, x) \in R \Rightarrow (z, y) \in S)\},$$

$$S / R = \{(x, y) \in U^2 : \forall z((y, z) \in R \Rightarrow (x, z) \in S)\},$$

$$I_U = \{(x, y) \in U^2 : x = y\}.$$

By $P(W)$ we denote the powerset of a set W. Then, $(P(U^2), \subseteq, \circ, \backslash, /, I_U)$ is a residuated monoid. It admits the meet operation $R \wedge S = R \cap S$ and the top $\top = U^2$. This frame will be called *the full relational frame* on U.

Let $T \subseteq U^2$. For $R, S \subseteq T$, one sets:

$$R \circ S = \{(x, y) \in T : \exists z((x, z) \in R \& (z, y) \in S)\},$$

$$R \backslash S = \{(x, y) \in T : \forall z((z, x) \in R \& (z, y) \in T \Rightarrow (z, y) \in S)\},$$

$$S / R = \{(x, y) \in T : \forall z((y, z) \in R \& (x, z) \in T \Rightarrow (x, z) \in S)\}.$$

Then, $(P(T), \subseteq, \circ, \backslash, /)$ is a residuated groupoid [28]. It admits the meet operation $R \wedge S = R \cap S$ and top $\top = T$. This frame will be called *the restricted relational frame* determined by T. If T is transitive, which means: $(x, y) \in T$ and $(y, z) \in T$ entail $(x, z) \in T$, then the operation $\circ$ is the same as for the full relational frame on U, and one can omit the conjunct $(z, y) \in T$ (resp. $(x, z) \in T$) in the definition of $R \backslash S$ (resp. S/R).

In [1], it has been shown that AL0 is complete with respect to full relational frames, and AL is complete with respect to restricted relational frames determined by irreflexive and transitive relations. Actually, the authors prove some representation theorems: (i) every residuated monoid is embeddable into a full relational frame, (ii) every residuated semigroup is embeddable into a restricted relational frame determined by an irreflexive and transitive relation. [28] shows that every residuated groupoid is embeddable into a restricted relational frame, which yields the completeness of NL with respect to restricted relational frames.

In [9], similar results were proved by applying labeled formulas in the sense of D. Gabbay and canonical models for arbitrary theories based on Lambek logics. (ii) was strengthened: every residuated semigroup is embeddable into a restricted relational frame determined by an irreflexive linear ordering. Also: every residuated monoid satisfying: $1 \leq ab$ entails $1 \leq a$ and $1 \leq b$, is embeddable into a restricted relational frame determined by a reflexive linear ordering. This yields the completeness of AL0, AL1 with respect to restricted relational frames determined by a reflexive linear ordering.

As a matter of fact, these representation theorems yield the strong completeness of the respective systems: the sequents provable in the system from a set of new axioms are precisely those which are true in all models over the respective frames such that all new axioms are true in the model. In [9], the way of reasoning was opposite: strong completeness was shown by means of a canonical model and used to prove the appropriate representation theorem.

Section 2 contains a complete proof of one representation theorem; we present the new method of partial representations in detail. In section 3 we briefly show how to adjust this method to different classes of structures.

2   A Representation Theorem

First, we give a new proof of the following representation theorem: every residuated semigroup is embeddable into a restricted relational frame determined by an irreflexive linear ordering. Then, we show how to modify the proof for the case of other algebras.

Let $(M, \leq, \cdot, \backslash, /)$ be a residuated semigroup. By *a partial representation* of M we mean a triple (U, T, f) such that U is a nonempty set, $T \subseteq U^2$ is an irreflexive linear ordering, and f is a mapping from M into $P(T)$, satisfying the following conditions:

(PR1) $f(a) \circ f(b) \subseteq f(ab)$,
(PR2) if $a \leq b$ then $f(a) \subseteq f(b)$,

for all $a, b \in M$. A partial representation of M is called *a representation* of M if it satisfies the following conditions:

(R1) $f(a) \circ f(b) = f(ab)$,
(R2) $f(a \backslash b) = f(a) \backslash f(b)$,
(R3) $f(a/b) = f(a)/f(b)$,
(R4) $a \leq b$ iff $f(a) \subseteq f(b)$,

for all $a, b \in M$. The operations $\circ$, $\backslash$, $/$ on relations are defined as for restricted relational frames.

Lemma 1. *Let (U, T, f) be a partial representation of M. Then, $f(a \backslash b) \subseteq f(a) \backslash f(b)$ and $f(a/b) \subseteq f(a)/f(b)$, for all $a, b \in M$.*

Proof. For $a, b \in M$, we have $a(a \backslash b) \leq b$, by $a \backslash b \leq a \backslash b$ and (RES). Then, $f(a) \circ f(a \backslash b) \subseteq f(b)$, by (PR1) and (PR2), hence $f(a \backslash b) \subseteq f(a) \backslash f(b)$, by (RES). The second inclusion can be shown in a dual way. $\square$

Let $D = (U, T, f)$ and $D' = (U', T', f')$ be partial representations of M. D' is called *a conservative extension* of D if $U \subseteq U'$, $T = T' \cap U^2$ and $f(a) = f'(a) \cap T$, for all $a \in M$; we write $D \sqsubseteq D'$ if D' is a conservative extension of D.

Let T be an irreflexive linear ordering on a nonempty set U, and let f be a mapping from M into $P(T)$. By *the closure* of f we mean a mapping $c(f)$ from M into $P(T)$, defined as follows: $(x, y) \in c(f)(a)$ iff there exist $x_0, \ldots, x_n \in U$ and $a_1, \ldots, a_n \in M$, $n > 0$, such that $x = x_0$, $y = x_n$, $(x_{i-1}, x_i) \in f(a_i)$, for all $0 < i \leq n$, and $a_1 \cdots a_n \leq a$.

Lemma 2. *$(U, T, c(f))$ is a partial representation of M, and $f(d) \subseteq c(f)(d)$, for all $d \in M$.*

Proof. (PR2) is obvious. We prove (PR1). Assume $(x, y) \in c(f)(a)$ and $(y, z) \in c(f)(b)$. There exist $x_0, \ldots, x_m$ and $a_1, \ldots, a_m$ such that $x = x_0$, $y = x_m$, $(x_{i-1}, x_i) \in f(a_i)$, for all $0 < i \leq m$, and $a_1 \cdots a_m \leq a$. There exist $y_0, \ldots, y_n$ and $b_1, \ldots, b_n$ such that $y = y_0$, $z = y_n$, $(y_{i-1}, y_i) \in f(b_i)$, for all $0 < i \leq n$, and $b_1 \cdots b_n \leq b$. By (MON), $a_1 \cdots a_m b_1 \cdots b_n \leq ab$. Then, the sequence $x_0, \ldots, x_m = y_0, \ldots, y_n$ witnesses $(x, z) \in c(f)(ab)$. Assume $(x, y) \in f(d)$. Then, x, y and d witness $(x, y) \in c(f)(d)$. $\square$

We describe three constructions of a partial representation of M which is a conservative extension of a given partial representation of M. Similar ideas have been employed in [9], following Kurtonina [16]. The latter papers, however, consider extensions of sets of labeled formulas, not models; [16] does not touch the representation problem.

Construction 1. Let (U, T, f) be a partial representation of M and $(x', y') \in f(ab)$. Let $u \notin U$. We set $U' = U \cup \{u\}$. T' equals T on U and contains all pairs (z, u), for $z \in U$ such that $(z, x') \in T$ or $z = x'$, and all pairs (u, z), for $z \in U$ such that $(x', z) \in T$. So, T' is an irreflexive linear ordering on U' in which u is positioned immediately after x'. Finally, we define $f'(a) = f(a) \cup \{(x', u)\}$, $f'(b) = f(b) \cup \{(u, y')\}$, and $f'(c) = f(c)$, for all $c \in M$ such that $c \neq a$ and $c \neq b$. So, f' maps M into $P(T')$, and $f'(d) \cap T = f(d)$, for all $d \in M$.

Lemma 3. $(U, T, f) \sqsubseteq (U', T', c(f'))$ and $(x', u) \in c(f')(a)$, $(u, y') \in c(f')(b)$.

Proof. By lemma 2, $(U', T', c(f'))$ is a partial representation of M. Clearly, $U \subseteq U'$ and $T = T' \cap U^2$. Also, for $d \in M$, $f(d) \subseteq f'(d)$ and $f'(d) \subseteq c(f')(d)$, which yields $f(d) \subseteq c(f')(d)$. Assume $(x, y) \in c(f')(d)$, where $(x, y) \in T$. Then, there exist $x_0, \ldots, x_n \in U'$ and $a_1, \ldots, a_n \in M$, $n > 0$, such that $x = x_0$, $y = x_n$, $(x_{i-1}, x_i) \in f'(a_i)$, for all $0 < i \leq n$, and $a_1 \cdots a_n \leq d$. Let k be the number of occurrences of u among $x_0, \ldots, x_n$. By induction on k, we prove $(x, y) \in f(d)$. (Actually, since $(u, u) \notin T$, then $k \leq 1$, but the inductive argument will be useful in variants of this proof, discussed later on.) Let $k = 0$. Then, (x, y) belongs to $f(a_1) \circ \cdots \circ f(a_n)$ which is contained in $f(d)$, by (PR1), (PR2). Assume the thesis holds for k. We prove it for $k + 1$. Let i be the least integer such that $x_i = u$. Clearly, $0 < i < n$. We have $(x_{i-1}, u) \in f'(a_i)$ and $(u, x_{i+1}) \in f'(a_{i+1})$. Then, $a_i = a$, $a_{i+1} = b$, and $x_{i-1} = x'$, $x_{i+1} = y'$. Since $(x', y') \in f(ab)$, then $(x, y) \in c(f')(d)$ can be witnessed by $x_0, \ldots, x_{i-2}, x', y', x_{i+2}, \ldots, x_n$ and $a_1 \cdots a_{i-1}aba_{i+2} \cdots a_n \leq d$, with u occurring k times in the former sequence. Consequently, $(x, y) \in f(d)$, by the induction hypothesis. We have shown $c(f')(d) \cap T = f(d)$. Since $(x', u) \in f'(a)$. $(u, y') \in f'(b)$. then the second part of the thesis is also true. $\qquad\square$

Construction 2. Let (U, T, f) be a partial representation of M, $(x'y') \in T$ and $(x', y') \notin f(a\backslash b)$. Let $u \notin U$. We set $U' = U \cup \{u\}$. T' equals T on U and contains all pairs (u, z), for $z \in U$. So, T' is an irreflexive linear ordering on U' in which u is the first element. Finally, we define $f'(a) = f(a) \cap \{(u, x')\}$ and $f'(c) = f(c)$, for all $c \neq a$. Then, f' maps M into $P(T')$, and $f'(d) \cap T = f(d)$, for all $d \in M$.

Lemma 4. $(U, T, f) \sqsubseteq (U', T', c(f'))$ and $(u, x') \in c(f')(a)$, $(u, y') \notin c(f')(b)$.

Proof. Again, $(U', T', c(f'))$ is a partial representation of M, $U \subseteq U'$, $T' \cap U^2 = T$ and $f(d) \subseteq c(f')(d)$, for all $d \in M$. Assume $(x, y) \in c(f')(d)$, $(x, y) \in T$. There exist $x_0, \ldots, x_n \in U'$ and $a_1, \ldots, a_n \in M$ which satisfy the same conditions as in the proof of lemma 3. Since no pair of the form (z, u) belongs to $f'(a_i)$, for $0 < i \leq n$, then the number k must be 0, and we argue as above, for the case $k = 0$. Consequently, $c(f')(d) \cap T = f(d)$. Clearly, $(u, x') \in c(f')(a)$. We show $(u, y') \notin c(f')(b)$. Suppose the contrary. There exist $x_0, \ldots, x_n$ and $a_1, \ldots, a_n$ such that $x_0 = u$, $x_n = y'$, and the remainder is the same as above (for $d = b$). By the same reason as above, x_0 is the only occurrence of u among $x_0, \ldots, x_n$. We have $(u, x_1) \in f'(a_1)$, hence $x_1 = x'$, $a_1 = a$. Since $x_n = y' \neq x'$, then $n > 1$. Further, $aa_2 \cdots a_n \leq b$, which yields $a_2 \cdots a_n \leq a\backslash b$, by (RES), and consequently, $(x', y') \in c(f')(a\backslash b)$. But $(x', y') \in T$, hence $(x', y') \in f(a\backslash b)$, by the first part of the lemma. This contradicts the assumption of construction 2. $\qquad\square$

Construction 3. This construction is dual to construction 2. Let (U, T, f) be a partial representation of M, $(x', y') \in T$ and $(x', y') \notin f(a/b)$. Let $u \notin U$. We set $U' = U \cup \{u\}$. T' equals T on U and contains all pairs (z, u), for $z \in U$. So, T' is an irreflexive linear ordering on U' in which u is the last element. Finally, we define $f'(b) = f(b) \cup \{(y', u)\}$ and $f'(c) = f(c)$, for all $c \neq b$. Then, f' maps

M into $P(T')$, and $f'(d) \cap T = f(d)$, for all $d \in M$. The following lemma is dual to lemma 4.

Lemma 5. $(U, T, f) \sqsubseteq (U', T', c(f'))$ and $(y', u) \in c(f')(b)$, $(x', u) \notin c(f')(a)$.

Let $D = (U, T, f)$, $D' = (U', T', f')$ be partial representations of M. We say that D' witnesses D, if $D \sqsubseteq D'$ and, for all $(x, y) \in T$ and $a, b \in M$, the following conditions hold true:

(W1) if $(x, y) \in f(ab)$, then $(x, y) \in f'(a) \circ f'(b)$,
(W2) if $(x, y) \notin f(a \backslash b)$, then $(x, y) \notin f'(a) \backslash f'(b)$,
(W3) if $(x, y) \notin f(a/b)$, then $(x, y) \notin f'(a)/f'(b)$.

Let $(D_\alpha)_{\alpha < \gamma}$, $\gamma > 0$, be a (possibly transfinite) sequence of partial representations of M such that $D_\alpha \sqsubseteq D_\beta$, for all $\alpha < \beta$. We define the limit $D = (U, T, f)$, by setting: $U = \bigcup_{\alpha < \gamma} U_\alpha$, $T = \bigcup_{\alpha < \gamma} T_\alpha$, and $f(a) = \bigcup_{\alpha < \gamma} f_\alpha(a)$, for all $a \in M$. The proof of the next lemma is easy.

Lemma 6. $D_\alpha \sqsubseteq D$, for all $\alpha < \gamma$.

Lemma 7. Let $(D_n)_{n < \omega}$ be an $\omega-$sequence of partial representations of M such that D_{n+1} witnesses D_n, for all $n < \omega$, and D_0 satisfies (R4). Let $D = (U, T, f)$ be the limit of this sequence. Then, D is a representation of M.

Proof. By lemma 6, $D_n \sqsubseteq D$, for all $n < \omega$, hence D is a partial representation of M. We denote $D_n = (U_n, T_n, f_n)$.

We show that D satisfies (R4). D satisfies (PR2). Assume $a \not\leq b$. Then, $f_0(a) \not\sqsubseteq f_0(b)$, since D_0 satisfies (R4). So, there exists $(x, y) \in f_0(a)$, $(x, y) \notin f_0(b)$. Since $D_0 \sqsubseteq D$, then $(x, y) \in f(a)$, $(x, y) \notin f(b)$, which yields $f(a) \not\sqsubseteq f(b)$.

We show that D satisfies (R1). D satisfies (PR1). Assume $(x, y) \in f(ab)$. Then, there exists $n < \omega$ such that $(x, y) \in f_n(ab)$. Since D_{n+1} witnesses D_n, then $(x, y) \in f_{n+1}(a) \circ f_{n+1}(b)$, which yields $(x, y) \in f(a) \circ f(b)$.

We show that D satisfies (R2). By lemma 1, $f(a \backslash b) \subseteq f(a) \backslash f(b)$. Assume $(x, y) \in T$, $(x, y) \notin f(a \backslash b)$. Then, there exists $n < \omega$ such that $(x, y) \in T_n$, $(x, y) \notin f_n(a \backslash b)$. Since D_{n+1} witnesses D_n, then $(x, y) \notin f_{n+1}(a) \backslash f_{n+1}(b)$. So, there exists $z \in U_{n+1}$ such that $(u, x) \in f_{n+1}(a)$, $(u, y) \notin f_{n+1}(b)$. Since $D_{n+1} \sqsubseteq D$, then $(u, x) \in f(a)$, $(x, y) \notin f(b)$, hence $(x, y) \notin f(a) \backslash f(b)$. In a dual way one shows that D satisfies (R3). $\square$

Lemma 8. Let $D = (U, T, f)$ be a partial representation of M. Then, there exists D' such that D' witnesses D.

Proof. The proof uses the axiom of choice. Let κ be the maximum of the cardinalities of U and M. First, assume that κ is infinite. The set $U^2 \times M^2 \times \{1, 2, 3\}$ is of cardinality κ. Let $(t_\alpha)_{\alpha < \kappa}$ be a sequence of all elements of this set. We shall define a chain $(D_\alpha)_{\alpha < \kappa}$ of partial representations of M such that $D_\alpha \sqsubseteq D_\beta$ whenever $\alpha < \beta$. We denote $D_\alpha = (U_\alpha, T_\alpha, f_\alpha)$. Put $D_0 = D$. Assume D_α, for $\alpha < \gamma$, have already been defined. If γ is a limit ordinal, we set D_γ to be the limit of the chain $(D_\alpha)_{\alpha < \gamma}$. By lemma 6, $D_\alpha \sqsubseteq D_\gamma$, for all $\alpha < \gamma$.

Let $\gamma = \beta + 1$. Take $t_\beta = (x', y', a, b, i)$. We consider four cases. (I) $i = 1$ and $(x', y') \in f_\beta(ab)$. We define $D_{\beta+1}$ by construction 1 with $U = U_\beta$, $T = T_\beta$, $f = f_\beta$, and $D_{\beta+1} = (U', T', c(f'))$. (II) $i = 2$ and $(x', y') \in T_\beta$, $(x', y') \notin f_\beta(a \backslash b)$. We define $D_{\beta+1}$ by construction 2. (III) $i = 3$ and $(x', y') \in T_\beta$, $(x, y) \notin f_\beta(a/b)$. We define $D_{\beta+1}$ by construction 3. (IV) cases (I), (II), (III) do not hold. We define $D_{\beta+1} = D_\beta$. By lemmas 3, 4, 5 $D_\beta \sqsubseteq D_{\beta+1}$, hence $D_\alpha \sqsubseteq D_\gamma$, for all $\alpha < \gamma$.

Let $D' = (U', T', g)$ be the limit of $(D_\alpha)_{\alpha < \kappa}$. By lemma 6, $D_\alpha \sqsubseteq D'$, for all $\alpha < \kappa$.

We show (W1). Assume $(x', y') \in T$, $(x', y') \in f(ab)$. Then, there exists $\beta < \kappa$ such that $t_\beta = (x', y', a, b.1)$. Case (I) holds and, by lemma 3, $(x', y') \in f_{\beta+1}(a) \circ f_{\beta+1}(b)$, which yields $(x', y') \in g(a) \circ g(b)$.

We show (W2). Assume $(x', y') \in T$, $(x', y') \notin f(a \backslash b)$. Then, there exists $\beta < \kappa$ such that $t_\beta = (x', y', a, b, 2)$. Since $D \sqsubseteq D_\beta$, then $(x', y') \in T_\beta$, $(x', y') \notin f_\beta(a \backslash b)$. So, case (II) holds and, by lemma 4, there is u such that $(u, x') \in f_{\beta+1}(a)$, $(u, y') \notin f_{\beta+1}(b)$. Since $D_{\beta+1} \sqsubseteq D'$, then $(u, x') \in g(a)$, $(u, y') \notin g(b)$, and consequently, $(x', y') \notin g(a) \backslash g(b)$. The proof of (W2) is dual.

If U and M are finite, then $U^2 \times M^2 \times \{1, 2, 3\}$ is also finite, and we consider finite sequences $(t_j)_{j < n}$ and $(D_j)_{j \leq n}$. Now, $D' = D_n$. Then, the proof goes as above. □

We are ready to prove:

Theorem 1. *For any residuated semigroup M, there exists a representation of M.*

Proof. Let M be a residuated semigroup. First, we define a partial representation of M which satisfies (R4). Here, we present one method. For every $a \in M$, choose different elements x_a, y_a. Let U consist of all x_a, y_a, for $a \in M$. Let T be an arbitrary irreflexive linear ordering on U which contains all pairs (x_a, y_a) (we use the axiom of choice). For $a \in M$, we define:

$$f(a) = \{(x_c, y_c) : c \leq a\}.$$

Take $D = (U, T, f)$. (PR2) is obvious. If $a \not\leq b$, then $(x_a, y_a) \in f(a)$, $(x_a, y_a) \notin f(b)$, hence (R4) is true. (PR1) also holds, since $f(a) \circ f(b) = \emptyset$, for all $a, b \in M$.

We form a chain $(D_n)_{n < \omega}$ of partial representations of M such that $D_0 = D$, and D_{n+1} witnesses D_n, for all $n < \omega$; the existence of such a chain follows from lemma 8. By lemma 7, the limit of this chain is a representation of M. □

Notice that, for a countable M, all D_n's are countable, hence in the resulting representation (U, T, f) the universe U is countable. Therefore, every countable residuated semigroup is embeddable into the restricted relational frame determined by a countable, irreflexive linear ordering T. Analogous facts are true for all variants, considered in the next section.

3 Variants

We shall briefly discuss different modifications of the above proof appropriate for other representation theorems.

Irreflexive Partial Orderings. In [1] it has been shown that every residuated semigroup is embeddable into the restricted relational frame determined by an irreflexive partial ordering. The proof uses the axiom of choice. By applying the above methods, the axiom of choice can be avoided. We redefine the notions of a partial representation and a representation, assuming T to be an irreflexive partial ordering on U. Lemma 8 can be proved without the axiom of choice. For every quadruple (x', y', a, b) such that $(x', y') \in f(ab)$, one introduces a new element u. For every quadruple (x', y', a, b) such that $(x', y') \notin f(a \backslash b)$, one introduces a new element v. For every quadruple (x', y', b, a) such that $(x', y') \notin f(b/a)$, one introduces a new element w. U' is U enriched with all new elements. T' is the smallest irreflexive partial ordering on U' which contains T and all pairs $(x', u), (u, y')$, for quadruples of the first kind and the corresponding new elements u, in which all new elements v are minimal elements and all new elements w are maximal elements. $f'(a)$ is $f(a)$ enriched with all pairs (x', u) such that there exists a quadruple (x', y', a, b) of the first kind, and u is the corresponding new element, all pairs (u, y') such that there exists a quadruple (x', y', b, a) of the first kind, and u is the corresponding new element, all pairs (v, x') such that there exists a quadruple (x', y', a, b) of the second kind, and v is the corresponding new element, and all pairs (y', w) such that there exists a quadruple (x', y', b, a) of the third kind, and w is the corresponding new element. We define $D' = (U', T', c(f'))$. The proof of $D \sqsubseteq D'$ joins arguments used in the proofs of lemmas 3, 4, 5, and similarly for (W1), (W2), (W3). Finally, in the proof of theorem 1, D_0 can be formed without the axiom of choice: T_0 consists of all pairs (x_a, y_a), for $a \in M$.

Residuated Groupoids. It has been shown in [28] that every residuated groupoid is embeddable into a restricted relational frame (now, T need not be transitive). The proof uses logical tools. Here, we show how to prove this theorem by the above methods. In the definitions of a partial representation and a representation, we assume T to be an arbitrary relation on U.

Since the product is nonassociative, $a_1 \cdots a_n \leq a$ is not meaningful, unless we fix a bracketing on $a_1 \cdots a_n$. Therefore, we say: $a_1 \cdots a_n \leq a$ with a bracketing p. If $a_1 \cdots a_n$ is bracketed by p, then the notion of a *constituent* of $a_1 \cdots a_n$ with respect to p is defined in a natural way. For example, the constituents of $((ab)(cd))$ are: $abcd$, ab, cd, a, b, c, d.

Let $(M, \leq, \cdot, \backslash, /)$ be a residuated groupoid, and let $T \subseteq U^2$. Let p be a bracketing on $a_1 \cdots a_n$. We say that a sequence $x_0, \ldots, x_n$ is *compatible* with p, T if, for every constituent $a_i \cdots a_j$ of $a_1 \cdots a_n$ with respect to p, it holds $(x_{i-1}, x_j) \in T$.

Let f be a mapping from M into $P(T)$. We define *the closure* of f as the mapping $c(f)$ from M into $P(T)$, satisfying: $(x, y) \in c(f)(a)$ iff there exist

$x_0, \ldots, x_n \in U$ and $a_1, \ldots, a_n \in M$, $n > 0$, such that $a_1 \cdots a_n \leq a$ with a bracketing p, the sequence $x_0, \ldots, x_n$ is compatible with p, T, $x_0 = x$, $x_n = y$, and $(x_{i-1}, x_i) \in f(a_i)$, for all $0 < i \leq n$.

Lemma 1 is true. The proof of lemma 2 can easily be modified. If p is the given bracketing on $a_1 \cdots a_m$, and q on $b_1 \cdots b_n$, then r on $a_1 \cdots a_m b_1 \cdots b_n$ is $((p)(q))$. Now, $a_1 \cdots a_m b_1 \cdots b_n \leq ab$ with bracketing r, and the sequence $x_0, \ldots, x_m = y_0, \ldots, y_n$ is compatible with r, T (notice that $(x, y) \in T$, by the definition of $\circ$ in restricted relational frames).

We assume that T is irreflexive in all partial representations (U, T, f) to be considered. In construction 1 we take $T' = T \cup \{(x', u), (u, y')\}$ and the remainder as above. In the proof of lemma 3, if u occurs among $x_0, \ldots, x_n$, it must be $u = x_i$, $0 < i < n$, $x_{i-1} = x'$, $x_{i+1} = y'$, and $a_i a_{i+1} = ab$ is a constituent of $a_1 \cdots a_n$. Assume it is not the case. Then, $a_i a_{i+1}$ is not a constituent of $a_1 \cdots a_n$. There exist k, l such that $a_k \cdots a_i$ and $a_{i+1} \cdots a_l$ are constituents of $a_1 \cdots a_n$, and either $k < i$, or $i+1 < l$. Consider the first case. Let k be the greatest integer less than i such that $a_k \cdots a_i$ is a constituent. Then, the bracketing is $((a_k \cdots a_{i-1}) a_i)$. By compatibility, $(x_{k-1}, u) \in T'$, hence $x_{k-1} = x'$. Since $x_{i-1} = x'$, we get $(x', x') \in T'$, which is impossible. In the second case, the reasoning is similar. Then, we can continue as in the proof of lemma 3.

In construction 2 we take $T' = T \cup \{(u, x'), (u, y')\}$ and the remainder as above. The proof of lemma 4 uses arguments similar to those from the preceding paragraph. In construction 3 we take $T' = T \cup \{(x', u), (y', u)\}$ and the remainder as above. The proof of lemma 5 is dual. Proofs of lemmas 6, 7, 8 can be copied. In the proof of theorem 1, T_0 consists of all pairs (x_a, y_a), for $a \in M$. Again, one can avoid the axiom of choice, by changing the proof of lemma 8.

Residuated Monoids. For $R, S \subseteq I_U$, we have $R \circ S = R \cap S$. In residuated monoids, $a \leq 1$ and $b \leq 1$ entails $ab \leq a$ and $ab \leq b$, hence $ab \leq a \wedge b$ (if $a \wedge b$ exists), but not necessarily $ab = a \wedge b$. Then, an embedding of a residuated monoid M into the full relational frame $P(U^2)$ is not required to satisfy $f(1) = I_U$. One stipulates a weaker condition:

$$(1) \quad 1 \leq a \text{ iff } I_U \subseteq f(a),$$

for all $a \in M$. This yields the completeness theorem for AL0 and AL1 in the usual sense: A is provable iff $1 \leq [A]$ in the Lindenbaum algebra iff $I_U \subseteq f([A])$ in the relational frame iff A is true the relational model.

In [1], it has been shown that every residuated monoid is embeddable into the full relational frame on some universe U. Again, this result can be proven by the above methods, and the axiom of choice can be eliminated. Now, we take $T = U^2$ in all partial representations (so, T can simply be dropped). Further, a partial representation of M must satisfy:

(PR3) if $1 \leq a$ then $I_U \subseteq f(a)$,

for all $a \in M$, and a representation of M must satisfy (1).

Lemma 1 is true. The closure of f is defined as above; one admits $n = 0$ if $1 \leq a$. This yields $I_U \subseteq c(f)(a)$ if $1 \leq a$. Lemma 2 can be shown as above. In constructions 1, 2, 3 we drop manipulations with T and preserve the remainder. Proofs of lemmas 3, 4, 5 can essentially be copied; one adds the case $n = 0$ which is easy. An additional construction is needed.

Construction 4. Let $D = (U, f)$ be a partial representation of M and $1 \not\leq a$. Let $u \notin U$. We set $U' = U \cup \{u\}$ and consider $D' = (U', c(f))$. We prove:

Lemma 9. $D \sqsubseteq D'$ and $I_{U'} \not\subseteq c(f)(a)$.

Proof. By lemma 2, D' is a partial representation of M. Assume $(x, y) \in c(f)(d)$, for $x, y \in U$. There exist $x_0, \ldots, x_n$ and $a_1, \ldots, a_n$ such that $x_0 = x$, $x_n = y$, $a_1 \ldots a_n \leq d$, and $(x_{i-1}, x_i) \in f(a_i)$, for $0 < i \leq n$. Assume $n > 0$. Since no pair from $f(a_i)$ contains u, we can proceed as in the proof of lemma 3 for $k = 0$. Assume $n = 0$. Then $1 \leq d$ and $x = y$. By (PR3) for D, $(x, y) \in f(d)$. Clearly, $(u, u) \notin c(f)(a)$, since $1 \not\leq a$. $\square$

The proof of lemma 6 can be copied. In the definition of witnessing, we add the condition:

(W4) if $1 \not\leq a$ then $I_{U'} \not\subseteq f'(a)$,

for all $a \in M$. Now, lemma 7 is easy. In the proof of lemma 8, the sequence $(t_\alpha)_{\alpha < \kappa}$ contains all elements of the set $(U^2 \times M^2) \cup M'$, where $M' = \{a \in M : 1 \not\leq a\}$. If $t_\beta = a$, $a \in M'$, then $D_{\beta+1}$ is formed out from D_β by construction 4. By conservativity, D' witnesses D. In the proof of theorem 1, we can define U and f as above and take $D_0 = (U, c(f))$ (one might also set $D_0 = (U, c(g))$, where $g(a) = \emptyset$, for all $a \in M$). The axiom of choice can be eliminated by changing the proof of lemma 8.

Reflexive Linear Orderings. Every residuated monoid in which $1 \leq ab$ holds only if $1 \leq a$ and $1 \leq b$ is embeddable into the restricted relational frame determined by a reflexive linear ordering T [9]. This can be shown by our methods, joining elements of the proofs for residuated semigroups and residuated monoids. We omit details.

Residuated Groupoids with Unit. Every residuated groupoid with unit is embeddable into the restricted relational frame determined by a reflexive relation T. This is shown by mixing the proofs for residuated groupoids and residuated monoids. Now, partial representations and representations are defined as for residuated groupoids, with T reflexive. Additionally, representations must satisfy (1), and partial representations must satisfy (PR1)-(PR3) and:

(PR4) if $1 \not\leq a$ then $f(a)$ is irreflexive,

for all $a \in M$. Now, we can proceed as in the proof for residuated groupoids; in the definition of $c(f)$ we add the possibility $n = 0$ if $1 \leq a$. In lemma 2, we must assume that f satisfies (PR4). In proofs of lemmas 3, 4, 5, we need some changes. Look at the proof that $D \sqsubseteq D'$, sketched in the fragment concerning residuated groupoids. We have rejected the possibility $(x', x') \in T'$, since T' is irreflexive. Now, $(x', x') \in T'$ is true, but we infer $(x', x') \in f(a_k \cdots a_{i-1})$ with the induced bracketing on $a_k \cdots a_{k-1}$, by (PR1), (PR2), and consequently, $1 \leq a_k \cdots a_{i-1}$, by (PR4). Therefore, $x_0, \ldots, x_{k-1}, x_i, \ldots, x_n$ and $a_1 \cdots a_{k-1} a_i \cdots a_n$ are shorter sequences, proving $(x, y) \in c(f')(d)$. Accordingly, arguments in lemmas 3, 4, 5 can be saved, if one always takes the least possible n. Clearly, (PR4) is preserved by these constructions. The remainder of the proof is more or less the same as for residuated groupoids. Construction 4 is not needed.

Residuated Semigroups with Meet. Assume that M admits meet. We stipulate that every partial representation (U, T, f) of M satisfies:

$$f(a \wedge b) = f(a) \cap f(b), \text{ for all } a, b \in M.$$

Lemma 1 is true. We preserve the definition of closure. Lemma 2 is true except for the claim that $c(f)$ satisfies the above equality. Lemmas 3, 4, 5 remain true. Let us consider lemma 3. We assume that $D = (U, T, f)$ is a partial representation of M, and we prove that $D' = (U', T', c(f'))$ is a partial representation of M. We must show:

$$c(f')(d \wedge e) = c(f')(d) \cap c(f')(e), \text{ for all } d, e \in M.$$

The inclusion $\subseteq$ is easy. We prove the converse inclusion. Let $(x, y) \in c(f')(d) \cap c(f')(e)$. Let $x \neq u$ and $y \neq u$. By conservativity, $(x, y) \in f(d) \cap f(e)$, which yields $(x, y) \in f(d \wedge e)$, hence $(x, y) \in c(f')(d \wedge e)$. Assume $x = u$. Then, $y \neq u$. There exist $x_0, \ldots, x_m$ and $a_1, \ldots, a_m$ such that $u = x_0$, $y = x_m$, $(x_{i-1}, x_i) \in f'(a_i)$, for all $0 < i \leq m$, and $a_1 \cdots a_m \leq d$. There exist $y_0, \ldots, y_n$ and $b_1, \ldots, b_n$ such that $u = y_0$, $y = y_n$, $(y_{i-1}, y_i) \in f'(b_i)$, for all $0 < i \leq n$, and $b_1 \cdots b_n \leq e$. Assume $m = 1$. Then, $y = x_1 = y'$, by construction 1, hence also $n = 1$ (otherwise, $(y', y') \in T'$, which is impossible). We have $a_1 = b_1 = b$, by construction 1, and $b \leq d$, $b \leq e$, which yields $b \leq d \wedge e$. Consequently, $(x, y) \in c(f')(d \wedge e)$. Assume $n = 1$. Then, $m = 1$, and we argue as above. Assume $m > 1$ and $n > 1$. We have $x_1 = y_1 = y'$ and $a_1 = b_1 = b$, as above. By (RES), $a_2 \cdots a_m \leq b \backslash d$ and $b_2 \cdots b_m \leq b \backslash e$. Thus, $(y', y) \in c(f')(b \backslash d)$ and $(y', y) \in c(f')(b \backslash e)$. By conservativity, $(y', y) \in f(b \backslash d)$ and $(y', y) \in f(b \backslash e)$, which yields $(y', y) \in f((b \backslash d) \wedge (b \backslash e))$. Now, $(b \backslash d) \wedge (b \backslash e) = b \backslash (d \wedge e)$ is true in residuated semigroups, hence $(y', y) \in f(b \backslash (d \wedge e))$. Consequently, $(x, y) \in c(f')(d \wedge e)$. For $y = u$, the argument is dual.

The partial representation D_0 in the proof of theorem 1 satisfies the above equality: $(x_c, y_c) \in f(a \wedge b)$ iff $c \leq a \wedge b$ iff $c \leq a$ and $c \leq b$ iff $(x_c, y_c) \in f(a) \cap f(b)$. It is easy to show that the limit of a chain of partial representations of M, satisfying the above equality, also satisfies this equality. Therefore, we can prove that every residuated semigroup admitting meet is embeddable into the restricted relational frame determined by an irreflexive linear ordering T, and the embedding f satisfies the equality (this theorem has been proven in [1] in a different way). Analogous results can be obtained for all other kinds of structures, considered in this paper.

Structures with Top. We cannot expect that $f(\top) = T$ or $f(\top) = U^2$ must be satisfied by all embeddings. Clearly, $f(\top)$ must contain all relations $f(a)$, for $a \in M$. We can define the restricted relational frame determined by $f(\top)$. Consider the case of residuated monoids. Since $\top\top \leq \top$ is true, then $f(\top)$ is a transitive relation. Then, for the relations $f(a)$, $a \in M$, the operations $\circ, \backslash, /$ for the full relational frame are the same as the corresponding operations for the restricted relational frame. Therefore, f can be treated as an embedding of M into the restricted relational frame which preserves top, and similarly for other kinds of structures.

MALL and Cyclic MALL. MALL is a propositional fragment of Classical Linear Logic [14]. Its formulas are formed out of propositional variables $p, q, r, \ldots$, negated propositional variables $p^\perp, q^\perp, r^\perp, \ldots$ and constants $0, 1, \perp, \top$ by means of binary operation symbols $\otimes, \oplus, \wedge, \vee$. (We use a notation following Troelstra.) Models are *phase spaces*, defined as follows. Let $(M, \cdot, 1)$ be a commutative monoid, and let $\perp^\star \subseteq M$. (Since formulas will be interpreted as subsets of M, $\perp^\star$ can be understood as an interpretation of a propositional constant which, however, does not explicitly appear in the language of MALL.) For $X, Y \subseteq M$, one defines:

$$X \circ Y = \{ab : a \in X \,\&\, b \in Y\}, \ X \to Y = \{a : \forall b(b \in X \Rightarrow ba \in Y)\}.$$

Then, $(P(M), \subseteq, \circ, \to, \{1\})$ is a commutative residuated monoid. Also $X\backslash Y = X \to Y$. One defines $X^\perp = X \to \perp^\star$. A set $X \subseteq M$ is called *a fact* if $X = X^{\perp\perp}$ (equivalently, $X = Y^\perp$, for some $Y \subseteq M$). A phase space is denoted $(M, \perp^\star)$.

A *model* over $(M, \perp^\star)$ is given by an assignment μ which assigns facts to propositional variables. It is defined for all formulas, by setting:

$$\mu(p^\perp) = \mu(p)^\perp, \ \mu(1) = (\perp^\star)^\perp, \ \mu(0) = \mu(1)^\perp, \ \mu(\top) = M, \ \mu(\perp) = M^\perp,$$

$$\mu(A \otimes B) = (\mu(A) \circ \mu(B))^{\perp\perp}, \ \mu(A \oplus B) = (\mu(A)^\perp \circ \mu(B)^\perp)^\perp,$$

$$\mu(A \wedge B) = \mu(A) \cap \mu(B), \ \mu(A \vee B) = (\mu(A)^\perp \cap \mu(B)^\perp)^\perp.$$

A is said to be true in the model if $1 \in \mu(A)$. As shown in [14], theorems of MALL are formulas true in all models over phase spaces.

From representation theorems, established above, it follows that commutative residuated monoids $(P(M), \subseteq, \circ, \to, \{1\})$ are embeddable into full relational frames, and the embeddings preserve meet and top. This yields a relational semantics for MALL. In this semantics, formulas are interpreted as binary relations, and linear implication $A \to B = A^\perp \oplus B$ is interpreted as residuation: $\mu(A \to B) = \mu(A)\backslash\mu(B)$. It is easy to see that all logical constants of MALL are definable in terms of relational residuation, top and meet plus the special constant $\perp^\star$.

Models of Cyclic MALL [30] are based on cyclic phase spaces. Now, $(M, \cdot, 1)$ is a monoid, and $\perp^\star \subseteq M$ satisfies: $ab \in \perp^\star$ entails $ba \in \perp^\star$. The remainder is as above. The embeddings described above yield a semantics for Cyclic MALL, based on full relational frames.

A Link to Pregroups. As a final result, we describe another, much simpler construction which works for right residuated monoids M, satisfying:

$$(2)\ a\backslash(bc) = (a\backslash b)c, \text{ for all } a, b, c \in M.$$

We define:

$$(3)\ f(a) = \{(x,y) \in M^2 : x \leq ay\}.$$

Assuming (2), we show that (M, f) is a representation of M. We prove (R1). Let $(x,y) \in f(a) \circ f(b)$. Then, $x \leq az$ and $z \leq by$, for some z, which yields $x \leq aby$, hence $(x,y) \in f(ab)$. Let $(x,y) \in f(ab)$. Then, $x \leq aby$, hence, for $z = by$, we have $(x,z) \in f(a)$ and $(z,y) \in f(b)$. We prove (R4). Clearly, $a \leq b$ entails $f(a) \subseteq f(b)$. Assume $a \nleq b$. Then, $(a,1) \in f(a)$, $(a,1) \notin f(b)$, hence $f(a) \nsubseteq f(b)$. We prove (R2). The inclusion $f(a\backslash b) \subseteq f(a)\backslash f(b)$ holds, by the same argument as in the proof of lemma 1. We prove the converse inclusion. Let $(x,y) \in f(a)\backslash f(b)$. Then, for any z, if $(z,x) \in f(a)$, then $(z,y) \in f(b)$. We have $(ax,x) \in f(a)$, so $ax \leq by$, which yields $x \leq a\backslash(by)$. By (2), $x \leq (a\backslash b)y$, hence $(x,y) \in f(a\backslash b)$.

We have $f(1) = \{(x,y) : x \leq y\}$. So, $I_M \subseteq f(1)$. Clearly, if $(x,y) \in f(a)$ and $x' \leq x$, $y \leq y'$, then $(x',y') \in f(a)$. Therefore, if $I_M \subseteq f(a)$, then $f(1) \subseteq f(a)$, which yields $1 \leq a$, by (R4). This shows (1).

Conversely, if (M, f) is a representation of M, where f is defined by (3), then M satisfies (2). Every right residuated monoid (even semigroup) satisfies $(a\backslash b)c \leq a\backslash(bc)$. Assume that f, defined by (3), is an embedding. We have: $(x,y) \in f(a)\backslash f(b)$ iff $z \leq ax$ entails $z \leq by$, for all z, iff $ax \leq by$ iff $x \leq a\backslash(by)$. Since $f(a)\backslash f(b) = f(a\backslash b)$, then $x \leq a\backslash(by)$ iff $x \leq (a\backslash b)y$, for all $x, y \in M$, which yields $a\backslash(by) = (a\backslash b)y$. We have shown that M satisfies (2), and we have proven:

Theorem 2. *For any right residuated monoid M, the mapping f defined by (3) is an embedding of M into the relational monoid $P(M^2)$ if, and only if, M satisfies (2).*

A similar result can be obtained for left residuated monoids. We notice that right (resp. left) residuated monoids satisfying (2) (resp. the dual of (2)) are precisely right (resp. left) pregroups in the sense of [8], structures closely related to pregroups in the sense of [19]. One may define $a^r = a\backslash 1$ (resp. $a^l = 1/a$). Then, $aa^r \leq 1$ and

$$1 \leq a\backslash a = a\backslash(1a) = a^r a,$$

and similarly for the dual operation. Conversely, in a right (resp. left) pregroup, one defines $a\backslash b = a^r b$ (resp. $a/b = ab^l$), and the resulting structure is a right (resp. left) residuated monoid, satisfying (2) (resp. its dual). Several results on relational and functional representations of (right, left) pregroups are proved in [7, 8].

References

1. Andréka, H., Mikulás, S.: Lambek Calculus and Its Relational Semantics. Completeness and Incompleteness. Journal of Logic, Language and Information **3.1** (1994) 1–37.

2. van Benthem, J.: Language in Action. Categories, Lambdas and Dynamic Logic. North Holland, Amsterdam, 1991.

3. Buszkowski, W.: Completeness Results for Lambek Syntactic Calculus. Zeitschrift für mathematische Logik und Grundlagen der Mathematik **32** (1986) 13–28.

4. Buszkowski, W.: Algebraic Structures in Categorial Grammar. Theoretical Computer Science **199** (1998) 1–24.

5. Buszkowski, W.: More on embeddings of residuated semigroups into algebras of relations (abstract). E. Orłowska and A. Szałas (eds.): Relational Methods in Logic, Algebra and Computer Science. Warsaw, 1998, 33–36.

6. Buszkowski, W.: Finite Models of Some Substructural Logics. Mathematical Logic Quarterly **48** (2002) 63–72.

7. Buszkowski, W.: Lambek Grammars Based on Pregroups. P. de Groote, G. Morrill and C. Retoré (eds.): Logical Aspects of Computational Linguistics. LNAI 2099, Springer, Berlin, 2001, 95–109.

8. Buszkowski, W.: Pregroups: Models and Grammars. H. de Swart (ed.): Relational Methods in Computer Science. LNCS 2561, Springer, Berlin, 2002, 35–49.

9. Buszkowski, W., Kołowska-Gawiejnowicz, M.: Representation of Residuated Semigroups in Some Algebras of Relations (The Method of Canonical Models). Fundamenta Informaticae **31** (1997) 1–12.

10. Dunn, J.M.: Partial gaggles applied to logics with restricted structural rules. [13], 63–108.

11. Došen, K.: Sequent Systems and Groupoid Models. Studia Logica **47** (1988) 353–386, **48** (1989) 41–65.

12. Došen, K.: A Brief Survey of Frames for The Lambek Calculus. Zeitschrift für mathematische Logik und Grundlagen der Mathematik **38** (1992) 179–187.

13. Došen, K., Schroeder-Heister, P. (eds.): Substructural Logics. Oxford University Press, Oxford, 1993.

14. Girard, J.Y.: Linear logic. Theoretical Computer Science **50** (1987) 1–102.

15. Hoare, C.A.R., Jifeng, H.: The weakest prespecification. Fundamenta Informaticae **9** (1986) 51–84, 217–262.

16. Kurtonina, N.: Frames and Labels. A Modal Analysis of Categorial Inference. Ph.D. Thesis, University of Utrecht, 1995.

17. Lambek, J.: The mathematics of sentence structure. American Mathematical Monthly **65** (1958) 154–170.

18. Lambek, J.: On the calculus of syntactic types. R. Jakobson (ed.): Structure of Language and Its Mathematical Aspects. AMS, Providence, 1961, 166–178.

19. Lambek, J.: Type Grammars Revisited. A. Lecomte, F. Lamarche and G. Perrier (eds.): Logical Aspects of Computational Linguistics. LNAI 1582, Springer, Berlin, 1999, 1–27.

20. MacCaull, W., Orłowska, E.: Correspondence Results for Relational Proof Systems with Application to the Lambek Calculus. Studia Logica **71** (2002) 389–414.

21. Moortgat, M.: Categorial Type Logics. J. van Benthem and A. ter Meulen (eds.): Handbook of Logic and Language. Elsevier, Amsterdam, 1997, 93–177.

22. Ono, H.: Semantics for Substructural Logics. [13], 259–291.

23. Orłowska, E.: Relational Interpretation of Modal Logics. H. Andréka, D. Monk and I. Nemeti (eds.): Algebraic Logic. North Holland, Amsterdam, 1988, 443–471.

24. Orłowska, E.: Relational proof system for relevant logic. Journal of Symbolic Logic **57** (1992) 1425–1440.

25. Pentus, M.: Models for The Lambek Calculus. Annals of Pure and Applied Logic **75** (1995) 179–213.

26. Pratt, V.: Action Logic and Pure Induction. J. van Eijck (ed.): Logics in AI. LNAI 478, Springer, Berlin, 1991, 97–120.
27. Restall, G.: An Introduction to Substructural Logics. Routledge, London, 2000.
28. Szczerba, M.: Relational Models for The Nonassociative Lambek Calculus. E. Orłowska and A. Szałas (eds.): Relational Methods for Computer Science Applications. Physica Verlag, Heidelberg, 2001, 149–159.
29. Venema, Y.: Tree models and (labelled) categorial grammars. Journal of Logic, Language and Information **5** (1996) 253–277.
30. Yetter, D.N.: Quantales and (non-commutative) linear logic. Journal of Symbolic Logic **55** (1996) 41–64.

Approximation Operators in Qualitative Data Analysis*

Ivo Düntsch and Günther Gediga**

Department of Computer Science
Brock University
St. Catharines, Ontario, L2S 3A1, Canada
{duentsch,gediga}@cosc.brocku.ca

Abstract. A large part of qualitative data analysis is concerned with approximations of sets on the basis of relational information. In this paper, we present various forms of set approximations via the unifying concept of modal–style operators. Two examples indicate the usefulness of the approach.

1   Introduction

In many instances it is not possible to describe a set precisely, owing to insufficient information or other sources of uncertainty. One way of handling this situation is to assign degrees of belief – probabilities, fuzzy membership assignments etc – to a statement such as "Object x is a member of set X". More cautious approaches consider intervals in which the relevant numerical functions lie, such as upper and lower probabilities or possibility theory. Such (approximations of) "point estimates" are problematic in many cases because the underlying model assumptions are often hard to fulfill. We have argued elsewhere [1] that qualitative tools often give comparative results under much less stringent assumptions. A frequently studied technique of qualitative set description is to determine a lower and an upper approximation of a set using non-numerical techniques. We will call a pair $\langle f,g \rangle$ of functions $2^U \to 2^U$ an *approximation pair* (on U) if

$$f(X) \subseteq X \subseteq g(X) \tag{1.1}$$

for all $X \subseteq U$. A *weak approximation pair* satisfies

$$f(X) \subseteq g(X). \tag{1.2}$$

This seems to be the weakest condition for a sensible concept of set approximation which is internal with respect to U. Stronger structural conditions require that f is an interior operator and g a closure operator, or that they are dual to each other (to be explained below).

A common mathematical basis of this type of set approximation are constructions associated with binary relations on the universe U. A frequently recurring theme is the

---

* Co-operation for this paper was supported by EU COST Action 274 "Theory and Applications of Relational Structures as Knowledge Instruments" (TARSKI), www.tarski.org. Ivo Düntsch gratefully acknowledges support from the National Sciences and Engineering Research Council of Canada.
** Equal authorship is implied

H. de Swart et al. (Eds.): TARSKI, LNCS 2929, pp. 214–230, 2003.

fact that each binary relation R on U gives rise to a "neighborhood" mapping $f_R : U \rightarrow 2^U$ via the assignment

$$x \mapsto \{y \in U : xRy\}. \tag{1.3}$$

Indeed, according to [2], it was already known to Tarski [3] in 1927 that there is a one-one correspondence between binary relations on U and mappings $f : 2^U \rightarrow 2^U$ which satisfy

$$f(\emptyset) = \emptyset,$$

$$f\left(\bigcup_{i \in I} X_i\right) = \bigcup_{i \in I} f(X_i).$$

This observation later formed the basis for the correspondence theory between Kripke frames and modal logics.

Examples are the approximation operators of rough set theory [4], various generalizations [5–8], the derivation operator of formal concept analysis [9], or the span-content operators of [10].

2 Operators and Relations

Much of the mathematical background of qualitative data analysis is concerned with set operators, relational structures, and the interplay among them: Closure and interior operators, (semi–) lattices, polarities, Galois correspondences, duality theory for Kripke frames, and Boolean algebras with operators. Most of the machinery has been developed in the first half of the 20th century, see for example [2, 11–14]. Many of the early results have been rediscovered by modal logicians (see [15] for a brief discussion), and re-rediscovered by the rough set community (see e.g. [16]). A good overview of the relevant correspondence results is given in [17], and, for rough set theory, in [18, 19].

For unexplained concepts we invite the reader to consult [20] for order and lattice theory, and [21] for Boolean algebras.

If $\langle A, +, \cdot, -, 0, 1 \rangle$ and $\langle B, +, \cdot, -, 0, 1 \rangle$ are Boolean algebras, and $f : A \rightarrow B$ is a mapping, then the *dual of f* is the function $f^\partial : A \rightarrow B$ defined by

$$f^\partial(a) = -f(-a). \tag{2.1}$$

Clearly, if f preserves $\cdot$, resp. $+$, then its dual preserves $+$, resp. $\cdot$.

If $\langle X, \leq \rangle$ and $\langle Y, \leq \rangle$ are partially ordered sets, a pair $\langle \psi, \varphi \rangle$ is called a *Galois connection* between X and Y, if $\psi : X \rightarrow Y$ and $\varphi : Y \rightarrow X$ are antitone (i.e. dually order preserving) mappings, and $x \leq x^{\psi\varphi}$, $y \leq y^{\varphi\psi}$ for all $x \in X$, $y \in Y$. $x \in X$ is called *Galois closed* with respect to $\langle \psi, \varphi \rangle$ if $x = x^{\psi\varphi}$.

A mapping $c : 2^U \rightarrow 2^U$ is called a *weak closure operator* if it satisfies

Cl1. $c(\emptyset) = \emptyset$,
Cl2. $X \subseteq c(X)$,
Cl3. $X \subseteq Y \Rightarrow c(X) \subseteq c(Y)$.

It is called a *closure operator* if it additionally satisfies

Cl4. $c(c(X)) = c(X)$.

Sets for which $c(X) = X$ are called *closed*. It is well known that the collection of all closed sets of a closure operator can be made into a complete lattice $\mathfrak{L}_c$ by setting

$$\bigvee\{X_i : i \in I\} = c\left(\bigcup\{X_i : i \in I\}\right), \tag{2.2}$$

$$\bigwedge\{X_i : i \in I\} = \bigcap\{X_i : i \in I\}. \tag{2.3}$$

Furthermore, each complete lattice is order isomorphic to the complete lattice of closed sets of some closure operator [14]. A closure operator is called *additive*, if it satisfies

Cl5. $c(X \cup Y) = c(X) \cup c(Y)$.

and *completely additive* if it distributes over arbitrary unions.

Note that Cl5 implies Cl3, but not vice versa: Suppose that U has at least three elements and define

$$c(X) = \begin{cases} X, & \text{if } |X| \leq 1, \\ U, & \text{otherwise.} \end{cases}$$

Then, c satisfies Cl1 – Cl4, but not Cl5. Additive closure operators are called "closure operators" in [11], where their corresponding lattices have been extensively studied. Each such $\mathfrak{L}_c$ is the lattice of closed sets of a topology; conversely, the topological closure operator satisfies Cl1 – Cl5. A nice survey of closure systems and related structures on finite sets can be found in [22].

We call c a *weak interior operator* if it satisfies

Int1. $i(U) = U$,
Int2. $i(X) \subseteq X$,
Int3. $X \subseteq Y \Rightarrow i(X) \subseteq i(Y)$.

and an *interior operator* if, additionally,

Int4. $i(i(X)) = i(X)$.

An interior operator is called *multiplicative* if it satisfies

Int5. $i(X \cap Y) = i(X) \cap i(Y)$,

and *completely multiplicative* if it distributes over arbitrary intersections. Clearly, the dual of a (weak, additive, completely additive) closure operator is a (weak, multiplicative, completely multiplicative) interior operator and vice versa.

If R is a relation between the elements of U and V, and $x \in U$, we let the *converse of R* be the relation $R^{\smile} = \{\langle y, x\rangle : xRy\}$. The *domain of R* is the set $\text{dom}(R) = \{x \in U : (\exists y \in V)xRy\}$, and $R(x)$ is the set $\{y \in V : xRy\}$; sometimes, $R(x)$ is called a *neighborhood of* x [23, 24].

As mentioned above, the assignment $x \mapsto R(x)$ defines a function $\overline{R} : U \to 2^V$. Conversely, given a function $f : U \to 2^V$, we can define a relation $S_f \subseteq U \times V$ by $x S_f y \iff y \in f(x)$. Clearly, $\overline{R_f} = f$, and $S_{\overline{R}} = R$.

R can be used to define several operators $2^U \to 2^V$:

$$\langle R \rangle(X) = \{b \in V : (\exists a \in X) a R b\}, \qquad \text{(Possibility operator)} \quad (2.4)$$

$$[R](X) = \{b \in V : (\forall a \in U)[a R b \Rightarrow a \in X]\}, \qquad \text{(Necessity operator)} \quad (2.5)$$

$$[[R]](X) = \{b \in V : (\forall a \in U))[a \in X \Rightarrow a R b]\}, \qquad \text{(Sufficiency operator)} \quad (2.6)$$

$$\langle\langle R \rangle\rangle(X) = \{b \in V : (\exists a \in U)[a \notin X \text{ and not } a R b]\}. \quad \text{(Dual sufficiency operator)} \quad (2.7)$$

The operators $(2.4) - (2.7)$ are generalizations of well known operators used in modal and algebraic logic, see e.g. [25, 26] and also [27, 10, 23].

Clearly, $\langle R \rangle$ and $[R]$, as well as $[[R]]$ and $\langle\langle R \rangle\rangle$ are dual to each other, and for each $\mathfrak{X} \subseteq 2^U$,

$$\langle R \rangle \left(\bigcup_{X \in \mathfrak{X}} X \right) = \bigcup_{X \in \mathfrak{X}} \langle R \rangle(X), \qquad (2.8)$$

$$[R] \left(\bigcap_{X \in \mathfrak{X}} X \right) = \bigcap_{X \in \mathfrak{X}} [R](X), \qquad (2.9)$$

$$[[R]] \left(\bigcup_{X \in \mathfrak{X}} X \right) = \bigcap_{X \in \mathfrak{X}} [[R]](X), \qquad (2.10)$$

$$\langle\langle R \rangle\rangle \left(\bigcap_{X \in \mathfrak{X}} X \right) = \bigcup_{X \in \mathfrak{X}} \langle\langle R \rangle\rangle(X). \qquad (2.11)$$

Furthermore, $\langle R \rangle$ and $[[R]]$ are, respectively, the existential and universal extension of the assignment $x \mapsto R(x)$ to subsets of U, since

$$\langle R \rangle(X) = \bigcup_{x \in X} R(x), \quad [[R]](X) = \bigcap_{x \in X} R(x).$$

It is also easily seen that

$$[[R]](X) = [(-R)](U \setminus X), \ \langle\langle R \rangle\rangle(X) = \langle\langle (-R) \rangle\rangle(U \setminus X). \qquad (2.12)$$

Here, $-R = \{\langle x, y \rangle \in U \times V : \langle x, y \rangle \notin R\}$ is the relational complement of R. The sufficiency operators $[[R]]$ and $[[R^\smile]]$ are intimately connected with Galois connections:

Proposition 1. *[14] The pair $\langle [[R]], [[R^\smile]] \rangle$ is Galois connection between $\langle 2^U, \subseteq \rangle$ and $\langle 2^V, \subseteq \rangle$, and each Galois connection between these sets has this form for some $R \subseteq U \times V$.*

The combined operators $[R^\smile]\langle R \rangle$ and $\langle R^\smile \rangle[R]$ will play a major role in our subsequent discussions. For these, we have

Proposition 2.   *1. $[R\check{}]\langle R\rangle$ is a closure operator on U.*
2. $[R\check{}]\langle R\rangle$ and $\langle R\check{}\rangle[R]$ are dual to each other.
3. $\langle R\check{}\rangle[R]$ is an interior operator on U.

Proof. 1.   Clearly, $[R\check{}]\langle R\rangle(\emptyset) = \emptyset$. For Cl2, let $x \in X \subseteq U$. Then, $R(x) \subseteq \langle R\rangle(X)$ by definition of $\langle R\rangle$, and hence, $x \in [R\check{}]\langle R\rangle(X)$ by (2.5).

Since both $[R\check{}]$ and $\langle R\rangle$ preserve $\subseteq$ by (2.8) and (2.9), so does $[R\check{}]\langle R\rangle$, and thus, it satisfies Cl3.

Let $Y \subseteq V$. Then,

$$q \in [R\check{}]\langle R\rangle[R\check{}](Y) \Rightarrow (\forall s \in V)[qRs \Rightarrow s \in \langle R\rangle[R\check{}](Y),$$
$$\Rightarrow (\forall s \in V)[qRs \Rightarrow (\exists p \in U)[pRs \wedge (\forall t \in V)[pRt \Rightarrow t \in Y]]],$$
$$\Rightarrow (\forall s \in V)[qRs \Rightarrow s \in Y],$$
$$\Rightarrow q \in [R\check{}](Y),$$

which implies Cl4.

2. First, note that $([R\check{}]\langle R\rangle)^{\partial}(X) = U \setminus [R\check{}]\langle R\rangle(U \setminus X)$. Now,

$$x \in U \setminus [R\check{}]\langle R\rangle(-X) \iff R(x) \not\subseteq \langle R\rangle(U \setminus X),$$
$$\iff (\exists z)[xRz \text{ and } z \notin \langle R\rangle(U \setminus X)],$$
$$\iff (\exists z)[xRz \text{ and } (\forall y)(y \notin X \Rightarrow y(-R)z)],$$
$$\iff (\exists z)[xRz \text{ and } (\forall y)(yRz \Rightarrow y \in X)],$$
$$\iff (\exists z)[xRz \text{ and } R\check{}(z) \subseteq X],$$
$$\iff (\exists z)[xRz \text{ and } z \in [R](X)],$$
$$\iff x \in \langle R\check{}\rangle[R](X).$$

3. follows from 1. and 2. by duality.   □

Observe that Proposition 2 is true for arbitrary R. The following result has been known in a different context for some time [9]:

Corollary 1 $[[R\check{}]][[R]]$ *is a closure operator.*

Proof. First,

$$[[R\check{}]][[R]](X) = [-R\check{}](V \setminus [[R]](X)), \qquad \text{by (2.12)}$$
$$= [-R\check{}](V \setminus ([-R](U \setminus X)), \qquad \text{by (2.12)}$$
$$= [-R\check{}]\langle -R\rangle(X), \qquad \text{by duality.}$$

The claim follows now from Proposition 2.   □

This result is also a direct consequence of Proposition 1; we have taken the route above to emphasize the connection with $[R\check{}]\langle R\rangle$.

To conclude this Section, let us consider the case where $U = V$. Correspondence theory [17] tells us that

$$X \subseteq \langle R\rangle(X) \iff R \text{ is reflexive,}, \qquad (2.13)$$
$$\langle R\rangle\langle R\rangle(X) \subseteq \langle R\rangle(X) \iff R \text{ is transitive,} \qquad (2.14)$$
$$X \subseteq [R]\langle R\rangle(X) \iff R \text{ is symmetric,} \qquad (2.15)$$

Note that $\langle R \rangle$ is a weak closure operator just in case R is reflexive, and that $\langle R \rangle$ is a completely additive closure operator, if, in addition, R is transitive. In this case, we denote the topology generated by the closed sets by τ_R, i.e.

$$\tau_R = \{U \setminus \langle R \rangle (Y) : Y \subseteq U\} = \{[R](X) : X \subseteq U\}. \tag{2.16}$$

Since these topologies (and related structures) have been considered in the present context (e.g. [23, 19, 24]), we recall some facts about their properties. In the sequel, let R be reflexive and transitive.

Proposition 3. *In τ_R, each $x \in U$ has a smallest open neighborhood.*

Proof. This result has appeared in a different form and context already in [28]. Let $x \in U$, and $\mathfrak{X}$ be the collection of all open neighborhoods of x. By (2.9) , $\bigcap \mathfrak{X}$ is open, and clearly, it is the smallest open set containing x. □

A topology with the property of Proposition 3 is called *principal*.

Proposition 4. *1. [28] The collection of all principal topologies on a set U can be made into a lattice which is anti–isomorphic to the lattice of all reflexive and transitive relations on U.*
2. [29] The following are equivalent:
   (a) R is an equivalence relation.
   (b) $(\forall X \subseteq U)[X \in \tau_R \Longleftrightarrow U \setminus X \in \tau_R]$.
   (c) τ_R is regular.

In this case, a basis for τ_R are the classes of the partition induced by R, and $R(x)$ is the smallest neighborhood of x.

3 Approximation Based on Homogenous Relations

In our first scenario, we consider a finite set U and various relations and operators on U. The basic idea is that objects are usually not considered in isolation, but are in some way related; these relationships are then used to obtain operators $2^U \rightarrow 2^U$ which can approximate subsets of U.

3.1 Equivalence Relations and Approximation Spaces

As a first example, consider the case of a spatial database. A frequently used key is the *minimum bounding box* which is the smallest aligned rectangle enclosing a spatial object. More generally, given a grid of rectangles (cells) in the plane, one can approximate a spatial object O from above by the smallest set of cells which cover O, and from below by the largest set of cells totally contained in O (Figure 1). This situation is easily captured in a relational setting: Suppose that U is an area in the plane which is covered by a grid of disjoint rectangles; some care must be taken that the boundary of adjacent cells belong to exactly one cell. We now define $R \subseteq U \times U$ by

$$xRy \Longleftrightarrow x \text{ and } y \text{ are in the same cell.}$$

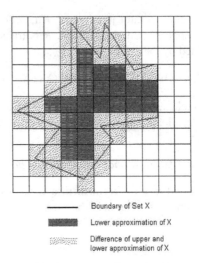

───── Boundary of Set X

████ Lower approximation of X

░░░░ Difference of upper and
lower approximation of X

Fig. 1. An approximated region

Clearly, R is an equivalence relation, and each cell corresponds to exactly one equivalence class of R. If X is a region in U, then the upper approximation $\overline{X}$ of X is the union of all classes of R which intersect X, and the lower approximation is the union of all classes of R totally contained in X. In other words,

$$\overline{X} = \{x : R(x) \cap X \neq \emptyset\} = \bigcup \{R(x) : x \in X\} = \langle R \rangle (X), \qquad (3.1)$$

$$\underline{X} = \{x : R(x) \subseteq X\} = \{x : R^{\smile}(x) \subseteq X\} = [R](X). \qquad (3.2)$$

More generally, an *approximation space* is a structure $\langle U, R \rangle$, where R is an equivalence relation on the set U [4]. An approximation space tells us the granularity of our knowledge about the world - we can distinguish objects only up to the equivalences classes of R, but not within the classes.

A *rough set* is a pair $\langle A, B \rangle$ such that

1. A and B are empty or a union of equivalence classes of R,
2. $A \subseteq B$,
3. If C is a singleton class contained in B, then $C \subseteq A$.

Rough sets are approximations of subsets of U in the following sense: Let $X \subseteq U$. Since we cannot distinguish within equivalence classes of R we can say with certainty that some $x \in U$ is a member of X just in case the whole class $R(x)$ is a subset of X. Similarly, we can be sure that $x \notin X$ only if the class $R(x)$ of x is disjoint to X. The rough set approximations are given by (3.1) and (3.2). The concepts agree well with the interpretation of these operators in modal logic: $x \in \underline{X}$ if x is certainly a member of X, and $x \in \overline{X}$, if x is possibly a member of X, according to the knowledge delivered by the granularity induced by R. The connection of the rough set approximation operators to modal S5 logic have first been observed by [30] and subsequently by many authors; overviews can be found in [31, 32].

Since R is reflexive, the upper approximation is a weak closure operator, and since R is also transitive, it is in fact a completely additive closure operator. Consequently, the lower approximation is a completely multiplicative interior operator. Furthermore, the properties of an equivalence relation imply that

$$\langle R \rangle = \langle R \rangle \langle R \rangle = [R]\langle R \rangle, \ [R] = [R][R] = \langle R \rangle [R]. \tag{3.3}$$

3.2 More General Relations

It was argued in [7] that for many applications the properties of an equivalence relation are too strong, and that relations with weaker properties should be considered when one wants to express some form of similarity. A first generalization is to require only that R is reflexive and symmetric, and relations with these properties have indeed be called *similarity relations*. This terminology may be somewhat misleading: While a a relation of similarity may be symmetric, calling a reflexive and symmetric relation a "similarity" may cause contextual problems: Suppose that we have agreed on what constitutes similarity and have expressed this by a reflexive and symmetric relation R. Let S be the relation $-R \cup 1'$. Since the complement of a symmetric relation is again symmetric, S is a similarity. Thus, if x is not similar to y according to R, we would say that x is similar to y, according to S.

To suppose that a relation of similarity is symmetric is also not always appropriate. Much of the similarity data used in Computer Science are expert judgments and it is quite reasonable to assume that experts judgments shows some bias. Indeed, the investigations of [33] show that similarities based on human judgment are often quite asymmetric, and we invite the reader to consult [34] for an example.

It has also been argued that non–reflexive relations can be interpreted as similarity [35], and the following example was given:

"We may discern persons by comparing photographs taken of them. But it may happen that we are unable to recognize that a same person appears in two different photographs".

A similar example is the famous experiment by [36]:

"The S[ubject]s of this experiment were exposed to pairs of aural Morse signals sent at a high tone speed. The signals of each pair were separated by a short temporal interval. The S[ubject]s were asked to indicate whether they thought the signals were the same (or different) by making the appropriate remark on an IBM True–False Answer sheet. Each S[ubject] was asked to respond in this fashion to 351 different pairs of Morse signals."

We interpret this as

$$xRy \iff x \text{ and } y \text{ are recognized as the same (person, signal).}$$

In both cases, the similarity of x and y is very much in the eye – or the ear – of the beholder, and not necessarily a property of x and y.

An equivalence relation R on U has the special property that

$$\bigcup\{R(x) : x \in X\} = \{x : R(x) \cap X \neq \emptyset\} = \bigcup\{R(x) : R(x) \cap X \neq \emptyset\}, \qquad (3.4)$$

and each is equal to $\overline{X}$. If $R \subseteq U \times U$ is arbitrary, then (3.4) need not be true. Thus, one needs to decide what constitutes a lower or upper approximation of X. In the literature, one finds many suggestions for such pairs; in (3.5) and (3.6) below, R is assumed to be reflexive:

$$R_*(X) = \{x \in U : R\breve{}(x) \subseteq X\}, \qquad R^*(X) = \bigcup\{R\breve{}(x) : x \in X\}, \qquad [8] \qquad (3.5)$$

$$R_{**}(X) = \bigcup\{R\breve{}(x) : R\breve{}(x) \subseteq X\} \quad R^{**}(X) = \bigcup\{R\breve{}(x) : R\breve{}(x) \cap X \neq \emptyset\} \quad [8] \qquad (3.6)$$

$$R_+(X) = \bigcup\{R(x) : R(x) \subseteq X\} \quad R^+(X) = \bigcup\{R(x) : R(x) \cap X \neq \emptyset\} \qquad [19] \qquad (3.7)$$

$$R^{\blacktriangledown}(X) = \{x \in U : R(x) \subseteq X\}, \qquad R^{\blacktriangle}(X) = \{x \in U : R(x) \cap X \neq \emptyset\} \qquad [35, 37, 16] \quad (3.8)$$

$$R_\circ(X) = \bigcup\{R\breve{}(x) : R\breve{}(x) \subseteq X\} \quad R^\circ(X) = \{x \in U : R(x) \subseteq \langle R \rangle(X)\} \qquad [10] \qquad (3.9)$$

All of these can be located within the modal operator framework:

$$R_*(X) = [R](X) \qquad\qquad R^*(X) = \langle R\breve{} \rangle(X),$$
$$R_{**}(X) = \langle R\breve{} \rangle[R](X) \qquad\qquad R^{**}(X) = \langle R\breve{} \rangle\langle R \rangle(X)$$
$$R_+(X) = \langle R \rangle[R\breve{}](X) \qquad\qquad R^+(X) = \langle R \rangle\langle R\breve{} \rangle(X)$$
$$R^{\blacktriangledown}(X) = [R\breve{}](X) \qquad\qquad R^{\blacktriangle}(X) = \langle R\breve{} \rangle(X)$$
$$R_\circ(X) = \langle R\breve{} \rangle[R](X) \qquad\qquad R^\circ(X) = [R\breve{}]\langle R \rangle(X)$$

We note without proof some inclusion properties among these relations:

Proposition 5.

$$(\forall X)[X \subseteq \langle R \rangle(X)] \Longleftrightarrow (\forall X)[[R](X) \subseteq X] \qquad \Longleftrightarrow R \text{ is reflexive} . \qquad (3.10)$$

$$(\forall X)[[R](X) \subseteq \langle R \rangle(X)] \Longleftrightarrow (\forall X)[X \subseteq \langle R \rangle\langle R\breve{} \rangle(X)] \Longleftrightarrow \operatorname{dom}(R\breve{}) = U. \quad (3.11)$$

$$(\forall X)[\langle R \rangle[R\breve{}](X) \subseteq \langle R \rangle\langle R\breve{} \rangle(X)]. \qquad (3.12)$$

It can be seen that that the only approximation pair for arbitrary R is $\langle R_\circ, R^\circ \rangle$; these functions are also dual to each other in the sense of (2.1) of Section 2 and a closure, respectively, an interior operator. All other pairs need extra conditions such as reflexivity or totality to satisfy the approximation conditions (1.1) or (1.2) on page 214. If R is reflexive, then $\langle R_\circ, R^\circ \rangle$ gives the tightest bounds for the approximation pairs $\langle f, g \rangle$ above in the sense that

$$f(X) \subseteq R_\circ(X) \subseteq X \subseteq R^\circ \subseteq g(X).$$

4   Approximation Based on Heterogeneous Relations

Another type of approximation arises when we have information about the properties of the elements of the domain. Such information may be given by an information system in the sense of [4], or, more generally, by a binary relation $R \subseteq U \times V$ connecting objects with properties. For this situation we have special names for some of the modal operators: If $X \subseteq U$ and $Y \subseteq V$, we say that

- $\langle R \rangle (X)$ is the *span of* X,
- $[R^{\backprime}](Y)$ is the *content of* Y,
- $[[R]](X)$ is the *intent of* X,
- $[[R^{\backprime}]](Y)$ is the *extent of* Y.

The span of X is the set of all properties which are related to some element of X, and the content of Y is the set of those objects which can be completely described by the properties in Y. The intent of X are those properties common to all elements of X, and the extent of Y is the set of all objects which possess all properties in Y. Extent and intent are the basic operators of formal concept analysis (FCA) [9]. Since we know from Corollary 1 that

$$[[R^{\backprime}]][[R]] = [-R^{\backprime}]\langle -R \rangle$$

the FCA operators are the content–span operator applied to $-R$, and it depends on the context which one is appropriate to use. We just mention a result from [27] which indicates in another way how the two closures differ:

Proposition 6. *For all* $x \in U, X \subseteq U,$

$$x \in [R^{\backprime}]\langle R \rangle (X) \Longleftrightarrow \bigcap \{R(y) : y \in X\} \subseteq R(x).$$
$$x \in [[R^{\backprime}]][[R]](X) \Longleftrightarrow R(x) \subseteq \bigcup \{R(y) : y \in X\}.$$

For an extensive algebraic and topological view of the connections between FCA and approximation spaces we refer the reader to [38].

In this Section we will investigate more closely the operators $[R^{\backprime}]\langle R \rangle$ and its dual $\langle R^{\backprime} \rangle [R]$. We know already that $[R^{\backprime}]\langle R \rangle$ is a closure operator, and $\langle R^{\backprime} \rangle [R]$ is an interior operator, so that they can serve as sensible approximations of $X \subseteq U$.

We first show that the approximation pair $\langle [R^{\backprime}]\langle R \rangle, \langle R^{\backprime} \rangle [R] \rangle$ are the original rough set approximation operators (3.1), (3.2) defined on page 220, derived from the standard data representation of rough set analysis: An *information system* is a structure

$$I = \langle U, \Omega, \{V_q : q \in \Omega\}, f \rangle, \tag{4.1}$$

where

- U is a finite set of objects.
- Ω is a finite set of *attributes*.
- For each $q \in \Omega$,
    - V_q is a set of *attribute values* of attribute q. In the sequel, we let $V = \bigcup_{q \in \Omega} V_q$.
- $f : U \times \Omega \to V$ is a function such that $f(x,q) \in V_q$ for all $x \in U$, $q \in \Omega$, called the *attribute* or *information function*. We interpret $f(x,q) = a$ as "Object x has the value a at attribute q".

Furthermore, if $Q = \{q_1, \dots, q_n\} \subseteq \Omega$ we lift f by setting

$$f_Q(x) = \langle f(x,q_1), \dots f(x,q_n) \rangle. \tag{4.2}$$

Let $R \subseteq U \times V_\Omega$ be the relational version of f_Ω, i.e.

$$xRt \iff f_\Omega = t.$$

Furthermore, let θ be the kernel of f_Ω, i.e.

$$x\theta y \iff f_\Omega(x) = f_\Omega(y).$$

The approximation space which we consider is $\langle U, \theta \rangle$.

Proposition 7. *Let $X \subseteq U$. Then,*

1. $[R^\smile]\langle R \rangle (X) = \overline{X}.$
2. $\langle R^\smile \rangle [R](X) = \underline{X}.$

Proof. We only show 1. since 2. follows immediately by duality.

$$
\begin{aligned}
z \in [R^\smile]\langle R \rangle(X) &\iff R(z) \subseteq \langle R \rangle(X), \\
&\iff f_\Omega(z) \in \langle R \rangle(X), \\
&\iff (\exists x \in X) x R f_\Omega(z), \\
&\iff (\exists x \in X) f_\Omega(x) = f_\Omega(z), \\
&\iff (\exists x \in X) x \theta z, \\
&\iff \theta z \cap X \neq \emptyset, \\
&\iff z \in \overline{X},
\end{aligned}
$$

which completes the proof.   □

4.1   Example: Student Assessment

Suppose S is a set of skills and Q is a set of problems, with which the skills in S should be tested. Let $R \subseteq Q \times S$ be a relation such that qRs is interpreted as

Skill s is necessary to solve q.   (4.3)

The skill set $R(q)$ is minimally sufficient to solve q.   (4.4)

This is an assignment given by an expert. The modal operators can be interpreted as follows: Let $P \subseteq Q$, $M \subseteq S$. Then,

$$s \in \langle R \rangle(P) \iff s \text{ is necessary to solve some problem in } P.$$
$$s \in [R](P) \iff s \text{ is necessary only for problems in } P.$$
$$s \in [[R]](P) \iff s \text{ is necessary for all problems in } P.$$
$$q \in \langle R^\smile \rangle(M) \iff \text{Some } s \in M \text{ is necessary to solve } q.$$
$$q \in [R^\smile](M) \iff q \text{ can be solved with the skills in } M.$$
$$q \in [[R^\smile]](M) \iff \text{All skills in } M \text{ are necessary to solve } q.$$

Suppose that P is a set of problems which student s has been able to solve; we are interested in the true state of knowledge of s. Let us suppose that the student has made no lucky guesses, i.e. we assume that s really possess all the skills required to solve the problems in P and possibly more; in this case, $\langle R \rangle (P)$ is a lower bound for the skills s has, and P is a lower bound for the set of problems s is able to solve. Now,

$$q \in [R^{\cdot}]\langle R \rangle (P) \iff R(q) \subseteq \langle R \rangle (P)$$
$$\iff \langle R \rangle (P) \text{ contains the skills sufficient to solve } q, \qquad \text{by (4.4).}$$

Thus, q should have been solved, since s has the skills to solve q; if $q \notin P$, it was due to a careless error. Therefore, q can be included in the true knowledge state of s.

For a more detailed description and a test theory based on skill functions we invite the reader to consult [27].

4.2   Example: Morse Data

Another example we shall look at is the famous Morse data collected by [36], a flag-ship of multidimensional scaling (MDS); the experiment has already been described on page 221. [39] describes the data using the dimensions

1. Length of the signal,
2. Distribution of dots and dashes in the signal, going from only dots to only dashes.

see Figure 2. The distances between the points in a plane spanned by these dimensions reflect (partially) the ordinal relation among the given proximities.

Fig. 2. MDS interpretation of the Morse data [39]

Table 1. Morse data

	a	b	c	d	e	f	g	h	i	j	k	l	m	n	o	p	q	r	s	t	u	v	w	x	y	z	*1	*2	*3	*4	*5	*6	*7	*8	*9	*0
A	92	4	6	13	3	14	10	13	46	5	22	3	25	34	6	6	9	35	23	6	37	13	17	12	7	3	2	7	5	5	8	6	5	6	2	3
B	5	84	37	31	5	28	17	21	5	19	34	40	6	10	12	22	25	16	18	2	18	34	8	84	30	42	12	17	14	40	32	74	43	17	4	4
C	4	38	87	17	4	29	13	7	11	19	24	35	14	3	9	51	34	24	14	6	6	11	14	32	82	38	13	15	31	14	10	30	28	24	18	12
D	8	62	17	88	7	23	40	36	9	13	81	56	8	7	9	27	9	45	29	6	17	20	27	40	15	33	3	9	6	11	9	19	8	10	5	6
E	6	13	14	6	97	2	4	4	17	1	5	6	4	4	5	1	5	10	7	67	3	3	2	5	6	5	4	3	5	3	5	2	4	2	3	3
F	4	51	33	19	2	90	10	29	5	33	16	50	7	6	10	42	12	35	14	2	21	27	25	19	27	13	8	16	47	25	26	24	21	5	5	5
G	9	18	27	38	1	14	90	6	5	22	33	16	14	13	62	52	23	21	5	3	15	14	32	21	23	39	15	14	5	10	4	10	17	23	20	11
H	3	45	23	25	9	32	8	87	10	10	9	29	5	8	8	14	8	17	37	4	36	59	9	33	14	11	3	9	15	43	70	35	17	4	3	3
I	64	7	7	13	10	8	6	12	93	3	5	16	13	30	7	3	5	19	35	16	10	5	8	2	5	7	2	5	8	9	6	8	5	2	4	5
J	7	9	38	9	2	24	18	5	4	85	22	31	8	3	21	63	47	11	2	7	9	9	22	32	28	67	66	33	15	7	11	28	29	26	23	
K	5	24	38	73	1	17	25	11	5	27	91	33	10	12	31	14	31	22	2	2	23	17	33	63	16	18	5	9	17	8	8	18	14	13	5	6
L	2	69	43	45	10	24	12	26	9	30	27	86	6	2	9	37	36	28	12	5	16	19	20	31	25	59	12	13	17	15	26	29	36	16	7	3
M	24	12	5	14	7	17	29	8	8	11	23	8	96	62	11	10	15	20	7	9	13	4	21	9	18	8	5	7	6	6	5	7	11	7	10	4
N	31	4	13	30	8	12	10	16	13	3	16	8	59	93	5	9	5	28	12	10	16	4	12	4	6	11	5	2	3	4	4	6	2	2	10	2
O	7	7	20	6	5	9	76	7	2	39	26	10	4	8	86	37	35	10	3	4	11	14	25	35	27	27	19	17	7	7	6	18	14	11	20	12
P	5	22	33	12	5	36	22	12	3	78	14	46	5	6	21	83	43	23	9	4	12	19	19	19	41	30	34	44	24	11	15	17	24	23	25	13
Q	8	20	38	11	4	15	10	5	2	27	23	26	7	6	22	51	91	11	2	3	6	14	12	37	50	63	34	32	17	12	9	27	40	58	37	24
R	13	14	16	23	5	34	26	15	7	12	21	33	14	12	12	29	8	87	16	2	23	23	62	14	12	13	7	10	13	4	7	12	7	9	1	2
S	17	24	5	30	11	26	5	59	16	3	13	10	5	11	2	15	59	72	14	4	3	9	11	12	36	42	87	16	21	27	9	10	6	7	8	2
T	13	10	1	5	46	3	6	6	14	6	14	7	6	5	6	11	4	4	7	96	8	5	4	2	2	6	5	5	3	3	3	8	7	6	14	6
U	14	29	12	32	4	32	11	34	21	7	44	32	11	13	6	20	12	40	51	6	93	57	34	17	9	11	6	6	16	34	10	9	9	7	4	3
V	5	17	24	16	9	29	6	39	5	11	26	43	4	1	9	17	10	17	11	6	32	92	17	57	35	10	10	14	28	79	44	36	25	10	1	5
W	9	21	30	22	9	36	25	15	4	25	29	18	15	6	26	20	25	61	12	4	19	20	86	22	25	22	10	22	19	16	5	9	11	6	3	7
X	7	64	45	19	3	28	11	6	1	35	50	42	10	8	24	32	61	10	12	3	12	17	21	91	48	26	12	20	24	27	16	57	29	16	17	6
Y	9	23	62	15	4	26	22	9	1	30	12	14	5	6	14	30	52	5	7	4	6	13	21	44	86	23	26	44	40	15	11	26	22	33	23	16
Z	3	46	45	18	2	22	17	10	7	23	21	51	11	2	15	59	72	14	4	3	9	11	12	36	42	87	16	21	27	9	10	25	66	47	15	15
1	2	5	10	3	3	5	13	4	2	29	5	14	9	7	14	30	28	9	4	2	3	12	14	17	19	22	84	63	13	8	10	8	19	32	57	55
2	7	14	22	5	4	20	13	3	25	26	9	14	2	3	17	37	28	6	5	3	6	10	11	17	30	13	62	89	54	20	5	14	20	21	16	11
3	3	8	21	5	4	32	6	12	2	27	23	26	7	6	36	39	19	9	7	6	4	16	6	22	25	12	18	64	86	31	23	41	16	17	8	10
4	6	19	19	12	8	25	14	16	7	21	13	19	3	3	2	17	29	11	9	3	17	55	8	37	24	3	5	26	44	89	42	44	32	10	3	3
5	8	45	15	14	2	45	4	67	7	14	4	41	2	0	4	13	7	9	27	2	14	45	7	45	10	10	14	30	69	90	42	24	10	6	5	
6	7	80	30	17	4	23	4	14	2	11	11	27	6	2	7	16	30	11	14	3	12	30	9	58	38	39	15	14	26	24	17	88	69	14	5	14
7	6	33	22	14	5	25	6	4	6	24	13	32	7	6	36	39	12	6	2	3	13	9	30	30	50	22	29	18	15	12	61	85	70	20	13	
8	3	23	40	6	3	15	15	6	2	33	10	14	3	6	14	12	45	2	6	4	6	7	5	24	35	50	42	29	16	16	9	30	60	89	61	26
9	3	14	23	3	1	6	14	5	2	30	6	7	16	11	10	31	32	5	6	7	6	3	8	11	21	24	57	39	9	12	4	11	42	56	91	78
0	9	3	11	2	5	7	14	4	5	30	8	3	2	3	25	21	29	2	3	4	5	3	2	12	15	20	50	26	9	11	5	22	17	52	81	94

Table 2. Distinguished sets

Stimulus (first position)	Stimulus (second position)
$X_1 = \{E,T\}$	$Y_1 = \{e,t\}$
$X_2 = \{A,I,M,N\}$	$Y_2 = \{a,i,m,n\}$
$X_3 = \{D,G,K,O,R,S,U,W\}$	$Y_3 = \{d,g,k,o,r,s,u,w\}$
$X_4 = \{B,C,F,H,J,L,P,Q,V,X,Y,Z\}$	$Y_4 = \{b,c,f,h,j,l,p,q,v,x,y,z\}$
$X_5 = \{0,1,2,3,4,5,6,7,8,9\}$	$Y_5 = \{*0,*1,*2,*3,*4,*5,*6,*7,*8,*9\}$

In the sequel we will present the re–analysis of the data given in [10] which uses the modal operators. Table 1 shows in each cell the percentage of subjects who regarded the two stimuli as the same. We use upper case letters for first stimuli and lower case letters for second stimuli; the numeric characters are prefixed by a $*$, if they occur as second stimuli. The matrix diagonal corresponds to pairs which are truly the same, the off-diagonal entries correspond to pairs which are truly different.

As the length of the signal is one of the dimension identified in [39] (and also in [40]), we are interested in the behavior of the modal–style operators on the sets

$$X_n = \{p : \text{The length of the Morse code for first stimulus } p \text{ is } n\},$$

$$Y_n = \{q : \text{The length of the Morse code for second stimulus } q \text{ is } n\}.$$

which are given in Table 2.

We are aiming at a description of similarity dependencies among these four sets of stimuli and their elements.

Let U be the set of first stimuli, V be the set of second stimuli, and pRq if a (fixed) subject regards them as the same. The operators can be interpreted as follows:

$q \in \langle R \rangle(X_n)$ $\Longleftrightarrow$ q was gauged to be the same as some first stimulus of length n.

$q \in [R](X_n)$ $\Longleftrightarrow$ q was gauged to be the same only as first stimuli of length n.

$q \in [[R]](X_n)$ $\Longleftrightarrow$ q was gauged to be the same to all first stimuli of length n, and possibly others.

$p \in [R^{\smile}]\langle R \rangle(X_n)$ $\Longleftrightarrow$ Every signal, which cannot be distinguished from p cannot be distinguished from some stimulus of length n.

$p \in \langle R^{\smile} \rangle[R](X_n)$ $\Longleftrightarrow$ Some signals, which cannot be distinguished from p were gauged to be the same only to stimuli of length n.

$p \in [[R^{\smile}]][[R]](X_n)$ $\Longleftrightarrow$ Whenever q cannot be distinguished from all stimuli of length n, then q cannot be distinguished from p.

In order to consider the aggregated data given in Table 1, we need to consider "cut–off" points, and set

$$R_s = \{\langle p,q \rangle : \text{At least } s\% \text{ of the subjects responded "same",}$$

$$\text{when } \langle p,q \rangle \text{ was presented}\}. \quad (4.5)$$

Observe that $R_s \subseteq R_t$ in case $t \leq s$. For first stimuli, the approximation operators now are interpreted as

$p \in [R_s^{\smile}]\langle R_t \rangle(X_n)$ $\Longleftrightarrow$ Every second stimulus which could not be distinguished from p by at least $s\%$ of all subjects could not be distinguished from some first stimulus of length n by at least $t\%$ of all subjects.

$p \in \langle R_s^{\smile} \rangle[R_t](X_n)$ $\Longleftrightarrow$ There is a second stimulus q such that at least $s\%$ of subjects gauged q to be the same as p, and at least $t\%$ of subjects gauged q to be the same only as stimuli of length n.

We have analyzed the data for various cut–points, and have found, among other results, that

- The signal length is the first determining factor for the discrimination of the stimuli, because:
  - Signals of length 1 or 2 are easy to discriminate from other stimuli.
  - Signals of length 3 are easy to discriminate from other stimuli, if they are located at the first position.
  - Signals of length 3 in the second position overlap with signals of length 4. Signals of length 4 overlap mainly with signals of length 5.
- The character of the impulses is of less effect because a signal must contain mainly short Morse impulses, and should contain at least 4 (first stimuli) or 3 (second stimuli) Morse impulses to be hard to discriminate.

We invite the reader to consult [10] for the details.

5 Conclusion and Outlook

In this paper we have explored various tools for set approximation based on a relational connection of two "universes" U and V, or a relational connection within one "universe" U. The intention was to find a proper extension of rough sets in case of any binary relation. There is a list of proposals for set approximations based on certain binary relations such as similarity relations or dominance relations. We have shown that all these proposals can be well expressed in terms of modal style operators and that the (new) operator $[R^{\smile}]\langle R\rangle$ (content-span operator) exhibits some kind of optimality because it gives the tightest bounds among the proposed operators based on a reflexive relation.

In case of any binary relation the content-span operator is interesting as well, because applying $[R^{\smile}]\langle R\rangle$ is a complementary approach to the one taken by formal concept analysis – with exactly the same expressive power.

Two examples from diverse application fields indicate that the operators $[R^{\smile}]\langle R\rangle$ are not only well suited for approximating sets, but that the resulting approximations offer meaningful interpretations. More applications, however, are needed to delineate the situations in which either of these can be applied.

References

1. Düntsch, I., Gediga, G.: Roughian – Rough Information Analysis. International Journal of Intelligent Systems **16** (2001) 121–147
2. Jónsson, B., Tarski, A.: Boolean algebras with operators I. American Journal of Mathematics **73** (1951) 891–939
3. Tarski, A.: Sur quelque propriét'es charactéristiques des images d'ensembles. Annales de la Societé Polonaise de Mathématique **6** (1927) 127–128
4. Pawlak, Z.: Rough sets. Internat. J. Comput. Inform. Sci. **11** (1982) 341–356
5. Wong, S., Wang, L., Yao, Y.: On modeling uncertainty with interval structures. Computational Intelligence **11** (1995) 406–426
6. Yao, Y.Y.: On generalizing Pawlak approximation operators. In Polkowski, L., Skowron, A., eds.: Proceedings of the 1st International Conference on Rough Sets and Current Trends in Computing (RSCTC-98). Volume 1424 of LNAI., Berlin, Springer (1998) 298–307
7. Słowiński, R., Vanderpooten, D.: Similarity relations as a basis for rough approximations. ICS Research Report 53, Polish Academy of Sciences (1995)
8. Słowiński, R., Vanderpooten, D.: A generalized definition of rough approximations based on similarity. IEEE Transactions on Knowledge and Data Engineering **12** (2000) 331–336
9. Wille, R.: Restructuring lattice theory: An approach based on hierarchies of concepts. In Rival, I., ed.: Ordered sets. Volume 83 of NATO Advanced Studies Institute. Reidel, Dordrecht (1982) 445–470
10. Düntsch, I., Gediga, G.: Modal–style operators in qualitative data analysis. In: Proceedings of the 2nd IEEE International Conference on Data Mining (ICDM-02). (2002) 155–162
11. McKinsey, J.C.C., Tarski, A.: The algebra of topology. Annals of Mathematics **45** (1944) 141–191
12. McKinsey, J.C.C., Tarski, A.: On closed elements in closure algebras. Annals of Mathematics **47** (1946) 122–162
13. Birkhoff, G.: Lattice Theory. 2 edn. Volume 25 of Am. Math. Soc. Colloquium Publications. AMS, Providence (1948)

14. Ore, O.: Theory of Graphs. Volume 38 of Am. Math. Soc. Colloquium Publications. AMS, Providence (1962)
15. Jónsson, B.: A survey of Boolean algebras with operators. In: Algebras and Orders. Volume 389 of NATO Adv. Sci. Inst. Ser. C, Math. Phys. Sci. Kluwer, Dordrecht (1993) 239–286
16. Yao, Y.Y.: Generalization of rough sets using modal logic. Intelligent Automation and Soft Computing **2** (1996) 103–120
17. van Benthem, J.: Correspondence theory. In Gabbay, D.M., Guenthner, F., eds.: Extensions of classical logic. Volume 2 of Handbook of Philosophical Logic. Reidel, Dordrecht (1984) 167–247
18. Yao, Y.Y.: Constructive and algebraic methods of the theory of rough sets. Information Sciences **109** (1998) 21–47
19. Yao, Y.Y.: Relational interpretations of neighborhood operators and rough set approximation operators. Information Sciences **111** (1998) 239–259
20. Davey, B.A., Priestley, H.A.: Introduction to Lattices and Order. Cambridge University Press (1990)
21. Koppelberg, S.: General Theory of Boolean Algebras. Volume 1 of Handbook on Boolean Algebras. North Holland (1989)
22. Caspard, N., Monjardet, B.: The lattices of closure systems, closure operators, and implicational systems on a finite set: a survey. Discrete Applied Mathematics **127** (2003) 241–269
23. Pagliani, P.: Concrete neighbourhood systems and formal pretopological spaces. Preprint (2002)
24. Lin, T.Y.: Granular computing on binary relations. I: Data mining and neighborhood systems. In Polkowski, L., Skowron, A., eds.: Rough sets in knowledge discovery, Vol. 1. Physica–Verlag, Heidelberg (1998) 107–121
25. Fitting, M.: Basic modal logic. In Gabbay, D.M., Hogger, C.J., Robinson, J.A., eds.: Logical foundations. Volume 1 of Handbook of Logic in Artificial Intelligence and Logic Programming. Clarendon Press, Oxford (1993) 368–448
26. Orłowska, E.: Information algebras. In Alagar, V.S., Nivat, M., eds.: Algebraic Methodology and Software Technology, 4th International Conference, AMAST '95, Montreal, Canada, Proceedings. Volume 639 of Lecture Notes in Computer Science., Springer–Verlag (1995) 50–65
27. Gediga, G., Düntsch, I.: Skill set analysis in knowledge structures. British Journal of. Mathematical and Statistical Psychology **55** (2002) 361–384
28. Steiner, A.: The lattice of topologies: Structure and complementation. Trans. Amer. Math. Soc **122** (1966) 379–398
29. Huebener, J.: Complementation in the lattice of regular topologies. Pacific J. Math. **41** (1972) 139–149
30. Orlowska, E.: Semantics of vague concepts. In Dorn, G., Weingartner, P., eds.: Foundations of Logic and Linguistics. Problems and Solutions. Selected contributions to the 7th Internat. Congress of Logic, Methodology, and Philosophy of Science, Salzburg 1983, London, New York, Plenum Press (1983) 465–482
31. Yao, Y.Y., Lin, T.Y.: Generalization of rough sets using modal logic. Intelligent Automation and Soft Computing **2** (1996) 103–120
32. Demri, S., Orłowska, E.: Incomplete Information: Structure, Inference, Complexity. EATCS Monographs in Theoretical Computer Science. Springer–Verlag, Heidelberg (2002)
33. Tversky, A.: Features of similarity. Psychological Review **84** (1977) 327–352
34. Johannesson, M.: Modelling asymmetric similarity with prominence. Cognitive Studies 55, Lund University (1997) Available from http://www.lucs.lu.se/Abstracts/LUCS_Studies/LUCS55.html, Dec 27, 2002.
35. Järvinen, J.: On the structure of rough approximations. TUCS Technical Report 447, Turku Centre for Computer Science, University of Turku (2002)

36. Rothkopf, E.Z.: A measure of stimulus similarity and errors in some paired-associate learning tasks. Journal of Experimental Psychology **53** (1957) 94–101
37. Yao, Y.: Two views of the theory of rough sets in finite universes. International Journal of Approximate Reasoning **15** (1996) 291–317
38. Pagliani, P.: From concept lattices to approximation spaces: Algebraic structures of some spaces of partial objects. Fundamenta Informaticae **18** (1993)
39. Shepard, R.N.: Analysis of proximities as a technique for the study of information processing in man. Human Factors **5** (1963) 33–48
40. Buja, A., Swayne, D.F.: Visualization methodology for multidimensional scaling. Preprint (2001)

Lattice–Based Relation Algebras and Their Representability[*]

Ivo Düntsch[1], Ewa Orłowska[2], and Anna Maria Radzikowska[3]

[1] Brock University
St. Catharines, Ontario, Canada, L2S 3A1
duentsch@cosc.brocku.ca
[2] National Institute of Telecommunications
Szachowa 1, 04–894 Warsaw, Poland
orlowska@itl.waw.pl
[3] Faculty of Mathematics and Information Science
Warsaw University of Technology
Plac Politechniki 1, 00–661 Warsaw, Poland
annrad@mini.pw.edu.pl

1 Introduction

The motivation for this paper comes from the following sources. First, one can observe that the two major concepts underlying the methods of reasoning with incomplete information are the concept of degree of truth of a piece of information and the concept of approximation of a set of information items. We shall refer to the theories employing the concept of degree of truth as to theories of fuzziness and to the theories employing the concept of approximation as to theories of roughness (see [6] for a survey). The algebraic structures relevant to these theories are residuated lattices ([7], [12], [13], [16], [17], [18], [22], [23]) and Boolean algebras with operators ([19], [21], [10], [11]), respectively. Residuated lattices provide an arithmetic of degrees of truth and Boolean algebras equipped with the appropriate operators provide a method of reasoning with approximately determined information. Both classes of algebras have a lattice structure as a basis. Second, both theories of fuzziness and theories of roughness develop generalizations of relation algebras to algebras of fuzzy relations [20] and algebras of rough relations ([4], [5], [9]), respectively. In both classes a lattice structure is a basis. Third, not necessarily distributive lattices with modal operators, which can be viewed as most elementary approximation operators, are recently developed in [24] (distributive lattices with operators are considered in [14] and [25]). With this background, our aim is to begin a systematic study of the classes of algebras that have the structure of a (not necessarily distributive) lattice and, moreover, in each class there are some operators added to the lattice which are relevant for binary relations. Our main interest is in developing relational representation theorems for the classes of lattices with operators

---

[*] The work was carried out in the framework of COST Action 274/TARSKI on *Theory and Applications of Relational Structures as Knowledge Instruments*.

H. de Swart et al. (Eds.): TARSKI, LNCS 2929, pp. 231–255, 2003.

under consideration. More precisely, we wish to guarantee that each algebra of our classes is isomorphic to an algebra of binary relations on a set. We prove the theorems of that form by suitably extending the Urquhart representation theorem for lattices ([26]) and the representation theorems presented in [1]. The classes defined in the paper are the parts which put together lead to what might be called lattice-based relation algebras. Our view is that these algebras would be the weakest structures relevant for binary relations. All the other algebras of binary relations considered in the literature would then be their signature and/or axiomatic extensions.

Throughout the paper we use the same symbol for denoting an algebra or a relational system and their universes.

2 Doubly Ordered Sets

In this section we recall the notions introduced in [25] and some of their properties.

Definition 1. *Let X be a non–empty set and let $\leqslant_1$ and $\leqslant_2$ be two quasi orderings in X. A structure $(X, \leqslant_1, \leqslant_2)$ is called a **doubly ordered set** iff for all $x, y \in X$, if $x \leqslant_1 y$ and $x \leqslant_2 y$ then $x = y$.*   □

Definition 2. *Let $(X, \leqslant_1, \leqslant_2)$ be a doubly ordered set. We say that $A \subseteq X$ is $\leqslant_1$–increasing (resp. $\leqslant_2$–increasing) whenever for all $x, y \in X$, if $x \in A$ and $x \leqslant_1 y$ (resp. $x \leqslant_2 y$), then $y \in A$.*   □

For a doubly ordered set $(X, \leqslant_1, \leqslant_2)$, we define two mappings $l, r : 2^X \to 2^X$ by: for every $A \subseteq X$,

$$l(A) = \{x \in X : (\forall y \in X)\ x \leqslant_1 y \Rightarrow y \notin A\} \tag{1}$$
$$r(A) = \{x \in X : (\forall y \in X)\ x \leqslant_2 y \Rightarrow y \notin A\}. \tag{2}$$

Observe that mappings l and r can be expressed in terms of modal operators as follows: $l(A) = [\leqslant_1](-A)$ and $r(A) = [\leqslant_2](-A)$, where $-$ is the Boolean complement and $[\leqslant_i]$, $i = 1, 2$, are the necessity operators determined by relations $\leqslant_i$. Consequently, r and l are intuitionistic–like negations.

Definition 3. *Given a doubly ordered set $(X, \leqslant_1, \leqslant_2)$, a subset $A \subseteq X$ is called l–stable (resp. r–stable) iff $l(r(A)) = A$ (resp. $r(l(A)) = A$).*   □

The family of all l-stable (resp. r–stable) subsets of X will be denoted by $L(X)$ (resp. $R(X)$).

Recall the following notion from e.g. [8]:

Definition 4. *Let $(X, \leqslant_1)$ and $(Y, \leqslant_2)$ be partially ordered sets and let f and g be mappings $f : X \to Y$, $g : Y \to X$. We say that f and g are **Galois connection** iff for all $x, y \in X$*

$$x \leqslant_1 g(y) \text{ iff } y \leqslant_2 f(x).$$ □

Lemma 1. *[24] For any doubly ordered set $(X, \leqslant_1, \leqslant_2)$ and for any $A \subseteq X$,*

(i) $l(A)$ *is $\leqslant_1$-increasing*

(ii) $r(A)$ *is $\leqslant_2$-increasing*

(iii) *if A is $\leqslant_1$-increasing, then $r(A) \in R(X)$*

(iv) *if A is $\leqslant_2$-increasing, then $l(A) \in L(X)$*

(v) *if $A \in L(X)$, then $r(A) \in R(X)$*

(vi) *if $A \in R(X)$, then $l(A) \in L(X)$*

(vii) *if $A, B \in L(X)$, then $r(A) \cap r(B) \in R(X)$.* ∎

Lemma 2. *[24] The family of $\leqslant_i$-increasing sets, $i = 1, 2$, forms a distributive lattice, where join and meet are union and intersection of sets.* ∎

Lemma 3. *[26] For every doubly ordered set $(X, \leqslant_1, \leqslant_2)$, the mappings l and r form a Galois connection between the lattice of $\leqslant_1$-increasing subsets of X and the lattice of $\leqslant_2$-increasing subsets of X.* ∎

In other words, Lemma 3 says that for any $A \in L(X)$ and for any $B \in R(X)$, $A \subseteq l(B)$ iff $B \subseteq r(A)$.

Lemma 4. *For every doubly ordered set $(X, \leqslant_1, \leqslant_2)$ and for every $A \subseteq X$,*

(i) $l(r(A)) \in L(X)$ *and* $r(l(A)) \in R(X)$

(ii) *if A is $\leqslant_1$-increasing, then $A \subseteq l(r(A))$*

(iii) *if A is $\leqslant_2$-increasing, then $A \subseteq r(l(A))$.*

Proof. Direct consequence of Lemmas 1 and 3. ∎

Lemma 4 immediately implies:

Corollary 1. *For every doubly ordered set $(X, \leqslant_1, \leqslant_2)$ and for every $A \subseteq X$,*

(i) *if $A \in L(X)$, then $A \subseteq l(r(A))$*

(ii) *if $A \in R(X)$, then $A \subseteq r(l(A))$.* ∎

Let $(X, \leqslant_1, \leqslant_2)$ be a doubly ordered set. Define two binary operations in 2^X: for all $A, B \subseteq X$,

$$A \sqcap B = A \cap B \tag{3}$$
$$A \sqcup B = l(r(A) \cap r(B)). \tag{4}$$

Observe that $\sqcup$ is defined from $\sqcap$ resembling a De Morgan law with two different negations.

Moreover, put

$$\mathbf{0} = \emptyset. \tag{5}$$
$$\mathbf{1} = X \tag{6}$$

Lemma 5. [26] *For any doubly ordered set* $(X, \leqslant_1, \leqslant_2)$, *the system* $(L(X), \sqcap, \sqcup, \mathbf{0}, \mathbf{1})$ *is a lattice.* ∎

Definition 5. *Let* $(X, \leqslant_1, \leqslant_2)$ *be a doubly ordered set. The lattice* $(L(X), \sqcap, \sqcup, \mathbf{0}, \mathbf{1})$ *is called the **complex algebra of** X.* □

3   Urquhart Representation of Lattices

In this paper we are interested in studying relationships between relational structures (frames) providing Kripke–style semantics of logics, and algebras based on lattices. Therefore, we do not assume any topological structure in the frames. As a result, we have a weaker form of the representation theorems than the original Urquhart result, which requires compactness.

Let $(W, \wedge, \vee, 0, 1)$ be a bounded lattice.

Definition 6. *A **filter-ideal pair** of a lattice* W *is a pair* $x = (x_1, x_2)$ *such that* x_1 *is a filter of* W, x_2 *is an ideal of* W *and* $x_1 \cap x_2 = \emptyset$. □

The family of all filter–ideal pairs of a lattice W will be denoted by $FIP(W)$.

Let us define the following two quasi ordering relations on $FIP(W)$: for any $(x_1, x_2), (y_1, y_2) \in FIP(W)$,

$$(x_1, x_2) \preccurlyeq_1 (y_1, y_2) \quad \text{iff} \quad x_1 \subseteq y_1 \tag{7}$$
$$(x_1, x_2) \preccurlyeq_2 (y_1, y_2) \quad \text{iff} \quad x_2 \subseteq y_2. \tag{8}$$

Next, define

$$(x_1, x_2) \preccurlyeq (y_1, y_2) \quad \text{iff} \quad (x_1, x_2) \preccurlyeq_1 (y_1, y_2) \ \& \ (x_1, x_2) \preccurlyeq_2 (y_1, y_2).$$

We say that $(x_1, x_2) \in FIP(W)$ is *maximal* iff it is maximal wrt $\preccurlyeq$. We will write $X(W)$ to denote the family of all maximal filter–ideal pairs of the lattice W.

Observe that $X(W)$ is a binary relation on 2^W.

Proposition 1. [26] *Let W be a bounded lattice. For any $(x_1, x_2) \in FIP(W)$ there exists $(y_1, y_2) \in X(W)$ such that $(x_1, y_1) \preccurlyeq (y_1, y_2)$.* ∎

For any $(x_1, x_2) \in FIP(W)$, the maximal filter–ideal pair (y_1, y_2) such that $(x_1, x_2) \preccurlyeq (y_1, y_2)$ will be referred to as an *extension* of (x_1, x_2).

Definition 7. *Let $(W, \wedge, \vee, 0, 1)$ be a bounded lattice. The **canonical frame of** W is the structure $(X(W), \preccurlyeq_1, \preccurlyeq_2)$.* □

Lemma 6. *For every bounded lattice W, its canonical frame $(X(W), \preccurlyeq_1, \preccurlyeq_2)$ is a doubly ordered set.* ∎

Consider the complex algebra $(L(X(W)), \sqcap, \sqcup, \mathbf{0}, \mathbf{1})$ of the canonical frame of a lattice $(W, \wedge, \vee, 0, 1)$. Observe that $L(X(W))$ is an algebra of subrelations of $X(W)$.

Let us define the mapping $h : W \to 2^{X(W)}$ as follows: for every $a \in W$,

$$h(a) = \{x \in X(W) : a \in x_1\}. \tag{9}$$

Theorem 1. [26] *For every lattice $(W, \wedge, \vee, 0, 1)$ the following assertions hold:*

    (i) *For every $a \in W$, $r(h(a)) = \{x \in X(W) : a \in x_2\}$*
    (ii) *$h(a)$ is l–stable for every $a \in W$*
    (iii) *h is a lattice embedding.*

Proof. By way of example we prove **(iii)**.

We show that h is injective. Assume that for some $a, b \in W$, $h(a) = h(b)$. It follows that for every $x \in X(W)$, $a \in x_1$ iff $b \in x_1$. In particular, if $x_1 = [a] = \{z \in W : a \leqslant z\}$, then clearly $a \in [a)$, and also by the assumption $b \in [a)$. Hence $a \leqslant b$. Similarly, if $x_1 = [b)$, then $b \leqslant a$. We conclude that $a = b$.
Now we show that h preserves the operations. By way of example we prove that $h(a) \sqcup h(b) = h(a \vee b)$. Indeed, for every $a, b \in W$,

$$
\begin{aligned}
&h(a) \sqcup h(b) \\
&= l(r(\{x \in X(W) : a \in x_1\}) \sqcap r(\{x \in X(W) : b \in x_1\})) \\
&= l(\{x \in X(W) : a \in x_2\} \sqcap \{x \in X(W) : b \in x_2\}) && \text{from (i)} \\
&= l(\{x \in X(W) : a \vee b \in x_2\}) && \text{since } x_2 \text{ is an ideal} \\
&= lr(\{x \in X(W) : a \vee b \in x_1\}) && \text{from (i)} \\
&= lr(h(a \vee b)) && \text{the definition of } h \\
&= h(a \vee b) && \text{from (ii).} && \blacksquare
\end{aligned}
$$

The following theorem is a weak version of the Urquhart result.

Theorem 2 (Representation theorem for lattices). *Every bounded lattice is isomorphic to a subalgebra of the complex algebra of its canonical frame.* ∎

4   LC Algebras

An LC algebra is a bounded lattice with an additional unary operator which is
an abstract counterpart of the relational converse.

Definition 8. *An **LC algebra** is a system* $(W, \wedge, \vee, 0, 1, {}^{\smile})$ *such that* $(W, \wedge, \vee,$
$0, 1)$ *is a bounded lattice and* ${}^{\smile}$ *is a unary operator on* W *such that for all*
$a, b \in W$,

  (C.1) $a^{\smile\smile} = a$
  (C.2) $(a \vee b)^{\smile} = a^{\smile} \vee b^{\smile}$.   □

For an LC algebra W and any $a \in W$, $a^{\smile}$ is called a *converse of a*.

The following lemma gives the basic properties of the converse operation.

Lemma 7. *Let* $(W, \wedge, \vee, 0, 1, {}^{\smile})$ *be an LC algebra. Then the following assertions
hold:*

  (i) $0^{\smile} = 0$, $1^{\smile} = 1$.
  (ii) *for all* $a, b \in W$, $a \leqslant b$ *implies* $a^{\smile} \leqslant b^{\smile}$
  (iii) *for all* $a, b \in W$, $(a \wedge b)^{\smile} = a^{\smile} \wedge b^{\smile}$.

Proof. The proof of **(i)** is similar to the one given in [3]. By Definition 8: $0^{\smile} = 0 \vee$
$0^{\smile} = 0^{\smile\smile} \vee 0^{\smile} = (0^{\smile} \vee 0)^{\smile} = 0^{\smile\smile} = 0$. Analogously, $1 = 1 \vee 1^{\smile} = 1^{\smile\smile} \vee$
$1^{\smile} = (1^{\smile} \vee 1)^{\smile} = 1^{\smile}$.

(ii) Assume that $a \leqslant b$. Then $a \vee b = b$, so $(a \vee b)^{\smile} = b^{\smile}$. By axiom **(C.2)**,
$a^{\smile} \vee b^{\smile} = b^{\smile}$, so $a^{\smile} \leqslant b^{\smile}$.

(iii) ($\leqslant$) Let $a, b \in W$. Since $a \wedge b \leqslant a$, by **(ii)** we get $(a \wedge b)^{\smile} \leqslant a^{\smile}$. Similarly,
$(a \wedge b)^{\smile} \leqslant b^{\smile}$, which yields $(a \wedge b)^{\smile} \leqslant a^{\smile} \wedge b^{\smile}$.
($\geqslant$) Since $a^{\smile} \wedge b^{\smile} \leqslant a^{\smile}$, we have $(a^{\smile} \wedge b^{\smile})^{\smile} \leqslant a^{\smile\smile} = a$ by **(ii)** and **(C.1)**.
Analogously, $(a^{\smile} \wedge b^{\smile})^{\smile} \leqslant b$, so $(a^{\smile} \wedge b^{\smile})^{\smile} \leqslant a \wedge b$. Applying again **(C.1)** and
(ii) we get $a^{\smile} \wedge b^{\smile} = (a^{\smile} \wedge b^{\smile})^{\smile\smile} \leqslant (a \wedge b)^{\smile}$. ■

Given an LC algebra $(W, \wedge, \vee, 0, 1, {}^{\smile})$, by a *filter* (resp. *ideal*) of W we mean a
filter (resp. ideal) of the underlying lattice $(W, \wedge, \vee, 0, 1)$.

For any $A \subseteq W$, by $A^{\smile}$ we will denote the set:

$$A^{\smile} = \{a^{\smile} \in W : a \in A\}. \tag{10}$$

We have the following:

Lemma 8. *Let* $(W, \wedge, \vee, 0, 1, {}^{\smile})$ *be an LC algebra. Then the following assertions
hold for all* $A, B \subseteq W$:

(i) $A^\smile = \{a \in W : a^\smile \in A\}$

(ii) $A^{\smile\smile} = A$

(iii) $A \subseteq B$ iff $A^\smile \subseteq B^\smile$

(iv) $(-A)^\smile = -(A^\smile)$

(v) $(A \cup B)^\smile = A^\smile \cup B^\smile$

(vi) $(A \cap B)^\smile = A^\smile \cap B^\smile$.

Proof. By way of example we show **(ii)** and **(iii)**.

(ii) Let $a \in W$. Then $a \in A^{\smile\smile}$ iff $a^\smile \in A^\smile$ iff $a^{\smile\smile} \in A$ iff $a \in A$.

(iii) ($\Rightarrow$) Let $A, B \subseteq W$ be such that $A \subseteq B$ and let $a \in A^\smile$. Hence, by definition (10), $a^\smile \in A$, which by assumption implies $a^\smile \in B$.

($\Leftarrow$) Assume that $A^\smile \subseteq B^\smile$ and $a \in A$. By **(C.1)**, $a^{\smile\smile} \in A$, so $a^\smile \in A^\smile$, which by assumption gives $a^\smile \in B^\smile$. It follows that $a^{\smile\smile} \in B$, and hence, $a \in B$. ∎

Lemma 9. *Let* $(W, \wedge, \vee, 0, 1, ^\smile)$ *be an LC algebra and let* $A \subseteq W$. *Then the following assertions hold:*

(i) *If* A *is a filter of* W, *then so is* $A^\smile$

(ii) *If* A *is an ideal of* W, *then so is* $A^\smile$.

Proof.

(i) Let A be a filter of W and $a, b \in W$ such that $a \leqslant b$ and $a \in A^\smile$. Then $a^\smile \in A$, and, by Lemma 7(ii) we also have $a^\smile \leqslant b^\smile$. This implies, $b^\smile \in A$, and thus, $b \in A^\smile$.

Let $a, b \in A^\smile$. This means that $a^\smile \in A$ and $b^\smile \in A$, so $a^\smile \wedge b^\smile \in A$ since A is a filter. By Lemma 7(iii), $a^\smile \wedge b^\smile = (a \wedge b)^\smile$, so that $(a \wedge b)^\smile \in A = A^{\smile\smile}$ by Lemma 8(ii). Then $(a \wedge b)^{\smile\smile} \in A^\smile$, or equivalently, $a \wedge b \in A^\smile$.

(ii) Let A be an ideal of W and let $a, b \in W$. Assume that $b \in A^\smile$ and $a \leqslant b$. Then $b^\smile \in A$ and by Lemma 7(ii), $a^\smile \leqslant b^\smile$. Hence $a^\smile \in A$, so $a \in A^\smile$.

Let $a, b \in A^\smile$. Then $a^\smile \in A$ and $b^\smile \in A$. Since A is an ideal, $a^\smile \vee b^\smile \in A$. By axiom **(C.1)**, $a^\smile \vee b^\smile = (a \vee b)^\smile$. Hence $(a \vee b)^\smile \in A$, so $a \vee b \in A^\smile$. ∎

4.1   LC Frames

Definition 9. *An **LC frame** is a relational system* $(X, \leqslant_1, \leqslant_2, C)$ *such that* $(X, \leqslant_1, \leqslant_2)$ *is a doubly ordered set and* C *is a mapping* $C : X \to X$ *satisfying the following conditions for all* $x, y \in X$:

(MC.1) $x \leqslant_1 y$ *implies* $C(x) \leqslant_1 C(y)$

(MC.2) $x \leqslant_2 y$ *implies* $C(x) \leqslant_2 C(y)$

(SC) $C(C(x)) = x$. □

Given an LC frame $(X, \leqslant_1, \leqslant_2, C)$ let us define a mapping $^\curlyvee : 2^X \to 2^X$ as follows: for every $A \subseteq X$,

$$A^\curlyvee = \{C(x) : x \in A\}. \tag{11}$$

The following two lemmas present some properties of $^\curlyvee$.

Lemma 10. *Let* $(X, \leqslant_1, \leqslant_2, C)$ *be an LC frame and let* $^\curlyvee$ *be defined by (11). Then for all* $A, B \subseteq X$,

- (i) $A^\curlyvee = \{x \in X : C(x) \in A\}$
- (ii) $A^{\curlyvee\curlyvee} = A$
- (iii) $A \subseteq B$ *implies* $A^\curlyvee \subseteq B^\curlyvee$
- (iv) $(A \cap B)^\curlyvee = A^\curlyvee \cap B^\curlyvee$.

Proof. By way of example we show **(ii)** and **(iv)**.

(ii) Let $x \in W$. By **(SC)**, **(i)** and the definition (11) we have the following equivalences: $x \in A$ iff $C(C(x)) \in A$ iff $C(x) \in A^\curlyvee$ iff $x \in A^{\curlyvee\curlyvee}$.

(iv)$(\subseteq)$ Since $A \cap B \subseteq A$, by **(iii)** it follows that $(A \cap B)^\curlyvee \subseteq A^\curlyvee$. Similarly, $(A \cap B)^\curlyvee \subseteq B^\curlyvee$. Then $(A \cap B)^\curlyvee \subseteq A^\curlyvee \cap B^\curlyvee$.

$(\supseteq)$ Since $A^\curlyvee \cap B^\curlyvee \subseteq A^\curlyvee$, from **(iii)** and **(ii)** it follows $(A^\curlyvee \cap B^\curlyvee)^\curlyvee \subseteq A^{\curlyvee\curlyvee} = A$. Also, $(A^\curlyvee \cap B^\curlyvee)^\curlyvee \subseteq B$. Then $(A^\curlyvee \cap B^\curlyvee)^\curlyvee \subseteq A \cap B$, so again by **(ii)** and **(iii)**, $A^\curlyvee \cap B^\curlyvee = (A^\curlyvee \cap B^\curlyvee)^{\curlyvee\curlyvee} \subseteq (A \cap B)^\curlyvee$. ∎

Lemma 11. *Let* $(X, \leqslant_1, \leqslant_2, C)$ *be an LC frame and let* $^\curlyvee$ *be defined by (11). Then for all* $A, B \subseteq X$,

- (i) $l(A^\curlyvee) = l(A)^\curlyvee$
- (ii) $r(A^\curlyvee) = r(A)^\curlyvee$.
- (iii) *if* A *is l–stable, then so is* $A^\curlyvee$.

Proof. By way of example we show **(i)** and **(iii)**.

(i) $(\subseteq)$ Let $x \notin l(A)^\curlyvee$. By the definition (11), this means that $C(x) \notin l(A)$, so there exists $y \in X$ such that **(i.1)** $C(x) \leqslant_1 y$, and **(i.2)** $y \in A$. By **(MC.1)**, **(i.1)** implies $C(C(x)) \leqslant_1 C(y)$, so by **(SC)** we get **(i.3)** $x \leqslant_1 C(y)$. Next, **(i.2)** and **(SC)** imply $C(C(y)) \in A$, whence $C(y) \in A^\curlyvee$, which together with **(i.3)** implies $x \notin l(A^\curlyvee)$.

$(\supseteq)$ can be proved in the similar way.

(iii) Let A be l–stable. By **(i)** and **(ii)**, $l(r(A^\curlyvee)) = l(r(A)^\curlyvee) = l(r(A))^\curlyvee = A^\curlyvee$, so $A^\curlyvee$ is l–stable. ∎

4.2 Complex Algebras of LC Frames

Definition 10. *Let $(X, \leqslant_1, \leqslant_2, C)$ be an LC frame. By the **complex algebra** of X we mean a structure $(L(X), \sqcap, \sqcup, \mathbf{0}, \mathbf{1}, {}^{\curlyvee})$ with the operations defined by (3), (4), (11) and the constants defined by (5) and (6).* □

Theorem 3. *The complex algebra of an LC frame is an LC algebra.*

Proof. From Lemma 10(ii), $A^{\curlyvee\curlyvee} = A$, so it suffices to show that $(A \sqcup B)^{\curlyvee} = A^{\curlyvee} \sqcup B^{\curlyvee}$. For every $x \in X$,

$$
\begin{array}{llll}
x \in (A \sqcup B)^{\curlyvee} & \text{iff} & x \in l(r(A) \cap r(B))^{\curlyvee} & \text{by the definition of } \sqcup \\
& \text{iff} & x \in l((r(A) \cap r(B))^{\curlyvee}) & \text{by Lemma 11(i)} \\
& \text{iff} & x \in l(r(A)^{\curlyvee} \cap r(B)^{\curlyvee}) & \text{by Lemma 10(iv)} \\
& \text{iff} & x \in l(r(A^{\curlyvee}) \cap r(B^{\curlyvee})) & \text{by Lemma 11(ii)} \\
& \text{iff} & x \in A^{\curlyvee} \sqcup B^{\curlyvee}. & \blacksquare
\end{array}
$$

4.3 Canonical Frames of LC Algebras

Let $(W, \wedge, \vee, 0, 1, {}^{\smile})$ be an LC algebra. As usual, $FIP(W)$ and $X(W)$ denote the family of all filter–ideal pairs (resp. maximal filter–ideal pairs) of W.

First, observe the following:

Lemma 12. $(x_1, x_2) \in FIP(W)$ *iff* $(x_1^{\smile}, x_2^{\smile}) \in FIP(W)$.

Proof. ($\Rightarrow$) Let $(x_1, x_2) \in FIP(W)$. From Lemma 9 it follows that $x_1^{\smile}$ is a filter of W and $x_2^{\smile}$ is an ideal of W. Note that $\emptyset^{\smile} = \emptyset$. Then, by Lemma 8(vi) we get that $x_1^{\smile} \cap x_2^{\smile} = \emptyset$, so $(x_1^{\smile}, x_2^{\smile}) \in FIP(W)$
($\Leftarrow$) Let $x_1, x_2 \subseteq W$ be such that $(x_1^{\smile}, x_2^{\smile}) \in FIP(W)$. Then, by Lemmas 8(ii) and 9, $x_1 = x_1^{\smile\smile}$ is a filter of W and $x_2 = x_2^{\smile\smile}$ is an ideal of W. Next, from Lemma 8(vi) we get $(x_1 \cap x_2)^{\smile} = \emptyset$, so $x_1 \cap x_2 = (x_1 \cap x_2)^{\smile\smile} = \emptyset$. Therefore we obtain $(x_1, x_2) \in FIP(W)$. $\blacksquare$

Let us now define a mapping $C^{\star} : FIP(W) \to FIP(W)$ as follows: for every $x \in FIP(W)$,

$$
C^{\star}(x) = (x_1^{\smile}, x_2^{\smile}). \tag{12}
$$

Lemma 13. *If x is a maximal filter–ideal pair of W, then so is $C^{\star}(x)$.*

Proof. Let $x = (x_1, x_2) \in FIP(W)$. Assume that $(x_1^{\smile}, x_2^{\smile})$ is not maximal. By Proposition 1, it can be extended to the maximal filter–ideal pair, say $y = (y_1, y_2)$. Then $x_1^{\smile} \subseteq y_1$, $x_2^{\smile} \subseteq y_2$ and $(x_1^{\smile}, x_2^{\smile}) \neq (y_1, y_2)$. By Lemma 8(ii) and 8(iii) we get $x_1 \subseteq y_1^{\smile}$, $x_2 \subseteq y_2^{\smile}$ and $(x_1, x_2) \neq (y_1^{\smile}, y_2^{\smile})$, which means that (x_1, x_2) is not a maximal filter–ideal pair. $\blacksquare$

Definition 11. *Let* $(W, \wedge, \vee, 0, 1, \smile)$ *be an LC algebra. A **canonical frame** of* W *is a structure* $(X(W), \preccurlyeq_1, \preccurlyeq_2, C^*)$, *where* $\preccurlyeq_1$, $\preccurlyeq_2$ *and* C^* *are defined by (7), (8) and (12), respectively.*   □

Theorem 4. *The canonical frame of an LC algebra is an LC frame.*

Proof. Let $x, y \in X(W)$ and assume that $x \preccurlyeq_1 y$. This means that $x_1 \subseteq y_1$. By Lemma 8(iii), $x_1^{\smile} \subseteq y_1^{\smile}$, so $C^*(x) \preccurlyeq_1 C^*(y)$. Hence **(MC.1)** holds. In the analogous way we can show that **(MC.2)** holds. Finally, let $x = (x_1, x_2) \in X(W)$. Then we have $C^*(C^*(x)) = (x_1^{\smile\smile}, x_2^{\smile\smile}) = (x_1, x_2) = x$ by Lemma 8(ii), so the condition **(SC)** also holds. ∎

4.4   Relational Representation of LC Algebras

In this section we conclude our discussion of LC algebras by showing their relational representability.

Theorem 5 (Representation theorem for LC algebras). *Every LC algebra is isomorphic to a subalgebra of the complex algebra of its canonical frame.*

Proof. Let $(W, \wedge, \vee, 0, 1, \smile)$ be an LC algebra, $(X(W), \preccurlyeq_1, \preccurlyeq_2, C^*)$ be the canonical frame of W and let $(L(X(W)), \sqcap, \sqcup, \mathbf{0}, \mathbf{1}, ^{\curlyvee})$ be the complex algebra of the canonical frame of W. By Theorems 3 and 4 it follows that $L(X(W))$ is an LC algebra, so it suffices to show that W is isomorphic to a subalgebra of $L(X(W))$. Let the mapping $h : W \rightarrow 2^{X(W)}$ be defined as in (9), i.e. $h(a) = \{x \in X(W) : a \in x_1\}$, $a \in W$. We show that for every $a \in W$, $h(a^{\smile}) = h(a)^{\curlyvee}$. For any $x \in X(W)$ and for any $a \in W$ we have: $x \in h(a^{\smile})$ iff $a^{\smile} \in x_1$ iff $a \in x_1^{\smile}$ iff $C^*(x) \in h(a)$ iff $x \in h(a)^{\curlyvee}$. Finally, from Theorem 1 h preserves the lattice operations and is injective. ∎

5   LP Algebras

LP algebras are a join of a not necessary distributive bounded lattice with a monoid. The monoid product operation is an abstract counterpart to the relational composition and the unit element of the monoid corresponds to the identity relation.

Definition 12. *An **LP algebra** is a structure* $(W, \wedge, \vee, 0, 1, \odot, 1')$ *such that* $(W, \wedge, \vee, 0, 1)$ *is a bounded lattice and for all* $a, b, c \in W$,

(P.1) $a \odot 1' = 1' \odot a = a$

(P.2) $a \odot (b \odot c) = (a \odot b) \odot c$

(P.3) $a \odot (b \vee c) = (a \odot b) \vee (a \odot c)$
(P.4) $(a \vee b) \odot c = (a \odot c) \vee (b \odot c)$. $\square$

Lemma 14. *Let* $(W, \wedge, \vee, 0, 1, \odot, 1')$ *be an LP algebra. For all* $a, b, c \in W$, *if* $a \leqslant b$, *then*

 (i) $c \odot a \leqslant c \odot b$
 (ii) $a \odot c \leqslant b \odot c$.

Proof.

(i) Let $a \leqslant b$. Then $a \vee b = b$, so $c \odot (a \vee b) = c \odot b$. Hence, by axiom **(P.3)**, we get $(c \odot a) \vee (c \odot b) = c \odot b$, which implies $c \odot a \leqslant c \odot b$.

(ii) This can be proved in an analogous way. ∎

5.1   LP Frames

In this section we follow the developments of Allwein and Dunn ([1]). However, the difference is that here we consider an abstract notion of an LP frame, not only the canonical frame of an LP algebra.

Definition 13. *An **LP frame** is a relational system* $(X, \leqslant_1, \leqslant_2, R, S, Q, I)$ *such that* $(X, \leqslant_1, \leqslant_2)$ *is a doubly ordered set,* R, S, Q *are ternary relations on* X *and* $I \subseteq X$ *is a unary relation on* X *such that the following conditions are satisfied: for all* $x, x', y, y', z, z' \in X$,

A. MONOTONICITY CONDITIONS
(MP.1) $R(x, y, z)$ & $x' \leqslant_1 x$ & $y' \leqslant_1 y$ & $z \leqslant_1 z' \Rightarrow R(x', y', z')$
(MP.2) $S(x, y, z)$ & $x \leqslant_2 x'$ & $y' \leqslant_1 y$ & $z' \leqslant_2 z \Rightarrow S(x', y', z')$
(MP.3) $Q(x, y, z)$ & $x' \leqslant_1 x$ & $y \leqslant_2 y'$ & $z' \leqslant_2 z \Rightarrow Q(x', y', z')$
(MP.4) $I(x)$ & $x \leqslant_1 x' \Rightarrow I(x')$

B. STABILITY CONDITIONS
(SP.1) $R(x, y, z) \Rightarrow \exists x'' \in X \ (x \leqslant_1 x''$ & $S(x'', y, z))$
(SP.2) $R(x, y, z) \Rightarrow \exists y'' \in X \ (y \leqslant_1 y''$ & $Q(x, y'', z))$
(SP.3) $S(x, y, z) \Rightarrow \exists z'' \in X \ (z \leqslant_2 z''$ & $R(x, y, z''))$
(SP.4) $Q(x, y, z) \Rightarrow \exists z'' \in X \ (z \leqslant_2 z''$ & $R(x, y, z''))$
(SP.5) $\exists u \in X(R(x, y, u)$ & $Q(x', u, z)) \Rightarrow \exists w \in X(R(x', x, w)$ & $S(w, y, z))$
(SP.6) $\exists u \in X(R(x, y, u)$ & $S(u, z, z')) \Rightarrow \exists w \in X(R(y, z, w)$ & $Q(x, w, z'))$
(SP.7) $I(x)$ & $(R(x, y, z)$ or $R(y, x, z)) \Rightarrow y \leqslant_1 z$
(SP.8) $\exists u \in X(I(u)$ & $S(u, x, x))$
(SP.9) $\exists u \in X(I(u)$ & $Q(x, u, x))$. $\square$

Lemma 15. *For every LP frame* $(X, \leqslant_1, \leqslant_2, R, S, Q, I)$ *and every l–stable subset* $A \subseteq X$ *it holds for all* $x, y, z \in X$:

$$I(x) \;\&\; (y \in A) \;\&\; (R(x, y, z) \text{ or } R(y, x, z)) \;\Rightarrow\; z \in A.$$

Proof.

Let $x, y, z \in X$ and assume that **(1)** $I(x)$ **(2)** $y \in A$ and **(3)** $R(x, y, z)$ or $R(y, x, z)$. By **(SP.7)**, we get from **(1)** and **(3)** that **(4)** $y \leqslant_1 z$. Since A is l–stable, by Lemma 1(i) it is $\leqslant_1$–increasing, so **(2)** and **(4)** imply $z \in A$. ∎

For an LP frame $(X, \leqslant_1, \leqslant_2, R, S, Q, I)$, let us define two mappings $\odot_Q, \odot_S$: $2^X \times 2^X \to 2^X$ by: for all $A, B \subseteq X$,

$$A \odot_Q B = \{z \in X : \forall x, y \in X \; (Q(x, y, z) \;\&\; x \in A \Rightarrow y \in r(B))\} \quad (13)$$
$$A \odot_S B = \{z \in X : \forall x, y \in X \; (S(x, y, z) \;\&\; y \in B \Rightarrow x \in r(A))\}. \quad (14)$$

Lemma 16. *For any* $A, B \subseteq X$,

(i) $A \odot_Q B$ *and* $A \odot_S B$ *are* $\leqslant_2$–*increasing*

(ii) *if* A *and* B *are l–stable, then* $A \odot_Q B$ *and* $A \odot_S B$ *are r–stable.*

Proof.

(i) Let $A, B \subseteq X$ and suppose that $A \odot_Q B$ is not $\leqslant_2$–increasing. Then, by Definition 2, there exist $x, y \in X$ such that **(i.1)** $x \in A \odot_Q B$, **(i.2)** $x \leqslant_2 y$ and **(i.3)** $y \notin A \odot_Q B$. By the definition (13), **(i.3)** means that there exist $u, w \in X$ such that **(i.4)** $Q(u, w, y)$, **(i.5)** $u \in A$ and **(i.6)** $w \notin r(B)$. However, by **(MP.3)**, **(i.2)** and **(i.4)** imply $Q(u, w, x)$, which together with **(i.5)** and **(i.6)** gives $x \notin A \odot_Q B$ – a contradiction with **(i.1)**.

In the similar way one can show that $A \odot_S B$ is $\leqslant_2$–increasing.

(ii) Let $A, B \subseteq X$ be l–stable sets. By (i), $A \odot_Q B$ is $\leqslant_2$–increasing. Therefore, by Lemma 4(iii), $A \odot_Q B \subseteq r(l(A \odot_Q B))$, so it suffices to show that $r(l(A \odot_Q B)) \subseteq A \odot_Q B$.

Let $z \in X$ and assume that $z \notin A \odot_Q B$. We show that **(ii.1)** $z \notin r(l(A \odot_Q B))$. By the definition (13) of $\odot_Q$, there exist $x, y \in X$ such that **(ii.2)** $Q(x, y, z)$, **(ii.3)** $x \in A$ and **(ii.4)** $y \notin r(B)$. From **(ii.4)**, there exists $y' \in X$ such that **(ii.5)** $y \leqslant_2 y'$ and **(ii.6)** $y' \in B$. By **(MP.3)**, **(ii.2)** and **(ii.5)** imply $Q(x, y', z)$, which by **(SP.4)** gives that there exists $z' \in X$ such that **(ii.7)** $z \leqslant_2 z'$ and **(ii.8)** $R(x, y', z')$.

Now we show that **(ii.9)** $z' \in l(A \odot_Q B)$. Let $z'' \in X$ and assume that **(ii.10)** $z' \leqslant_1 z''$. Hence, by **(MP.1)**, from **(ii.8)** we get that $R(x, y', z'')$, which by **(SP.2)** implies that there exists $y'' \in X$ such that **(ii.11)** $y' \leqslant_1 y''$ and **(ii.12)** $Q(x, y'', z'')$. Next, since B is l–stable, it is $\leqslant_1$–increasing, so from **(ii.6)** and **(ii.11)**, $y'' \in B$. It follows that **(ii.13)** $y'' \notin r(B)$. Furthermore, **(ii.12)**, **(ii.13)** and **(ii.3)** imply **(ii.14)** $z'' \notin A \odot_Q B$. Since z'' is an arbitrary element satisfying **(ii.10)**, we obtain that for any $z'' \in X$, $z' \leqslant_1 z''$ implies $z'' \notin A \odot_Q B$, so **(ii.9)** follows.

Proceeding in the similar way we can show that $A \odot_S B$ is r–stable. ∎

Lemma 17. *For every LP frame* $(X, \leqslant_1, \leqslant_2, R, S, C, I)$ *and for all l–stable subsets* $A, B \subseteq X$, $A \odot_s B = A \odot_Q B$.

Proof. $(\subseteq)$ Let $z \in X$ and assume that $z \notin A \odot_Q B$. We show that $z \notin A \odot_s B$. By assumption, there exist $x, y \in X$ such that **(1)** $Q(x, y, z)$, **(2)** $x \in A$ and **(3)** $y \notin r(B)$. From **(3)**, there is $y' \in X$ such that **(4)** $y \leqslant_2 y'$ and **(5)** $y' \in B$. Next, from **(1)** and **(4)** we get by **(MP.3)** that $Q(x, y', z)$, which by **(SP.4)** implies that there exists $z' \in X$ such that **(6)** $z \leqslant_2 z'$ and **(7)** $R(x, y', z')$. Furthermore, applying **(SP.1)**, we get from **(7)** that there exists $x' \in X$ such that **(8)** $x \leqslant_1 x'$ and **(9)** $S(x', y', z')$. From **(2)**, **(8)** and the assumption that A is l–stable we get that $x' \notin r(A)$, which together with **(5)** and **(9)** gives by the definition (14) that **(10)** $z' \notin A \odot_s B$. From Lemma 16(i), $A \odot_s B$ is $\leqslant_2$–increasing. Hence, by **(6)** and **(10)** we get $z \notin A \odot_s B$.

The proof of $(\supseteq)$ is similar. ∎

Following [1] and [2] we introduce the two relations S and Q in the LP frames, although in view of Lemma 17 one could argue that we only need S or Q. Keeping an extra relation makes the proofs a bit easier.

Let us define a mapping $\otimes : 2^X \times 2^X \to 2^X$ as follows: for all $A, B \subseteq X$,

$$A \otimes B = l(A \odot_Q B). \tag{15}$$

Lemma 18. *Let* $A, B \subseteq X$. *If* A *and* B *are l–stable sets, then so is* $A \otimes B$.

Proof. By Lemma 16(ii), $A \odot_Q B$ is r–stable, so from Lemma 1(vi) we get that $l(A \odot_Q B)$ is l–stable. ∎

Given an LP frame $(X, \leqslant_1, \leqslant_2, R, S, Q, I)$, let us define

$$\mathbf{1'} = l(r(I)) \tag{16}$$

Lemma 19. $\mathbf{1'}$ *is l–stable.*

Proof. Follows from Lemma 1(ii) and (iv). ∎

5.2   Complex Algebras of LP Frames

Definition 14. *Let* $(X, \leqslant_1, \leqslant_2, R, S, Q, I)$ *be an LP frame. A **complex algebra** of* X *is a structure* $(L(X), \sqcap, \sqcup, \mathbf{0}, \mathbf{1}, \otimes, \mathbf{1'})$ *with operations defined by (3)–(6), (15) and (16).* □

Our aim now is to show that complex algebras of LP frames are LP algebras. To this end, we show that any complex algebra of an LP frame satisfies axioms **(P.1)–(P.4)**.

First, we prove that axiom **(P.1)** is satisfied in $L(X)$.

Lemma 20. *Let $(L(X), \sqcap, \sqcup, \mathbf{0}, \mathbf{1}, \otimes, \mathbf{1}')$ be a complex algebra of an LP frame X. Then for every $A \in L(X)$,*

(i) $\mathbf{1}' \otimes A = A$

(ii) $A \otimes \mathbf{1}' = A$.

Proof.

(i) Let $A \in L(X)$. We show that $\mathbf{1}' \odot_Q A = r(A)$. Then we will have $l(\mathbf{1}' \odot_Q A) = l(r(A))$, which by the assumption and the definition (15) gives the result.

($\subseteq$) Let $z \in X$ and assume that $z \notin r(A)$. Then there exists $w \in X$ such that **(i.1)** $z \leqslant_2 w$ and **(i.2)** $w \in A$. From **(SP.8)**, there exists $x \in X$ such that **(i.3)** $I(x)$ and **(i.4)** $S(x, w, w)$. Next, from **(MP.4)**, I is $\leqslant_1$–increasing, so by Lemma 4(ii), $I \subseteq l(r(I))$, which together with **(i.3)** implies $x \in l(r(I))$, or equivalently, by the definition (16), **(i.5)** $x \in \mathbf{1}'$. Also, by **(MP.2)**, **(i.1)** and **(i.4)** imply $S(x, w, z)$. Then, in view of **(SP.3)**, there exists $z' \in X$ such that **(i.6)** $z \leqslant_2 z'$ and **(i.7)** $R(x, w, z')$. Furthermore, **(i.7)** implies by **(SP.2)** that there exists $y \in X$ such that **(i.8)** $w \leqslant_1 y$ and **(i.9)** $Q(x, y, z')$. From **(i.6)** and **(i.9)** we get by **(MP.3)** that **(i.10)** $Q(x, y, z)$. Since A is l–stable, by Lemma 1(i) it is $\leqslant_1$–increasing. Then **(i.2)** and **(i.8)** imply $y \in A$, so $y \notin r(A)$, which together with **(i.5)** and **(i.10)** gives $z \notin \mathbf{1}' \odot_Q A$ by the definition (13).

($\supseteq$) Let $z \in X$ and assume that $z \notin \mathbf{1}' \odot_Q A$. We show that $z \notin r(A)$. By assumption, for some $x, y \in X$ we have: **(i.11)** $Q(x, y, z)$, **(i.12)** $x \in \mathbf{1}'$ and **(i.13)** $y \notin r(A)$. From **(i.13)**, there exists $y' \in X$ such that **(i.14)** $y \leqslant_2 y'$ and **(i.15)** $y' \in A$. By **(MP.3)**, from **(i.11)** and **(i.14)** it follows that $Q(x, y', z)$, which by **(SP.4)** implies that there exists $z' \in X$ such that **(i.16)** $z \leqslant_2 z'$ and **(i.17)** $R(x, y', z')$. It suffices to show that $z' \in A$. Then, by **(i.16)** and the definition (2), $z \notin r(A)$.

Suppose that $z' \notin A$. By l–stability of A, this means that $z' \notin l(r(A))$, i.e. there exists $z'' \in X$ such that **(i.18)** $z' \leqslant_1 z''$ and **(i.19)** $z'' \in r(A)$. Applying **(MP.1)**, from **(i.17)** and **(i.18)**, $R(x, y', z'')$, which by **(SP.1)** implies that there exists $x' \in X$ such that **(i.20)** $x \leqslant_1 x'$ and **(i.21)** $S(x', y', z'')$. From **(i.12)** we have: $x \in \mathbf{1}' = l(r(I))$. Hence, by **(i.20)** and the definition (1), $x' \notin r(I)$, so there exists $x'' \in X$ such that **(i.22)** $x' \leqslant_2 x''$ and **(i.23)** $x'' \in I$. By **(MP.2)**, **(i.21)** and **(i.22)** imply $S(x'', y', z'')$, which by **(SP.3)** gives that there exists $w \in X$ such that **(i.24)** $z'' \leqslant_2 w$ and **(i.25)** $R(x'', y', w)$. Now, applying Lemma 15, **(i.25)**, **(i.23)** and **(i.15)** imply $w \in A$, which together with **(i.24)** gives $z'' \notin r(A)$, which contradicts **(i.19)**.

(ii) Let $A \in L(X)$. As before, we will show that $A \odot_Q \mathbf{1}' = r(A)$. Hence we will have $A \otimes \mathbf{1}' = l(A \odot_Q \mathbf{1}') = r(l(A)) = A$.

($\subseteq$) Let $z \in X$ and assume that $z \notin r(A)$. By the definition (2) this means that there exists $x \in X$ such that **(ii.1)** $z \leqslant_2 x$ and **(ii.2)** $x \in A$. From **(SP.9)**, there exists $y \in X$ such that **(ii.3)** $y \in I$ and **(ii.4)** $Q(x, y, x)$. From **(MP.4)**, I is $\leqslant_1$–increasing, so by Lemma 4(ii), $I \subseteq l(r(I)) = \mathbf{1}'$. Then from **(ii.3)** it follows that $y \in \mathbf{1}'$, whence **(ii.5)** $y \notin r(\mathbf{1}')$. Next, by **(MP.3)**, **(ii.1)** and **(ii.4)** imply **(ii.6)**

$Q(x, y, z)$. Therefore, there exist $x, y \in X$ such that **(ii.2)**, **(ii.5)** and **(ii.6)** hold, which by the definition (13) means that $z \notin A \odot_Q 1'$.

($\supseteq$) We have to show that $r(A) \subseteq A \odot_Q 1'$. Assume that $z \notin A \odot_Q 1'$. By the definition (13), there exist $x, y \in X$ such that **(ii.7)** $Q(x, y, z)$, **(ii.8)** $x \in A$ and **(ii.9)** $y \notin r(1')$. From **(ii.9)**, there exists $y' \in X$ such that **(ii.10)** $y \leqslant_2 y'$ and **(ii.11)** $y' \in 1'$. By **(MP.3)**, **(ii.7)** and **(ii.10)** imply $Q(x, y', z)$, from which, by **(SP.4)**, it follows that there exists $z' \in X$ such that **(ii.12)** $z \leqslant_2 z'$ and $R(x, y', z')$. Proceeding in the similar way as in the proof of ($\supseteq$) in **(i)**, we can show that $z' \in A$, which together with **(ii.12)** implies $z \notin r(A)$. ∎

The following lemma states that **(P.2)** is satisfied in $L(X)$.

Lemma 21. *Let* $(L(X), \sqcap, \sqcup, 0, 1, \otimes, 1')$ *be the complex algebra of an LP frame* X. *Then for all* $A, B, C \in L(X)$, *the following equality holds:*

$$A \otimes (B \otimes C) = (A \otimes B) \otimes C.$$

Proof. Let $A, B, C \subseteq X$ be l–stable sets. In view of Lemma 17 it suffices to show that $A \odot_Q (B \otimes C) = (A \otimes B) \odot_s C$.

($\subseteq$) Let $z \in X$ and assume that $z \notin (A \otimes B) \odot_s C$. By the definition (14), this means that there exist $x, y \in X$ such that **(1)** $S(x, y, z)$, **(2)** $y \in C$ and **(3)** $x \notin r(A \otimes B)$. But $r(A \otimes B) = r(l(A \odot_Q B))$ by the definition (15). Furthermore, by Lemma 17, $r(l(A \odot_Q B)) = r(l(A \odot_s B))$. Moreover, $A \odot_s B$ is r–stable by Lemma 16(ii), so $r(l(A \odot_s B)) = A \odot_s B$. Then $r(A \otimes B) = A \odot_s B$. Hence, from **(3)** we get $x \notin A \odot_s B$, so there exist $u, v \in X$ such that **(4)** $S(u, v, x)$, **(5)** $v \in B$ and **(6)** $u \notin r(A)$. From **(6)**, there exists $u' \in X$ such that **(7)** $u \leqslant_2 u'$ and **(8)** $u' \in A$. By the monotonicity condition **(MP.2)**, **(4)** and **(7)** imply $S(u', v, x)$, which by **(SP.3)** gives that there exists $x' \in X$ such that **(9)** $x \leqslant_2 x'$ and **(10)** $R(u', v, x')$. Applying again **(MP.2)**, from **(1)** and **(9)** we get that $S(x', y, z)$. Therefore, there exists $x' \in X$ such that $R(u'v, x')$ and $S(x', y, z)$. By **(SP.6)**, this implies that there exists $w \in X$ such that **(11)** $R(v, y, w)$ and **(12)** $Q(u', w, z)$. Next, by **(SP.1)**, **(11)** implies that there exists $v' \in X$ such that **(13)** $v \leqslant_1 v'$ and **(14)** $S(v', y, w)$. Since B is l–stable, it is $\leqslant_1$–increasing. Then **(5)** and **(13)** imply $v' \in B$, so **(15)** $v' \notin r(B)$. We have then obtained that there exist $y, v' \in X$ such that **(2)**, **(14)** and **(15)** hold, which by the definition (14) means that **(16)** $w \notin B \odot_s C$. From Lemma 16, $B \odot_s C$ is r–stable, so $B \odot_s C = r(l(B \odot_s C)) = r(l(B \odot_Q C)) = r(B \otimes C)$ by Lemma 17. Hence **(16)** implies **(17)** $w \notin r(B \otimes C)$. Then we finally get that there exist $u', w \in X$ such that **(8)**, **(12)** and **(17)** hold. By the definition (13), this means that $z \notin A \odot_Q (B \otimes C)$.

Proceeding in the similar way, and using **(SP.5)**, ($\supseteq$) can be proved. ∎

Finally, we show that axioms **(P.3)** and **(P.4)** are satisfied in complex algebras of LP frames.

Lemma 22. *Let* $(L(X), \sqcap, \sqcup, 0, 1, \otimes, 1')$ *be the complex algebra of an LP frame* X. *Then for all* $A, B, C \in L(X)$,

(i) $A \otimes (B \sqcup C) = (A \otimes B) \sqcup (A \otimes C)$
(ii) $(B \sqcup C) \otimes A = (B \otimes A) \sqcup (C \otimes A)$.

Proof.

(i) Let $A, B, C \subseteq X$ be l–stable sets. First we show that (i.1) $A \odot_Q (B \sqcup C) = (A \odot_Q B) \cap (A \odot_Q C)$. For every $x \in X$ the following equivalences hold:

$$x \in (A \odot_Q B) \cap (A \odot_Q C)$$
$$\text{iff } x \in A \odot_Q B \ \& \ x \in A \odot_Q C$$
$$\text{iff } [\forall y, z \in X \ (Q(y, z, x) \ \& \ y \in A \Rightarrow z \in r(B))]$$
$$\& \ [\forall y, z \in X \ (Q(y, z, x) \ \& \ y \in A \Rightarrow z \in r(C))]$$
$$\text{iff } \forall y, z \in X \ (Q(y, z, x) \ \& \ y \in A \Rightarrow z \in r(B) \cap r(C)).$$

By Lemma 1(vii), $r(B) \cap r(C)$ is r–stable, so $r(B) \cap r(C) = r(l(r(B) \cap r(C)))$. Hence, by the definition (4) we get

$$x \in (A \odot_Q B) \cap (A \odot_Q C)$$
$$\text{iff } \forall y, z, \in X \ (Q(y, z, x) \ \& \ y \in A \Rightarrow z \in r(l(r(B) \cap r(C))))$$
$$\text{iff } x \in A \odot_Q l(r(B) \cap r(C))$$
$$\text{iff } x \in A \odot_Q (B \sqcup C).$$

So (i.1) holds. Hence (i.2) $l(A \odot_Q (B \sqcup C)) = l((A \odot_Q B) \cap (A \odot_Q C))$. Note that (i.3) $l(A \odot_Q (B \sqcup C)) = A \otimes (B \sqcup C)$. Next, by Lemma 16(ii), $A \odot_Q B$ is r–stable, so $r(l(A \odot_Q B)) = A \odot_Q B$. Using again the definition (4), we get

$$l((A \odot_Q B) \cap (A \odot_Q C)) = l(r(l(A \odot_Q B)) \cap r(l(A \odot_Q C)))$$
$$= l(A \odot_Q B) \sqcup l(A \odot_Q C)$$
$$= (A \otimes B) \sqcup (A \otimes C).$$

Hence, by (i.2) and (i.3) we get the required result.

(ii) can be proved in the similar way. ∎

From Lemmas 20, 21 and 22 we get:

Theorem 6. *The complex algebra of an LP frame is an LP algebra.* ∎

5.3   Canonical Frames of LP Algebras

Let $(W, \wedge, \vee, 0, 1, \odot, 1')$ be an LP algebra. As before, by a *filter* (resp. *ideal*) of W we mean a filter (resp. ideal) of the underlying lattice $(W, \wedge, \vee, 0, 1)$. We will write $X(W)$ to denote the family of all maximal filter–ideal pairs of the lattice reduct of W.

Let us define the following ternary relations on $X(W)$: for all $x, y, z \in X(W)$,

$$R^\star(x, y, z) \quad \text{iff} \quad (\forall a, b \in W)\ a \in x_1\ \&\ b \in y_1 \Rightarrow a \odot b \in z_1 \tag{17}$$

$$S^\star(x, y, z) \quad \text{iff} \quad (\forall a, b \in W)\ a \odot b \in z_2\ \&\ b \in y_1 \Rightarrow a \in x_2 \tag{18}$$

$$Q^\star(x, y, z) \quad \text{iff} \quad (\forall a, b \in W)\ a \odot b \in z_2\ \&\ a \in x_1 \Rightarrow b \in y_2 \tag{19}$$

Moreover, let

$$I^\star = \{x \in X(W) : 1' \in x_1\}. \tag{20}$$

Definition 15. *Let an LP algebra* $(W, \wedge, \vee, 0, 1, \odot, 1')$ *be given. The structure* $(X(W), \preccurlyeq_1, \preccurlyeq_2, R^\star, S^\star, Q^\star, I^\star)$ *is called a **canonical frame of** W.*   □

The following two lemmas can be proved as in [1].

Lemma 23. *Let* $(X(W), \preccurlyeq_1, \preccurlyeq_2, R^\star, S^\star, Q^\star, I^\star)$ *be the canonical frame of an LP algebra. Then* $R^\star$, $S^\star$, $Q^\star$ *and* $I^\star$ *satisfy monotonicity conditions* **(MP.1)**– **(MP.4)** *of Definition 13.* ∎

Lemma 24. *Let* $(X(W), \preccurlyeq_1, \preccurlyeq_2, R^\star, S^\star, Q^\star, I^\star)$ *be the canonical frame of an LP algebra. Then* $R^\star$, $S^\star$, $Q^\star$ *and* $I^\star$ *satisfy stability conditions* **(SP.1)**–**(SP.9)** *of Definition 13.* ∎

Lemmas 23 and 24 imply the following theorem:

Theorem 7. *The canonical frame of an LP algebra is an LP frame.* ∎

5.4   Relational Representation for LP Algebras

Let $(W, \wedge, \vee, 0, 1, \odot, 1')$ be an LP algebra, $(X(W), \preccurlyeq_1, \preccurlyeq_2, R^\star, S^\star, Q^\star, I^\star)$ be its canonical frame and let $(L(X(W)), \sqcap, \sqcup, \mathbf{0}, \mathbf{1}, \otimes, \mathbf{1}')$ be the complex algebra of $X(W)$. Let the mapping $h : W \to 2^{X(W)}$ be defined as in (9), i.e. for every $a \in W$,

$$h(a) = \{x \in X(W) : a \in x_1\}.$$

Our aim is to show that W is isomorphic to a subalgebra of $L(X(W))$.

To begin, we introduce the following auxiliary notation. Given an LP algebra, for any $A, B \subseteq W$ denote

$$A \odot B = \{a \odot b : a \in A\ \&\ b \in B\}.$$

First, note the following:

Lemma 25. *Let an LP algebra* $(W, \wedge, \vee, 0, 1, \odot, 1')$ *be given and let* F *and* I *be a filter and an ideal of* W, *respectively. Then*

$$U = \{a \in W : (\{a\} \odot F) \cap I \neq \emptyset\}$$
$$V = \{a \in W : (F \odot \{a\}) \cap I \neq \emptyset\}$$

are ideals of W.

Proof. We show that U is an ideal of W. Let **(1)** $a \in U$ and **(2)** $b \leqslant a$. From the definition of U, **(1)** implies that there is $c \in F$ such that **(3)** $a \odot c \in I$. By Lemma 14(i), **(2)** implies **(4)** $b \odot c \leqslant a \odot c$. Since I is a ideal, **(3)** and **(4)** give **(5)** $b \odot c \in I$. So for some $c \in F$, **(5)** holds, which gives $b \in U$.

Let $a, b \in U$. We shall show that $a \vee b \in U$. By assumption, there exist $c, d \in F$ such that **(6)** $a \odot c \in I$ and **(7)** $b \odot d \in I$. Since $c \wedge d \leqslant c$ and $c \wedge d \leqslant d$, by Lemma 14(i) we have $a \odot (c \wedge d) \leqslant a \odot c$ and $b \odot (c \wedge d) \leqslant b \odot d$, so by **(6)** and **(7)** we get $a \odot (c \wedge d) \in I$ and $b \odot (c \wedge d) \in I$. Then $(a \odot (c \wedge d)) \vee (b \odot (c \wedge d)) \in I$, since I is an ideal. From axiom **(P.4)**, $(a \odot (c \wedge d)) \vee (b \odot (c \wedge d)) = (a \vee b) \odot (c \wedge d)$, so we get $(a \vee b) \odot (c \wedge d) \in I$. Since F is a filter, $c \wedge d \in F$. We have shown then that for some $c' = c \wedge d \in F$, $(a \vee b) \odot c' \in I$, which by the definition of U implies that $a \vee b \in U$.

Proceeding in the similar way we can show that V is an ideal. ∎

Theorem 8 (Representation theorem for LP algebras). *Every LP algebra is isomorphic to a subalgebra of the complex algebra of its canonical frame.*

Proof. In view of Theorem 1 it suffices to show that

(i) $h(1') = \mathbf{1}'$
(ii) $h(a \odot b) = h(a) \otimes h(b)$.

(i) Note that by (20), $I^* = h(1')$, so from Theorem 1(ii), it is an l–stable set. Also, $\mathbf{1}' = l(r(I^*))$ by (16). Hence $\mathbf{1}' = I^*$, i.e. $h(1') = \mathbf{1}'$.

(ii) ($\subseteq$) Let $a, b \in W$, $z \in X(W)$ and assume that $z \in h(a \odot b)$. Then it holds **(ii.1)** $a \odot b \in z_1$. We have to show that $z \in h(a) \otimes h(b)$. By the definition (15) and Lemma 17, this means that for any $w \in X(W)$, if $z \preccurlyeq_1 w$, then $w \notin h(a) \odot_s h(b)$. Assume that $z \preccurlyeq_1 w$, i.e. **(ii.2)** $z_1 \subseteq w_1$. We will show that **(ii.3)** $w \notin h(a) \odot_s h(b)$.

Let $[a)$ be the filter generated by a, i.e. $[a) = \{e \in W : a \leqslant e\}$. Define

$$U = \{c \in W : ([a) \odot \{c\}) \cap w_2 \neq \emptyset\}.$$

By Lemma 25, U is an ideal. We show that **(ii.4)** $b \notin U$. Suppose that $b \in U$. Then there exists $e \in [a)$ such that **(ii.5)** $e \odot b \in w_2$. Since $e \in [a)$, $a \leqslant e$, so $a \odot b \leqslant e \odot b$ by Lemma 14(ii). Then, since w_2 is an ideal, by **(ii.5)** we get $a \odot b \in w_2$, so $a \odot b \notin w_1$, which by **(ii.2)** gives $a \odot b \notin z_1$ – a contradiction with **(ii.1)**.

Then $([b), U) \in FIP(W)$. Let (y_1, y_2) be its extension to the maximal filter–ideal pair. Hence **(ii.6)** $[b) \subseteq y_1$ and **(ii.7)** $U \subseteq y_2$. From **(ii.6)**, $b \in y_1$, so **(ii.8)** $y \in h(b)$.

Next, let us consider the set

$$V = \{c \in W : (\{c\} \odot y_1) \cap w_2 \neq \emptyset\}.$$

By Lemma 25, V is an ideal. We show that (ii.9) $a \notin V$. Suppose that $a \in V$. Then there exists $e \in L$ such that (ii.10) $e \in y_1$ and (ii.11) $a \odot e \in w_2$. By the definition of U, (ii.11) means that $e \in U$, so by (ii.7), $e \in y_2$. Whence $e \notin y_1$, which contradicts (ii.10). So (ii.9) was proved. Then $([a), V) \in FIP(W)$. Let (x_1, x_2) be its extension to the maximal filter–ideal pair. Then we have (ii.12) $[a) \subseteq x_1$ and (ii.13) $V \subseteq x_2$. From (ii.12), $a \in x_1$, so $x \in h(a)$, which implies (ii.14) $x \notin r(h(a))$. Finally, we show that (ii.15) $S^\star(x, y, w)$. By the definition (18) of $S^\star$, this means that for all $c, d \in W$, $c \notin x_2$ and $d \in y_1$ imply $c \odot d \notin w_2$. Let $c, d \in W$ be such that $c \notin x_2$ and $d \in y_1$. It suffices to show that $c \odot d \notin w_2$. Indeed, $c \notin x_2$ implies by (ii.13) that $c \notin V$. From the definition of V, this gives that for any $e \in y_1$, $c \odot e \notin w_2$, so in particular $c \odot d \notin w_2$. Therefore, for some $x, y \in X(W)$, (ii.8), (ii.14) and (ii.15) hold – by the definition (14) of $\odot_s$ (ii.3) follows.

($\supseteq$) Let $a, b \in L$ and let $z \in X(W)$. Assume that $z \in h(a) \otimes h(b)$. By the definition (15) this is equivalent to (ii.16) $z \in l(h(a) \odot_Q h(b))$. Let $y \in X(W)$ be such that (ii.17) $z_1 \subseteq y_1$. Then from (ii.16), $y \notin h(a) \odot_Q h(b)$, which means that there exist $x, w \in X(W)$ such that (ii.18) $Q^\star(x, w, y)$, (ii.19) $x \in h(a)$ and (ii.20) $w \notin r(h(b))$. From (ii.19), $a \in x_1$. Next, by Theorem 1(i), (ii.20) implies (ii.21) $b \notin w_2$. By the definition (19), (ii.18) gives that for all $a, b \in W$, $a \odot b \in y_2$ and $a \in x_1$ imply $b \in w_2$. Hence, from (ii.19) and (ii.21), $a \odot b \notin y_2$. Applying again Theorem 1(i), we obtain $y \notin r(h(a \odot b))$, which together with (ii.17) gives $z \in l(r(h(a \odot b)))$. Since by Theorem 1(ii), $h(a \odot b)$ is l–stable, we finally get $z \in h(a \odot b)$. ∎

6   LCP Algebras

LCP algebras are meant to be relation algebras based on arbitrary bounded lattices. Their axioms consist of the axioms (C.1) and (C.2) of converse, the axioms (P.1),...,(P.4) of product and an additional axiom which tells us how converse and product are related with each other. In the axiomatization of classical relation algebras, this is done by postulating that converse distributes over composition and also by the axiom

$$a \odot - (a^\smile \odot - b) \leqslant b, \tag{21}$$

where $-x$ is the Boolean complement of x. It is well known that (21) is equivalent to de Morgan's *Theorem K*, one form of which states that

$$a \odot b \leqslant -c \text{ iff } a^\smile \odot c \leqslant -b \text{ iff } c \odot b^\smile \leqslant -a. \tag{22}$$

In our present setting, we do not have complementation as a distinguished operation, and thus, we cannot use (21). It may be argued that one could use the complement free version of (22), namely,

$$(a \odot b) \wedge c = 0 \text{ iff } (a^\smile \odot c) \wedge b = 0 \text{ iff } (c \odot b^\smile) \wedge a = 0. \tag{23}$$

However, it is not quite clear whether this is useful because of the following: Suppose that W is an LCP algebra (formally defined below), and let $m \notin W$. Set $W' = W \cup \{m\}$, and extend ordering and the operations of W over W' by

$$m \leqslant x, \quad m^{\smile} = m, \quad m \odot x = x \odot m = m$$

for all $x \in W'$. In other words, we are adding a new smallest element to W.

Lemma 26.

1. W' is an LCP algebra which satisfies (23)
2. If $\tau = \sigma$ is an equation of LCP algebras not containing 0, then

$$W \models \tau = \sigma \quad \text{iff} \quad W' \models \tau = \sigma.$$

Proof. The idea is to show that the equational classes generated by W and W' are the same when 0 is omitted in the signature. Suppose that $Eq(W)$ and $Eq(W')$ are these classes. It suffices to show that $W \in Eq(W')$ and $W' \in Eq(W)$. The first claim follows from the fact that W is a subalgebra of W'. The second claim can be seen as follows. Let $f : W' \to W \times W$ be defined by

$$f(x) = \begin{cases} \langle x, 1 \rangle, & \text{if } x \neq m, \\ \langle 0, 0 \rangle, & \text{if } x = m. \end{cases}$$

Then, f is an injective homomorphism, showing that $W' \in ISP(W)$. ∎

It may also be worthy to note that (21) is equivalent to the statement

$$-(a^{\smile} \odot - b) \text{ is the largest } x \text{ with } a \odot x \leqslant b,$$

i.e. $-(a^{\smile} \odot -b)$ is the (right) residuum of $\odot$. Hence, a more coherent way would be to introduce residua as e.g. in [23] as distinguished operators.

Let us formally define LCP algebras.

Definition 16. *An **LCP algebra** is a system* $(W, \wedge, \vee, 0, 1, {}^{\smile}, \odot, 1')$ *such that* $(W, \wedge, \vee, 0, 1, {}^{\smile})$ *is an LC algebra,* $(W, \wedge, \vee, 0, 1, \odot, 1')$ *is an LP algebra and for all* $a, b \in W$ *the following holds:*

    (CP) $(a \odot b)^{\smile} = b^{\smile} \odot a^{\smile}.$ □

Note that

Lemma 27. *For every LCP algebra* $(W, \wedge, \vee, 0, 1, {}^{\smile}, \odot, 1')$, $1'$ *is an equivalence element.*

Proof. We have to show that $1'$ is transitive and symmetric. Transitivity follows from $1' \odot 1' = 1'$, and symmetry can be shown as follows:

$$
\begin{aligned}
1'^\smile &= 1'^\smile \odot 1', && \text{by (P.1)} \\
&= (1'^\smile \odot 1')^{\smile\smile}, && \text{by (C.1)} \\
&= (1'^\smile \odot 1')^\smile, && \text{by (CP) and (C.1)} \\
&= 1'^{\smile\smile}, && \text{by (P.1)} \\
&= 1', && \text{by (C.1).} \quad \blacksquare
\end{aligned}
$$

Example 1. Let $(W, \wedge, \vee, 0, 1, \smile, 1')$ be a system such that $W = \{0, a, b, c, d, e, 1\}$, $1' = b$ and the operations $\smile$ and $\odot$ are given in Tables 1 and 2 below, respectively.

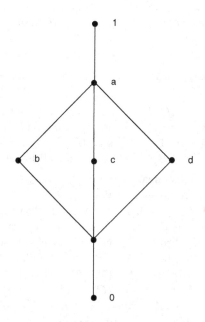

Fig. 1. An example.

Table 1. Converse $\smile$ of W

x	$x^\smile$
0	0
a	a
b	b
c	d
d	c
e	e
1	1

Table 2. Product $\odot$ of W

$\odot$	0	a	b	c	d	e	1
0	0	0	0	0	0	0	0
a	0	a	a	c	d	e	1
b	0	a	b	c	d	e	1
c	0	c	c	c	e	e	c
d	0	d	d	0	d	0	d
e	0	e	e	0	e	0	e
1	0	1	1	c	d	e	1

By straightforward verification it is easy to check that W is an LCP algebra. Note that the underlying lattice is not distributive and product of W is non–commutative $(c \odot e = e \neq 0 = e \odot c)$. $\qquad\Box$

As before, we start our discussion on LCP algebras by defining *LCP frames*.

Definition 17. *An **LCP frame** is a system $(X, \leqslant_1, \leqslant_2, C, R, S, Q, I)$ such that $(X, \leqslant_1, \leqslant_2, C)$ is an LC frame, $(X, \leqslant_1, \leqslant_2, R, S, Q, I)$ is an LP frame and moreover, for all $x, y, z \in X$,*

$\qquad$ **(SCP)** $Q(x, y, z) = S(C(y), C(x), C(z))$. $\qquad\Box$

Definition 18. *Let $(X, \preccurlyeq_1, \preccurlyeq_2, C, R, S, Q, I)$ be an LCP frame. A **complex algebra of** X is a structure $(L(X), \sqcap, \sqcup, 0, 1, ^{\curlyvee}, \otimes, 1')$ with operations defined by (3)–(6), (11), (15) and (16).* $\qquad\Box$

Theorem 9. *The complex algebra of an LCP frame is an LCP algebra.*

Proof. Let $(X, \preccurlyeq_1, \preccurlyeq_2, C, R, S, Q, I)$ be an LCP frame and let $(L(X), \sqcap, \sqcup, 0, 1, ^{\curlyvee}, \otimes, 1')$ be its complex algebra. By Theorems 4 and 6, $(L(X), \sqcap, \sqcup, 0, 1, ^{\curlyvee})$ is an LC algebra and $(L(X), \sqcap, \sqcup, 0, 1, \otimes, 1')$ is an LP algebra. It suffices to show that **(CP)** holds in $L(X)$, i.e. $(A \otimes B)^{\curlyvee} = B^{\curlyvee} \otimes A^{\curlyvee}$ for all l–stable sets $A, B \subseteq X$.

($\subseteq$) Let $z \in (A \otimes B)^{\curlyvee}$. This means that $C(z) \in A \otimes B$, or equivalently, **(1)** $C(z) \in l(A \odot_Q B)$. Let $z' \in X$ be such that **(2)** $C(z) \leqslant_1 z'$. By **(MC.1)** and **(SC)** this is equivalent to **(3)** $z \leqslant_1 C(z')$. From **(1)** and **(2)** we get $z' \notin A \odot_Q B$. By the definition (13) this means that there exist $x, y \in X$ such that **(4)** $Q(x, y, z')$, **(5)** $x \in A$ and **(6)** $y \notin r(B)$. From **(6)**, there exists $y' \in X$ such that **(7)** $y \leqslant_2 y'$ and **(8)** $y' \in B$. Next, **(5)** implies **(9)** $C(x) \in A^{\curlyvee}$. Similarly, from **(8)** it follows $C(y') \in B^{\curlyvee}$, so **(10)** $C(y') \notin r(B^{\curlyvee})$. By **(MP.3)**, **(4)** and **(7)** imply $Q(x, y', z')$, which by **(SCP)** is equivalent to **(11)** $S(C(y'), C(x), C(z'))$. Hence, by **(9)** and **(10)** we get that $C(z') \notin B^{\curlyvee} \odot_S A^{\curlyvee}$. Since A and B are l–stable, by Lemma 11(iii) $A^{\curlyvee}$ and $B^{\curlyvee}$ are also l–stable. Hence, from Lemma 17, $B^{\curlyvee} \odot_S A^{\curlyvee} = B^{\curlyvee} \odot_Q A^{\curlyvee}$ which together with **(3)** gives $z \in l(B^{\curlyvee} \odot_Q A^{\curlyvee})$, that is $z \in B^{\curlyvee} \otimes A^{\curlyvee}$.

The proof of ($\supseteq$) is similar. $\qquad\blacksquare$

Let $(W, \wedge, \vee, 0, 1, ^{\smallsmile}, \odot, 1')$ be an LCP algebra. As before, by a *filter* (*ideal*) of W we mean a filter (ideal) of the underlying lattice $(W, \wedge, \vee, 0, 1)$. We will write $X(W)$ to denote the family of all maximal filter–ideal pairs of the lattice reduct of W.

Definition 19. *Let $(W, \wedge, \vee, 0, 1, ^{\smallsmile}, \odot, 1')$ be an LCP algebra. A **canonical frame of** W is a structure $(X(W), \preccurlyeq_1, \preccurlyeq_2, C^{\star}, R^{\star}, S^{\star}, Q^{\star}, I^{\star})$ such that $\preccurlyeq_1$ and $\preccurlyeq_2$ are defined by (7) and (8), respectively, and relations $C^{\star}$, $R^{\star}$, $S^{\star}$, $Q^{\star}$ and $I^{\star}$ are defined by (12) and (17)–(20), respectively.* $\qquad\Box$

Theorem 10. *The canonical frame of an LCP algebra is an LCP frame.*

Proof. Let $(X, \wedge, \vee, 0, 1, \smile, \odot, 1')$ be an LCP algebra and let $(X(W), \preccurlyeq_1, \preccurlyeq_2,$ $C^\star, R^\star, S^\star, Q^\star, I^\star)$ be its canonical frame. By Theorem 4, $(X(W), \preccurlyeq_1, \preccurlyeq_2, C^\star)$ is an LC frame. Next, by Theorem 7, $(X(W), \preccurlyeq_1, \preccurlyeq_2, R^\star, S^\star, Q^\star, I^\star)$ is an LP frame. It suffices to show that **(SCP)** holds.

Let $x, y, z \in X(W)$. We have the following equivalences:

$S^\star(C^\star(y), C^\star(x), C^\star(z))$
iff $(\forall a, b \in W) \ a \odot b \in z_2^\smile \ \& \ b \in x_1^\smile \Rightarrow a \in y_2^\smile$ by (18)
iff $(\forall a, b \in W) \ (a \odot b)^\smile \in z_2 \ \& \ b^\smile \in x_1 \Rightarrow a^\smile \in y_2$ by (10)
iff $(\forall a, b \in W) \ b^\smile \odot a^\smile \in z_2 \ \& \ b^\smile \in x_1 \Rightarrow a^\smile \in y_2$ by axiom **(CP)**
iff $Q^\star(x, y, z)$ by (19). ∎

We conclude the paper by stating the representability of LCP algebras.

Theorem 11. *Every LCP algebra is isomorphic to a subalgebra of the complex algebra of its canonical frame.*

Proof. Follows from Theorems 5, 8, 9 and 10. ∎

7 Conclusions

In this paper we have studied not necessarily distributive lattices with the operators which are the abstract counterparts to the converse and composition of binary relations. On the algebraic side, we have presented relational representation theorems for these classes of algebras. The representation theorems are obtained by suitable extensions of the Urquhart representation theorem for lattices [26]. However, here we stress the relational aspect of representability, and we omit the topological aspect. On the logical side, with every class of algebras studied in the paper we have associated an appropriate class of frames. These frames constitute a basis of a Kripke-style semantics for the logics whose algebraic semantics is determined by the classes of algebras presented in the paper. The representation theorems would enable us to prove completeness of the logics. For a detailed elaboration of the respective relational logics one can follow the developments in [1] and [24].

The further work is planned on the class of relation algebras based on double residuated lattices presented in [23]. The signature of such algebras extends the signature of LCP algebras with the residuation operators determined by the product, and with the sum and its dual residua. In these algebras complement operations are definable in terms of residua. Therefore the axioms of this class of algebras should include some counterparts to the De Morgan theorem K.

An interesting axiomatic extension of the presented algebras would be LCP with the modular law relating product and converse operations. These problems will be discussed in the ongoing paper.

References

1. Allwein, G and Dunn, J. M. (1993). Kripke models for linear logic. *J. Symb. Logic*, **58**, 514–545.
2. Allwein, G. (2003). Private communication.
3. Chin, L. and Tarski, A. (1951). Distributive and modular laws in the arithmetic of relation algebras. *University of California Publications in Mathematics* **1**, 341–384.
4. Comer, S. (1991). An algebraic approach to the approximation of information. *Fundamenta Informaticae* **14**, 492–502.
5. Comer, S. (1993). On connections between information systems, rough sets and algebraic logic. In: Rasiowa, H. and Rauszer, C. (eds), *Algebraic Methods in Logic, Algebra and Computer Science*, Banach Center Publications, Warsaw.
6. Demri, S. and Orłowska, E. (2002). *Incomplete Iformation: Structure, Inference, Complexity*. EATCS Monographs in Theoretical Computer Science, Springer.
7. Dilworth, R. P. and Ward, N. (1939). Residuated lattices. *Transactions of the American Mathematical Society* **45**, 335–354.
8. Dunn, J. M. (2001). Gaggle theory: an abstraction of Galois connections and residuation, with application to nagation, implication and various logical operators. *Lecture Notes in Artificial Intelligence* **478**, Springer–Verlag, 31–51.
9. Düntsch, I. (1994). Rough relation algebras. *Fundamenta Informaticae* **21**, 321–331.
10. Düntsch, I. and Orłowska, E. (2001). Beyond modalities: sufficiency and mixed algebras. In: Orłowska, E. and Szałas, A. (eds) *Relational Methods for Computer Science Applications*, Physica–Verlag, Heidelberg, 263–285.
11. Düntsch, I. and Orłowska, E. (2002). Boolean algebras arising from information systems. *Proceedings of the Tarski Centenary Conference*, Banach Centre, Warsaw, June 2001. To appear.
12. Esteva. F. and Godo, L. (2001). Monoidal t–norm based logic: towards a logic for left–continuous t–norms. *Fuzzy Sets and Systems* **124**, 271–288.
13. Esteva, F. and Godo, L. (2001). On complete residuated many-valued logics with t-norm conjunction. *Proceedings of the 31st International Symposium on Multiple-Valued Logic*, Warsaw, Poland, 81–86.
14. Gehrke, M. and Jónsson, B. (1994). Bounded distributive lattices with operators. *Mathematica Japonica* **40**, No 2, 207–215.
15. Grätzer, G. (1978). *General Lattice Theory*. Basel Birkhäuser.
16. Hajek, P. (1998). *Metamathematics of Fuzzy Logic*, Kluwer, Dordrecht.
17. Höhle, U. (1996). Commutative, residuated l-monoids. In [18], 53–106.
18. Höhle, U. and Klement, U. P. (eds) (1996). *Non–Classical Logics and their Applications to Fuzzy Subsets*, Kluwer, Dordrecht.
19. Jónsson, B. and Tarski, A. (1951) Boolean algebras with operators. Part I. *American Journal of Mathematics* **73**, 891–936.
20. Kawahara, Y. and Furusawa, H. (1999) An algebraic formalization of fuzzy relations. *Fuzzy Sets and Systems* **101**, 125–135
21. Orłowska, E. (1995). Information algebras. *Lecture Notes in Computer Science*. **639**, Proceedings of *AMAST'95*, Montreal, Canada, 50–65.
22. Orłowska, E. and Radzikowska, A. M. (2001). Information relations and opeators based on double residuated lattices. *Procedings of the 6th International Workshop on Relational Methods in Computer Science RelMiCS'01*, Oisterwijk, Netherlands, 185–199.

23. Orłowska, E. and Radzikowska, A. M. (2002). Double residuated lattices and their applications. *Relational Methods in Computer Science*, H. C. M. de Swart (ed), Lecture Notes in Computer Science **2561**, Springer–Verlag, Heidelberg, 171–189.

24. Orłowska, E. and Vakarelov, D. (2003). Lattices with modal operators and lattice–based modal logic. Proceedings of the 12th International Congress of Logic, Methodology and Philosophy of Science, Oviedo, August 2003, to appear. Abstract in the Volume of Abstracts, 22–23.

25. Sofronie-Stokkermans, V. (2000). Duality and canonical extensions of bounded distributive lattices with operators, and applications to the semantics of non-classical logics. Studia Logica **64**, Part I pp. 93-122, Part II, 151–172.

26. Urquhart, A. (1978). A topological representation theorem for lattices. Algebra Universalis **8**, 45–58.

Binary Multirelations

Ingrid Rewitzky

Department of Mathematics and Applied Mathematics,
University of Cape Town, South Africa
`rewitzky@maths.uct.ac.za`

Abstract. Relational models for imperative programming languages
provide a representation of commands in terms of binary input-output
relations over states. Various relational models have arisen from modelling decisions on the distinction between angelic- and demonic nondeterminism, and have been shown to be isomorphic to disjunctive- or
conjunctive predicate transformer semantics. For commands with both
angelic- and demonic nondeterminism it is known that monotone unary
operators provide a predicate transformer semantics but there is no conventional relational model. In this paper we propose a novel relational
representation, in terms of binary multirelations, for such commands.
Then we show that binary multirelations and monotone unary operators
are intertranslatable.

1   Introduction

Recent extensions [Mor88,BaW89,Nel89,BaW92] of Dijkstra's guarded command
language [Dij75,Dij76] allow commands with both angelic- and demonic nondeterminism. A predicate transformer semantics of such extended languages
can easily be defined in terms of monotone unary operators. However, no relational model has been provided. Since known relational models capture either
angelic- or demonic nondeterminism, and binary relations are Jónsson/Tarski
dual to either disjunctive- or conjunctive predicate transformers [ReB95,BrR01],
we already know that there are no conventional binary relations corresponding to monotone predicate transformers. In this paper we propose the notion
of binary multirelation as a relational representation of commands with both
angelic- and demonic nondeterminism. The idea for the multirelational representation arose from the games of choice used for discussing nondeterminism in
[Hes94,BaW95,Hoa96,BaW98], as follows. Execution of a command with angelic-
and demonic nondeterminism may be viewed as a two-player two-step game of
choice. We may refer colloquially to the players as 'the user' and 'the machine'.
The game is lost by the player who is faced with the choice from an empty set.
It is assumed that the user will take any available opportunity to win, while the
machine assumes the role of a devil's advocate who is trying to make sure the
user loses. The user makes the first move by selecting a set of possible winning
positions, and the machine makes the second move by selecting the actual final
position from this set. Thus a loss for the user is the empty family, and the user

H. de Swart et al. (Eds.): TARSKI, LNCS 2929, pp. 256–271, 2003.

has the opportunity of winning the game whenever the family of choices contains the empty set. From each starting position the user may have the choice of more than one set of possible winning positions, therefore the game determines a binary relation between starting positions and sets of possible winning positions. Such binary relations from a set to a set of sets, may be seen as generalising multifunctions analogous to the way binary relations from a set to (in general another) set generalise functions. Accordingly, we will call them *binary multirelations*. Keeping this game-theoretic intuition in our minds, but without going into the depths of game theory, we proceed as follows. In Section 2 we introduce the notion of *binary multirelation*, and define multirelation forming operations on binary multirelations in terms of the usual set-theoretic operations and their power operations [Bri93]. Using results from lattice theory [DaP90] we then investigate the algebraic structure of different families of binary multirelations. In Section 3 we observe that certain up-closed binary multirelations correspond to binary relations and hence, by the Jónsson/Tarski translations [JoT51,JoT52], correspond to certain monotone unary operators. This suggests that there should be a bijective correspondence between up-closed binary multirelations and monotone unary operators. We show in Section 4 that this is the case. As a consequence we generalise the factorisation result [GMM94] for monotone unary operators which lies at the heart of the simulation rule technique for developing or verifying an implementation against its more abstract specification.

2 An Algebra of Binary Multirelations

As an example to motivate the definition of the notion of binary multirelation consider the following program α, that allows both angelic and demonic choice

$$((x := x - 1 \sqcap x := x + 1); x := x - 1) \sqcup (x := x + 1; (x := x - 1 \sqcap x := x + 1)),$$

where $\sqcap$ denotes angelic choice, $\sqcup$ denotes demonic choice, ; denotes sequential composition. During an execution of α, the user is given the choice between

$$\alpha_1 : \quad (x := x - 1 \sqcap x := x + 1); x := x - 1$$
$$\alpha_2 : \quad x := x + 1; (x := x - 1 \sqcap x := x + 1),$$

where both α_1 and α_2 in turn provide the machine with the choice of two statements to execute. Suppose the state space is the set of integers $\mathbb{Z}$. Then α_1 and α_2 and may be represented respectively by the following binary relations

$$r_{\alpha_1} = \{(i, i - 2) \mid i \in \mathbb{Z}\} \cup \{(i, i) \mid i \in \mathbb{Z}\}$$
$$r_{\alpha_2} = \{(i, i) \mid i \in \mathbb{Z}\} \cup \{(i, i + 2) \mid i \in \mathbb{Z}\}.$$

To capture the angelic choice between α_1 and α_2, we use pairs

$$(i, r_{\alpha_1}(i)) \quad \text{and} \quad (i, r_{\alpha_2}(i)) \qquad \text{for } i \in \mathbb{Z},$$

where $r_{\alpha_n}(i) = \{j \in \mathbb{Z} \mid (i,j) \in r_{\alpha_n}\}$ $(n = 0, 1)$. Hence, α may be represented as the binary relation $R_\alpha \subseteq \mathbb{Z} \times \mathcal{P}(\mathbb{Z})$ given by:

$$R_\alpha = \{(i, \{i - 2, i\}) \mid i \in \mathbb{Z}\} \cup \{(i, \{i, i + 2\}) \mid i \in \mathbb{Z}\}.$$

More generally, let $\mathbf{S}$ be the set of all states regarded as infinite vectors of values of a countable collection of program variables. A program α, that allows both demonic and angelic choice, may be represented as a binary relation $R_\alpha \subseteq \mathbf{S} \times \mathcal{P}(\mathbf{S})$, the idea being that for $s \in \mathbf{S}$ and $Q \in \mathcal{P}(\mathbf{S})$,

$sR_\alpha Q$ iff   program α, when started in state s,
is guaranteed to terminate in a state in which Q holds,
and, every state in Q is a possible outcome of α.

For each $s \in \mathbf{S}$,

$$R_\alpha(s) = \{Q \in \mathcal{P}(\mathbf{S}) \mid sR_\alpha Q\}$$

captures the angelic choices available to the user, and for each $Q \in R_\alpha(s)$,

$$\{t \in \mathbf{S} \mid t \in Q\}$$

captures the demonic choices available to the machine. We now formalise this representation in terms of the notion of a binary multirelation.

Definition 1. *Let X and Y be sets. A binary multirelation is a subset of the Cartesian product $X \times \mathcal{P}(Y)$, that is, a set of ordered pairs (x, Q) where $x \in X$ and $Q \subseteq Y$. Mostly we will deal with the case of $X = Y = \mathbf{S}$.*

As subsets of $\mathbf{S} \times \mathcal{P}(\mathbf{S})$, binary multirelations can be partially ordered by superset inclusion $\supseteq$ where

$$R \supseteq S \quad \text{iff} \quad (\forall s \in \mathbf{S})[R(s) \supseteq S(s)],$$

with the intuition that program S is 'better' than program R if the user has less choice in S than in R. Since the image set of any $s \in \mathbf{S}$ under any binary multirelation is an element of $\mathcal{P}(\mathcal{P}(\mathbf{S}))$ the pointwise extensions of the *power operations* [Bri93,BrR01] of the operations on the powerset Boolean algebra $\mathcal{P}(\mathcal{S}) = (\mathcal{P}(\mathbf{S}), \cup, \cap, ^-, \emptyset, \mathbf{S})$ give rise to further operations on binary multirelations. A pre-order on binary multirelations is given by the *upper power order*, in the sense of [Bri93], of set inclusion:

$$R \subseteq^+ S \quad \text{iff} \quad (\forall s \in \mathbf{S})(\forall P \in R(s))(\exists Q \in S(s))[Q \subseteq P]$$

with the intuition that program S is 'better' than program R if Q always allows the user to choose an equal or smaller set than P does. For any $X \subseteq \mathbf{S}$, we define the upclosure of X by $\uparrow X = \{y \in \mathbf{S} \mid (\exists x \in X)[x \leq y]\}$. Then

Lemma 1. *For any binary multirelations $R, S \subseteq \mathbf{S} \times \mathcal{P}(\mathbf{S})$,*

$$R \subseteq^+ S \quad \text{iff} \quad (\forall s \in \mathbf{S})[\uparrow R(s) \subseteq \uparrow S(s)].$$

Proof. By definition of $\subseteq^+$, $R \subseteq^+ S$ iff $(\forall s \in \mathbf{S})[R(s) \subseteq \uparrow S(s)]$. Now, for any $s \in \mathbf{S}$,

$$R \subseteq^+ S \text{ and } P \in \uparrow R(s) \Rightarrow (\forall x \in \mathbf{S})[R(x) \subseteq \uparrow S(x)] \text{ and } (\exists P' \in R(s))[P' \subseteq P]$$
$$\text{by definition of } \subseteq^+ \text{ and } \uparrow(\cdot)$$
$$\Rightarrow (\exists Q \in S(s))[Q \subseteq P]$$
$$\text{first order logic}$$
$$\Rightarrow P \in \uparrow S(s)$$
$$\text{by definition of } \uparrow(\cdot). \qquad \square$$

The usual set-theoretic operations of *union* ($\cup$), *intersection* ($\cap$) and *complement* ($^-$) with respect to $\mathbf{S}$ can be used to define operations of union, intersection, complement for binary multirelations, as follows. For any two binary multirelations R and S, their *power union* is defined by

$$R \cup^+ S = \{(s, Q_1 \cup Q_2) \mid sRQ_1 \text{ and } sSQ_2\},$$

and may be interpreted as a program in which the user is asked for two independent choices of programs to be executed and then the machine nondeterministically decides which to execute. Similarly, the *power intersection* of two binary multirelations R and S is defined by

$$R \cap^+ S = \{(s, Q_1 \cap Q_2) \mid sRQ_1 \text{ and } sSQ_2\},$$

and may be interpreted as a program in which the user is asked for two independent choices of programs to be executed and then the program terminates in a state in which both the chosen programs may terminate. For any binary multirelation R, we may also define the *power negation* of R by

$$\overline{R}^+ = \{(s, \overline{Q}) \mid sRQ\},$$

and interpret it as a program in which the user chooses a program not to be executed by the machine. Combining set complementation and power negation, we can define a notion of the *dual* of a binary multirelation R by

$$R^o = \overline{\overline{R}^+} = \{(s, \overline{Q}) \mid (s, Q) \notin R\}.$$

Note that the converse of a binary multirelation is, in general, not again a binary multirelation. With respect to the ordering $\subseteq^+$, the smallest (or worst) multirelation is the *universal* multirelation T^+ given by

$$T^+ = \{(s, Q) \mid s \in \mathbf{S} \text{ and } Q \in \mathcal{P}(\mathbf{S})\}$$

and which may be interpreted as a program which is always enabled. The largest (or best) is the *null* multirelation $\perp^+$ given by

$$\perp^+ = \{(s, \emptyset) \mid s \in \mathbf{S}\}$$

and which may be interpreted as a program which is never enabled. The *identity* multirelation I^+ is given by

$$I^+ = \{(s, \{s\}) \mid s \in \mathbf{S}\}.$$

Multirelational composition may be defined as follows: for any two binary multirelations R and S,

$$R; S = \{(s, Q) \mid (\exists Y)[sRY \text{ and } Y \subseteq \{y \mid ySQ\}]\}.$$

Note that relational composition for relations $r, s \subseteq \mathbf{S} \times \mathbf{S}$, namely

$$r; s = \{(x, z) \mid (\exists y)[xry \text{ and } ysz]\} = \{(x, z) \mid (\exists y)[xry \text{ and } y \in \{u \mid usz\}]\}$$

corresponds to multirelational composition. Furthermore, composition for monotone predicate transformers will be shown to correspond to multirelational composition on page 269. Binary multirelations have many interesting and useful properties. Here are some of them.

Definition 2. *Let $R \subseteq \mathbf{S} \times \mathcal{P}(\mathbf{S})$ be a binary multirelation. Then*

(a) *R is proper if, for each $s \in \mathbf{S}, R(s) \neq \emptyset$.*
(b) *R is total if, for each $s \in \mathbf{S}, \emptyset \notin R(s)$.*
(c) *R is up-closed if, for each $s \in \mathbf{S}, R(s)$ is an up-set with respect to set inclusion $\subseteq$ (that is, $Q \in R(s)$ and $Q \subseteq Q_1$ imply $Q_1 \in R(s)$).*
(d) *R is multiplicative (or an intersection structure) if, for each $s \in \mathbf{S}$ and any non-empty indexed set $\{Q_i\}_{i \in I}$ of subsets of $\mathbf{S}$,*

$$\bigcap \{Q_i \mid Q_i \in R(s)\} \in R(s).$$

(e) *R is additive if, for each $s \in \mathbf{S}$ and any non-empty indexed set $Q_{i_{i \in I}}$ of subsets of $\mathbf{S}, \cup_{i \in I} Q_i \in R(s)$ implies $Q_i \in R(s)$ for some $i \in I$.*

The next theorem provides some distributivity properties binary multirelational composition; the proofs are easy and left for the reader.

Theorem 1. *For any binary multirelations R, S, T,*

(a) *$\top^+; R = \top^+$. $R; \top^+ = \top^+$, if R is proper.*
(b) *$S \subseteq T$ implies $S \cap R \subseteq T \cap R$ and $S \cup R \subseteq T \cup R$. $S \subseteq T$ implies $S; R \subseteq T; R$ and $R; S \subseteq R; T$.*
(c) *$(S \cap T); R \subseteq S; R \cap T; R$ $R; (S \cap T) = R; S \cap R; T$, if R is multiplicative.*
(d) *$(S \cup T); R = S; R \cup T; R$. $R; (S \cup T) = R; S \cup R; T$, if R is additive.* □

In the remainder of this section we investigate the lattice-theoretic properties of different families of binary multirelations. Before doing so, we recall some lattice-theoretic notions for complete lattices. [Note that these definitions are included for ease of reference; further details may be found in, e.g., [BeS71,GHK80,DaP90] and in the original sources referenced in these standard texts on lattice theory.]

[**Interlude**] A complete sublattice of the lattice of all subsets of a set is known as a *complete ring of sets*. Let $(L, \vee, \wedge)$ be a complete lattice. An element x of L is *completely join-irreducible* if, for any non-empty indexed set $\{y_i\}_{i \in I}$ of elements of L,

$$x = \bigvee_{i \in I} y_i \quad \text{implies} \quad x = y_i \quad \text{for some } i \in I.$$

A subset X of L is *join-dense* if every element of L is a join of elements from X. A *completely meet-irreducible* element and a *meet-dense* subset of L are defined dually. An element k of L is *finite* (or *compact*) if, for every subset X of L,

$$k \leq \bigvee X \quad \text{implies} \quad k \leq \bigvee Y \quad \text{for some finite subset } Y \text{ of } X.$$

A complete lattice L is said to be *algebraic* if, for each $a \in L$,

$$a = \bigvee \{k \in L \mid k \leq a \quad \text{and} \quad k \text{ is finite}\}.$$

A family $\mathcal{X}$ of subsets of a set X ordered by inclusion and closed under arbitrary intersections of non-empty subfamilies is called an *intersection structure*; if in addition $X \in \mathcal{X}$ then $\mathcal{X}$ is called a *topped intersection structure*. [**End of Interlude**] Two subfamilies of the family of up-closed binary multirelations have interesting descriptions in terms of filters in $\mathcal{P}(\mathbf{S})$, namely, the subfamily of proper multiplicative up-closed binary multirelations and the subfamily of total additive up-closed binary multirelations. Take any proper multiplicative up-closed binary multirelation R. Then for each $s \in \mathbf{S}, R(s)$ is a filter in $\mathcal{P}(\mathbf{S})$ closed under arbitrary intersections and hence a principal filter in $\mathcal{P}(\mathbf{S})$ ([BeS71], p 107). That is, for each $s \in \mathbf{S}$ there is some $Y_s \in \mathcal{P}(S)$ such that:

$$R(s) \;=\; \{Q \in \mathcal{P}(\mathbf{S}) \mid Y_s \subseteq Q\}.$$

The set Y_s may be viewed as the set of outcomes of a program α, and then the set $R(s)$, being generated by Y_s, represents a program in which *there is no angelic choice* (i.e. the user is offered no choice). This means that, under our assumption of demonic nondeterminism for choices made by the machine, these multirelations capture demonic nondeterminism and hence we will refer to them as *demonic multirelations*. If the set Y_s generating $R(s)$ is finite then we call R a *finite demonic multirelation*. Dually, for any total additive up-closed binary multirelation R and any $s \in \mathbf{S}$, there is some $Y_s \subseteq \mathbf{S}$ such that:

$$\begin{aligned}
R(s) &= \{Q \in \mathcal{P}(\mathbf{S}) \mid Y_s \not\subseteq \overline{Q}\} \\
&= \{Q \in \mathcal{P}(\mathbf{S}) \mid (\exists t \in Y_s)[t \in Q]\} \\
&= \{Q \in \mathcal{P}(\mathbf{S}) \mid Q \in \bigcup_{t \in Y_s} \uparrow\!\{t\}\}.
\end{aligned}$$

Each singleton $\{t\}$ may be viewed as the set of outcomes of a deterministic program, and the set, being generated by the set of these singletons, then represents a program in which *there is no demonic choice* (i.e. the machine is offered no choice). This means that, under our assumption of angelic nondeterminism for angelic choice, these multirelations capture angelic nondeterminism and hence we will refer to them as *angelic multirelations*. A binary multirelation R which is both angelic and demonic is such that for each $s \in \mathbf{S}, R(s) = \{Q \in \mathcal{P}(\mathbf{S}) \mid t \in Q\}$, for some $t \in \mathbf{S}$, that is, for each $s \in \mathbf{S}, R(s)$ is a principal ultrafilter in $\mathcal{P}(\mathbf{S})$. These binary multirelations then represent programs in which neither the machine nor the user is offered a choice, and hence we refer to them as *deterministic multirelations*. Let $\mathbf{R}$ denote the family of all binary multirelations $R \subseteq \mathbf{S} \times \mathcal{P}(\mathbf{S}), \mathbf{R}_u$ the family of all up-closed multirelations, $\mathbf{R}_a$ the family of all angelic multirelations, $\mathbf{R}_d$ the family of all demonic multirelations, and $\mathbf{R}_t$ the family of all total multiplicative up-closed multirelations. In Theorem 2 we note that the family $\mathbf{R}_u$ of up-closed binary multirelations over $\mathbf{S}$ has a very rich lattice-theoretic structure inherited from the lattice of up-closed sets of the powerset Boolean algebra $\mathcal{P}(\mathcal{S}) = (\mathcal{P}(\mathbf{S}), \cup, \cap, ^-, \emptyset, \mathcal{P}(\mathbf{S}))$, and that subfamilies of $\mathbf{R}_u$ inherit some of this structure.

Theorem 2.

(a) *The structures* $(\mathbf{R}, \cap^+, \top^+)$ *and* $(\mathbf{R}, \cup^+, \perp^+)$ *are commutative monoids. The structures* $(\mathbf{R}_u, ;, I_u^+), (\mathbf{R}_a, ;, I_u^+)$ *and* $(\mathbf{R}_d, ;, I_u^+)$ *are monoids, where* $I_u^+ = \{(s, Q) \mid s \in Q\}$.

(b) *The family of up-closed binary multirelations* $\mathbf{R}_u$ *is a complete ring of sets in which*

$$\bigwedge \{R_i \mid i \in I\} = \bigcap_{i \in I} R_i = \overset{+}{\bigcup_{i \in I}} R_i \text{ and } \bigvee \{R_i \mid i \in I\} = \bigcup_{i \in I} R_i.$$

The bottom element is $\perp^+$ *and the top element is* $\top^+$. *The finite elements are the finite joins of demonic multirelations in* $\mathbf{R}_d$. *The completely join-irreducible elements are the demonic multirelations and the completely meet-irreducible elements are the angelic multirelations.*

(c) *The family of demonic multirelations* $\mathbf{R}_d$ *is a complete algebraic lattice in which joins of directed subfamilies of* $\{R_i\}_{i \in I}$ *and non-empty meets are as in* $\mathbf{R}_u$. *In general, joins are given by*

$$\bigvee \{R_i \mid i \in I\} = \overset{+}{\bigcap_{i \in I}} R_i = \bigcap \{R \in \mathbf{R}_d \mid \bigcup_{i \in I} R_i \subseteq R\}.$$

The bottom element is $\perp^+$ *and the top element is* $\top^+$. *The finite elements are the finite demonic multirelations.*

(d) *The family of angelic multirelations* $\mathbf{R}_a$ *is a complete lattice in which joins are as in* $\mathbf{R}_u$ *and meets are given by*

$$\bigwedge \{R_i \mid i \in I\} = \overset{+}{\bigcup_{i \in I}} R_i = \bigcup \{R \in \mathbf{R}_a \mid \bigcap_{i \in I} R_i \subseteq R\}.$$

The bottom element is $\perp^+$ *and the top element is* $\top_{\emptyset}^+$.

(e) *The family of total, multiplicative up-closed multirelations* $\mathbf{R}_t$ *is a complete meet-semilattice in which meets are as in* $\mathbf{R}_u$, *and the maximal elements are the deterministic multirelations.*

Proof.

(a) That $(\mathbf{R}, \sqcap^+, \top^+)$ and $(\mathbf{R}, \sqcup^+, \bot^+)$ are commutative monoids follows from the result by Gautam [Gau57] that a (non-trivial) equation involving the operations of an algebra remains valid of the corresponding power operations iff the individual variables in the equation occur only once on each side of the equation.

(b) Let $\mathcal{U}(\mathcal{P}(S))$ denote the lattice of up-closed subsets of $\mathcal{P}(S)$ ordered by set-inclusion. The mapping $\phi : \mathbf{R}_u \to \mathcal{P}(\mathcal{U}(\mathcal{P}(\mathbf{S})))$ given by $R \mapsto \{R(s) \mid s \in \mathbf{S}\}$ is an isomorphism, and hence $\mathbf{R}_u$ is a complete lattice with meets and joins as given above. Moreover, by ([Dav79], Proposition 1.1(i) and (ii)), $\mathcal{U}(\mathcal{P}(\mathbf{S}))$ is a complete ring of sets and hence so is $\mathbf{R}_u$. Take any demonic multirelation R. Let $\{R_i\}_{i \in I}$ be a non-empty family of up-closed multirelations such that $R = \bigcup_i R_i$. We need to show that $R = R_i$ for some $i \in I$. Take any $s \in \mathbf{S}$. Then $R(s) \neq \emptyset$ since R is proper, and hence there is some $Q \in R(s)$. By multiplicativity of $R(s), \bigcap\{Q \mid Q \in R(s)\} \in R(s)$, so $\bigcap\{Q \mid Q \in R(s)\} \in \bigcup_{i \in I} R_i(s)$ and hence $\bigcap\{Q \mid Q \in R(s)\} \in R_i(s)$ for some $i \in I$. For each $Q \in R(s), \bigcap\{Q \mid Q \in R(s)\} \subseteq Q$ and hence, since R_i is up-closed, $Q \in R_i(s)$. Therefore, since $s \in \mathbf{S}$ was arbitrary, $R \subseteq R_i$ for some $i \in I$. By definition of union $R_i \subseteq \bigcup_{i \in I} R_i$ so $R_i \subseteq R$. This shows that R is completely join-irreducible in $\mathbf{R}_u$. Dually, any angelic multirelation is completely meet-irreducible in $\mathbf{R}_u$. Since the demonic multirelations are the completely join-irreducible elements of $\mathbf{R}_u$, it follows by ([Dav79], Proposition 1.1(v)) that they are join-dense. Therefore the finite elements of $\mathbf{R}_u$ are finite joins of demonic multirelations.

(c) The family $\mathbf{R}_d$ contains $\top^+$ and is closed under non-empty meets since each multirelation $R \in \mathbf{R}_d$ is multiplicative. That is, $\bigcap_{i \in I} R_i \in \mathbf{R}_d$ for every non-empty indexed subfamily $\{R_i\}_{i \in I} \subseteq \mathbf{R}_d$. Any subfamily of $\mathbf{R}_d$ has at least one upper bound in $\mathbf{R}_d$, namely $\top^+$, and hence, by ([DaP90], Lemma 2.15), has a least upper bound in $\mathbf{R}_d$ is a topped intersection structure and hence, by ([DaP90], Corollary 2.17), a complete lattice with meets and joins as given above. The family $\mathbf{R}_d$ is also a topped algebraic intersection structure since it is closed under unions of directed subfamilies and is hence, by ([DaP90], Theorem 3.26(a)), an algebraic lattice in which the join of any directed subfamily is given by set union. By the characterisation of the finite elements of a topped intersection structure ([DaP90], Lemma 3.25), the finite elements of $\mathbf{R}_d$ are the finite demonic multirelations.

(d) Since any angelic multirelation is the dual of a demonic multirelation, (d) follows by dualising the proof for (c).

(e) The family $\mathbf{R}_t$ is closed under all meets since each multirelation $R \in \mathbf{R}_t$ is total and multiplicative. However, the multirelation $\top^+$ is not total so, in general, least upper bounds of subfamilies of $\mathbf{R}_t$ do not exist. Therefore, $\mathbf{R}_t$ os a complete meet-semilattice with meets as in $\mathbf{R}_u$. Take any deterministic multirelation R. Suppose there is some multirelation $R_1 \in \mathbf{R}_t$ such that $R \subseteq R_1$ and $R_1 \not\subseteq R$. Then for some $x \in \mathbf{S}$ there is some $Q \subseteq \mathbf{S}$ such that $Q \in R_1(x)$ and $Q \notin R(x)$. Since $R(x)$ is a principal prime filter in $\mathcal{P}(\mathbf{S}), \overline{Q} \in R(x) \subseteq R_1(x)$ and hence, by multiplicativity of $R_1, \emptyset = Q \cap \overline{Q} \in R_1(x)$ which contradicts the totality of R_1. This shows that the deterministic multirelations are maximal in $\mathbf{R}_t$.   □

3   Binary Multirelations and Binary Relations

The purpose of this rather technical section is to motivate the translations used in Section 4. For this we observe that binary relations may be viewed as demonic binary multirelations or as angelic binary multirelations. Then we extend the Jónsson/Tarski [JoT51,JoT52] translation between binary relations and meet-homomorphisms to a translation between demonic binary multirelations and meet homomorphisms. Dually, we obtain a translation between angelic binary multirelations and join-homomorphisms. Observing that these two translations are the same we obtain a translation between binary multirelations and monotone unary operators. Any binary relation r over the state space $\mathbf{S}$ may be viewed as an angelic binary multirelation $R_r \subseteq \mathbf{S} \times \mathcal{P}(\mathbf{S})$ given by:

$$sR_rQ \quad \text{iff} \quad (\exists t)[srt \text{ and } t \in Q], \qquad \text{for any } s \in \mathbf{S} \text{ and any } Q \in \mathcal{P}(\mathbf{S}),$$

and as a demonic binary multirelation $(R_r)^d \subseteq \mathbf{S} \times \mathcal{P}(\mathbf{S})$ given by:

$$s(R_r)^dQ \quad \text{iff} \quad (\forall t)[srt \Rightarrow t \in Q], \qquad \text{for any } s \in \mathbf{S} \text{ and any } Q \in \mathcal{P}(\mathbf{S}).$$

It is easy to check that R_r is an angelic binary multirelation and $(R_r)^d$ a demonic binary multirelation. To each demonic binary multirelation $R \subseteq \mathbf{S} \times \mathcal{P}(\mathbf{S})$ there corresponds some binary relation $r_R \subseteq \mathbf{S}^2$ given by $r_R(s) = \bigcap\{Q \mid Q \in R(s)\}$ for any $s \in \mathbf{S}$. Dually, if R is an angelic binary multirelation then the corresponding binary relation $r_R \subseteq \mathbf{S}^2$ is given by $r_R(s) = \bigcap\{Q \mid \overline{Q} \notin R(s)\}$ for any $s \in \mathbf{S}$. From any binary relation $r \subseteq \mathbf{S}^2$ we may define, using the techniques of Jónsson/Tarski, a join-homomorphism $g_r : \mathcal{P}(\mathbf{S}) \to \mathcal{P}(\mathbf{S})$ by

$$g_r(Q) = \{s \mid (\exists t)[srt \text{ and } t \in Q]\}, \qquad \text{for any } Q \in \mathcal{P}(\mathbf{S}),$$

and dually, a meet-homomorphism $(g_r)^d : \mathcal{P}(\mathbf{S}) \to \mathcal{P}(\mathbf{S})$ by

$$(g_r)^d(Q) = \{s \mid (\forall t)[srt \Rightarrow t \in Q]\}, \qquad \text{for any } Q \in \mathcal{P}(\mathbf{S}).$$

There is a pleasant connection between the angelic binary multirelation R_r and the join-homomorphism g_r of a binary relation $r \subseteq \mathbf{S}^2$, and between its demonic binary multirelation $(R_r)^d$ and its meet-homomorphism $(g_r)^d$. Namely,

Theorem 3. *For any binary relation $r \subseteq \mathbf{S}^2$ and any $Q \in \mathcal{P}(\mathbf{S})$,*

$$g_r(Q) = \{s \in \mathbf{S} \mid sR_rQ\} \quad \text{and} \quad (g_r)^d(Q) = \{s \in \mathbf{S} \mid s(R_r)^dQ\}.$$

On the other hand, consider any Boolean algebra with meet-homomorphisms $\mathcal{B} = (B, \vee, \wedge, ^-, 0, 1, \{g_i \mid i \in I\})$. Let $\mathcal{F}(B)$ denote the set of all prime filters of B and $\mathcal{F}_\circ(B)$ the set of all its filters. For any meet-homomorphism $g : B \to B$, the binary relation $r_g \subseteq \mathcal{F}(B)^2$ is given by

$$X r_g Y \quad \text{iff} \quad g^{-1}(X) \subseteq Y, \qquad \text{for any } X, Y \in \mathcal{F}(B).$$

The corresponding binary multirelation is then a subset of $\mathcal{F}(B) \times \mathcal{P}(\mathcal{F}(B))$. As a Corollary of the Ultrafilter Theorem for Boolean algebras ([DaP90], p 192) we have that every proper filter $Y \in \mathcal{F}_\circ(B)$ is the intersection of prime filters containing it, and hence determines and is determined by a set of prime filters, namely $A_Y = \{F \in \mathcal{F}(B) \mid Y \subseteq F\}$. Therefore, there is a bijective correspondence between $\mathcal{P}(\mathcal{F}(B))$ and $\mathcal{F}_\circ(B)$, and hence the demonic multirelation $R_{r_g} \subseteq \mathcal{F}(B) \times \mathcal{P}(\mathcal{F}(B))$ defined from r_g is, by the definition on page 264, given by

$$X R_{r_g} A_Y \quad \text{iff} \quad (\forall Z \in \mathcal{F}(B))[g^{-1}(X) \subseteq Z \Rightarrow Y \subseteq Z]$$

for any $X \in \mathcal{F}(B)$ and any $Y \in \mathcal{F}_\circ(B)$.

Theorem 4. *For any meet-homomorphism $g : B \to B$,*

$$X R_{r_g} A_Y \quad \text{iff} \quad Y \subseteq g^{-1}(X), \qquad \text{where } X \in \mathcal{F}(B) \text{ and } Y \in \mathcal{F}_\circ(B).$$

Proof. Take any $X \in \mathcal{F}(B)$ and any $Y \in \mathcal{F}_\circ(B)$. We need to show that

$$Y \subseteq g^{-1}(X) \quad \text{iff} \quad (\forall Z \in \mathcal{F}(B))[g^{-1}(X) \subseteq Z \Rightarrow Y \subseteq Z].$$

The left to right direction follows from the transitivity of the inclusion relation. For the right to left direction we prove the contrapositive, namely,

$$\text{if } Y \not\subseteq g^{-1}(X) \text{ then } (\exists Z \in \mathcal{F}(B))[g^{-1}(X) \subseteq Z \text{ and } Y \not\subseteq Z].$$

Suppose $Y \not\subseteq g^{-1}(X)$. Then there is some $y \in Y$ such that $y \notin g^{-1}(X)$. Since g is a meet-homomorphism it is monotone, so $\overline{g^{-1}(X)}$ is a down-set and hence $\overline{g^{-1}(X)}$ contains the principal ideal $\downarrow y$. Also $g^{-1}(X)$ is a filter such that $g^{-1}(X) \cap \downarrow y = \emptyset$. By the Ultrafilter Theorem for Boolean algebras [DaP90] there is some prime filter $Z \in \mathcal{F}(B)$ such that $g^{-1}(X) \subseteq Z$ and $\downarrow y \cap Z = \emptyset$ which means $g^{-1}(X) \subseteq Z$ and $Y \not\subseteq Z$ as required. $\qquad\square$

Dually, for any join-homomorphism $g : B \to B$, the binary relation $r_g \subseteq \mathcal{F}(B)^2$ is given by

$$X r_g Y \quad \text{iff} \quad Y \subseteq g^{-1}(X), \qquad \text{for any } X, Y \in \mathcal{F}(B),$$

and the angelic multirelation $R_{r_g} \subseteq \mathcal{F}(B) \times \mathcal{P}(\mathcal{F}(B))$ obtained from the binary relation r_g is, by the definition on page 264, given by:

$$X R_{r_g} A_Y \quad \text{iff} \quad (\exists Z \in \mathcal{F}(B))[Z \subseteq g^{-1}(X) \text{ and } Y \subseteq Z]$$

for any $X \in \mathcal{F}(B)$ and any $Y \in \mathcal{F}_\circ(B)$.

Theorem 5. *For any join-homomorphism* $g : B \to B$,

$$X R_{r_g} A_Y \quad \text{iff} \quad Y \subseteq g^{-1}(X), \qquad \text{where } X \in \mathcal{F}(B) \text{ and } Y \in \mathcal{F}_o(B).$$

Therefore, by Theorems 4 and 5, the definition of a binary multirelation from a meet-homomorphism is the same as that from a join-homomorphism, and therefore can be used more generally for any monotone unary operator.

4   Binary Multirelations and Monotone Unary Operators

In this section we establish a bijective correspondence between binary multirelational structures (i.e. a set endowed with some binary multirelations) and Boolean algebras with monotone unary operators.

Definition 3. *A binary multirelational structure* $\mathcal{S} = (\mathbf{S}, \{R_i \mid i \in I\})$ *is such that* $\mathbf{S}$ *is a set and* $\{R_i \mid i \in I\}$ *is a collection of up-closed binary multirelations over* $\mathbf{S}$.

For an up-closed binary multirelation $R \subseteq \mathbf{S} \times \mathcal{P}(\mathbf{S})$, a mapping g_R over $\mathcal{P}(\mathbf{S})$ is defined by

$$g_R(Q) = \{s \in \mathbf{S} \mid sRQ\}.$$

Since R is up-closed, g_R is monotone. By translating each up-closed binary multirelation R in a binary multirelational structure into a monotone unary operator g_R we obtain a certain kind of Boolean algebra with operators. Since we wish to establish a bijective correspondence we define these in general.

Definition 4. *A Boolean algebra with monotone unary operators*

$$\mathcal{B} = (B, \vee, \wedge, ^-, 0, 1, \{g_i \mid i \in I\})$$

is such that $\mathcal{B} = (B, \vee, \wedge, ^-, 0, 1)$ *is a Boolean algebra and* $\{g_i \mid i \in I\}$ *is a collection of monotone unary operators over* B.

In order to establish a bijective correspondence between Boolean algebras with monotone unary operators and binary multirelational structures, we must show how each gives rise to and can be recovered from the other. From above we have,

Theorem 6. *Given any binary multirelational structure* $\mathcal{S} = (\mathbf{S}, \{R_i \mid i \in I\})$ *its power algebra* $\mathcal{P}(\mathcal{S}) = (\mathcal{P}(\mathbf{S}), \cup, \cap, ^-, \emptyset, \mathbf{S}, \{g_{R_i} \mid i \in I\})$ *is a Boolean algebra with monotone unary operators.*

Next we show that any Boolean algebra with monotone unary operators in turn gives rise to a binary multirelational structure by invoking the basic Stone representation [Sto37]. That is, we represent the elements of the Boolean algebra as subsets of some universal set, namely the set of all prime filters, and then define binary multirelations over this universe. Let $\mathcal{B} = (B, \vee, \wedge, ^-, 0, 1, \{g_i \mid i \in I\})$ be a Boolean algebra with monotone unary operators, and let $\mathcal{F}(B)$ be the set of all prime filters in $\mathcal{B}$ considered as a Boolean algebra, and $\mathcal{F}_o(B)$ the set of all its filters. For each monotone operator $g : B \to B$, the binary multirelation $R_g \subseteq \mathcal{F}(B) \times \mathcal{P}(\mathcal{F}(B))$ is defined by

$$X R_g A_Y \quad \text{iff} \quad Y \subseteq g^{-1}(X), \qquad \text{for any } X \in \mathcal{F}(B) \text{ and any } Y \in \mathcal{F}_o(B)$$

where $A_Y = \{F \in \mathcal{F}(B) \mid Y \subseteq F\}$.

Theorem 7. *Given any Boolean algebra with monotone unary operators*

$$\mathcal{B} = (B, \vee, \wedge, ^-, 0, 1, \{g_i \mid i \in I\})$$

its prime filter structure $(\mathcal{F}(B), \{R_{g_i} \mid i \in I\})$ is a binary multirelational structure.

Proof. We need to show that for each monotone operator $g : B \to B$, the binary multirelation $R_g \subseteq \mathcal{F}(B) \times \mathcal{P}(\mathcal{F}(B))$ is up-closed. Take any $X \in \mathcal{F}(B)$ and $Y, Y_1 \in \mathcal{F}_o(B)$ such that $X R_g A_Y$ and $A_Y \subseteq A_{Y_1}$. Then $Y \subseteq g^{-1}(X)$ and

$$Y = \bigcap \{Q \in \mathcal{F}(B) \mid Y \subseteq Q\} \supseteq \bigcap \{Q \in \mathcal{F}(B) \mid Y_1 \subseteq Q\} = Y_1,$$

so $Y_1 \subseteq g^{-1}(X)$ and hence $A_{Y_1} \in R_g(X)$. □

Thus every binary multirelational structure gives rise to a Boolean algebra with monotone unary operators, and conversely. The next two theorems show that each can also be recovered from the other. Let $\mathcal{B} = (B, \vee, \wedge, ^-, 0, 1, \{g_i \mid i \in I\})$ be a Boolean algebra with monotone unary operators. Then the Stone [Sto37] mapping $h : B \to \mathcal{P}(\mathcal{F}(B))$, given by $h(a) = \{F \in \mathcal{F}(B) \mid a \in F\}$, is an embedding of the Boolean algebra $\mathcal{B} = (B, \vee, \wedge, ^-, 0, 1)$ into the powerset Boolean algebra

$$\mathcal{P}(\mathcal{F}(\mathcal{B})) = (\mathcal{P}(\mathcal{F}(B)), \cup, \cap, ^-, \emptyset, \mathcal{F}(B)).$$

Now we need to show that h preserves monotone unary operators over B.

Theorem 8. *Any Boolean algebra with monotone unary operators is isomorphic to a subalgebra of the Boolean algebra with unary operators of its underlying binary multirelational structure.*

Proof. Consider any monotone operator $g : B \to B$ and any $a \in B$. We have to show that $h(g(a)) = g_{R_g}(h(a))$.

$$
\begin{aligned}
g_{R_g}(h(a)) \quad &= \quad g_{R_g}(A_{\uparrow a}) \\
&\qquad \text{since } \uparrow a \in \mathcal{F}(B) \text{ and for } F \in \mathcal{F}(B), a \in F \text{ iff } \uparrow a \in F \\
&= \quad \{F \in \mathcal{F}(B) \mid A_{\uparrow a} \in R_g(F)\} \\
&\qquad \text{by definition on page 266 of } g_{R_g} \text{ from } R_g \\
&= \quad \{F \in \mathcal{F}(B) \mid \uparrow a \subseteq (g)^{-1}(F)\} \\
&\qquad \text{by definition on page 267 of } R_g \text{ from } g \\
&= \quad \{F \in \mathcal{F}(B) \mid a \in (g)^{-1}(F)\} \\
&\qquad \text{since } (g)^{-1}(F) \text{ is up} - \text{closed}
\end{aligned}
$$

$$= \quad \{F \in \mathcal{F}(B) \mid g(a) \in F\}$$
$$\text{by definition of } (\cdot)^{-1}$$
$$= \quad h(g(a))$$
$$\text{by definition of } h. \hspace{2cm} \square$$

Consider any binary multirelational structure $\mathcal{S} = (\mathbf{S}, \{R_i \mid i \in I\})$. The powerset $\mathcal{P}(\mathbf{S})$ of $\mathbf{S}$ endowed with the mappings g_R yields the Boolean algebra with monotone unary operators $\mathcal{P}(\mathcal{S}) = (\mathcal{P}(\mathbf{S}), \cup, \cap, ^-, \emptyset, \mathbf{S}, \{g_{R_i} \mid i \in I\})$. Forming the prime filter structure of this yields a binary multirelational structure which contains an isomorphic copy of the original binary multirelational structure. Each of the original up-closed binary multirelations R over $\mathbf{S}$ gives rise to a monotone unary operator $g_R : \mathcal{P}(\mathbf{S}) \to \mathcal{P}(\mathbf{S})$, which in turn gives rise to an up-closed binary multirelation R_{g_R} over $\mathcal{F}(\mathcal{P}(\mathbf{S}))$. There is a bijective correspondence between the elements of $\mathbf{S}$ and certain prime filters in $\mathcal{P}(\mathbf{S})$, namely the principal prime filters under the mapping $a \mapsto k(a) = \{A \subseteq \mathbf{S} \mid a \in A\}$. An extension of this mapping provides a bijective correspondence between subsets of $\mathbf{S}$ and principal filters, namely $Y \mapsto k(Y) = \{A \subseteq \mathbf{S} \mid Y \subseteq A\}$. We need to show that this mapping preserves structure.

Theorem 9. *Any binary multirelational structure is isomorphic to a substructure of the prime filter binary multirelational structure of its Boolean algebra with monotone unary operators.*

Proof. Consider any up-closed binary multirelation $R \subseteq \mathbf{S} \times \mathcal{P}(\mathbf{S})$. For $x \in \mathbf{S}$ and $Y \subseteq \mathbf{S}$, we show that

$$k(x)R_{g_R}A_{k(Y)} \quad \text{iff} \quad xRY, \quad \text{where } A_{k(Y)} = \{Q \in \mathcal{F}(\mathcal{P}(\mathbf{S})) \mid k(Y) \subseteq Q\}.$$

$$
\begin{aligned}
k(x)R_{g_R}A_{k(Y)} \quad &\text{iff} \quad && k(Y) \subseteq (g_R)^{-1}(k(x)) \\
& && \text{by definition on page 267 of } R_{g_R} \text{ from } g_R \\
&\text{iff} \quad && \{Z \in \mathcal{P}(\mathbf{S}) \mid Y \subseteq Z\} \subseteq (g_R)^{-1}(k(x)) \\
& && \text{by definition of } k(Y) \\
&\text{iff} \quad && (\forall Z \in \mathcal{P}(\mathbf{S}))[Y \subseteq Z \Rightarrow g_R(Z) \in k(x)] \\
& && \text{by definition of } \subseteq \\
&\text{iff} \quad && (\forall Z \in \mathcal{P}(\mathbf{S}))[Y \subseteq Z \Rightarrow x \in g_R(Z)] \\
& && \text{by definition of } k(x) \\
&\text{iff} \quad && (\forall Z \in \mathcal{P}(\mathbf{S}))[Y \subseteq Z \Rightarrow Z \in R(x)] \\
& && \text{by definition on page 266 of } g_R \text{ from } R \\
&\text{iff} \quad && Y \in R(x) \\
& && \text{since } R(x) \text{ is an up}-\text{set} \\
&\text{iff} \quad && xRY. \hspace{2cm} \square
\end{aligned}
$$

As a consequence of these bijective correspondences we are able to translate between properties of up-closed binary multirelations and properties of monotone unary operators.

Theorem 10. *Let $g : B \to B$ be a monotone unary operator. Then properties of g translate into properties of R_g as (i) and (ii) below. Conversely, let $R \subseteq \mathbf{S} \times \mathcal{P}(\mathbf{S})$ be an up-closed binary multirelation. Then the properties of R translate into properties of g_R as (ii) to (i) below.*

(a) (i) *g is 0-preserving*   (ii) *R is total;*
(b) (i) *g is 1-preserving*   (ii) *R is proper;*
(c) (i) *g is meet-preserving*   (ii) *R is multiplicative;*
(d) (i) *g is join-preserving*   (ii) *R is additive.*

Moreover, we are able to justify the definition on page 260 of composition of binary multirelations. Let R and S be up-closed binary multirelations over $\mathbf{S}$. For $x \in \mathbf{S}$ and $Q \in \mathcal{P}(\mathbf{S})$,

$$x \in (g_R \circ g_S)(Q) \quad \text{iff} \quad x \in g_R(g_S(Q))$$
$$\text{definition of functional composition}$$
$$\text{iff} \quad g_S(Q) \in \{V \mid xRV\}$$
$$\text{definition on page 266 of } g_R \text{ from } R$$
$$\text{iff} \quad (\exists Y)[xRY \text{ and } Y \subseteq g_S(Q)]$$
$$\text{since } R \text{ is up} - \text{closed}$$
$$\text{iff} \quad (\exists Y)[xRY \text{ and } (\forall y \in Y)[ySQ]]$$
$$\text{by definition of } \subseteq \text{ and } g_S \text{ from } S.$$

Another consequence of the bijective correspondences is a decomposition of any monotone unary operator into a meet- and join-homomorphism. For this it is important to note that any binary multirelation $R \subseteq \mathbf{S} \times \mathcal{P}(\mathbf{S})$ can also be viewed as a binary relation of the form $R \subseteq \mathbf{S} \times T$ (with $T = \mathcal{P}(\mathbf{S})$) and as such gives rise to a join-homomorphism $R^\uparrow : \mathcal{P}(\mathcal{P}(\mathbf{S})) \to \mathcal{P}(\mathbf{S})$ where

$$R^\uparrow(\mathbf{Q}) = \{s \mid (\exists Q)[sRQ \text{ and } Q \in \mathbf{Q}]\}, \quad \text{for any } \mathbf{Q} \in \mathcal{P}(\mathcal{P}(\mathbf{S})),$$

and a meet-homomorphism $R^{\uparrow^d} : \mathcal{P}(\mathcal{P}(\mathbf{S})) \to \mathcal{P}(\mathbf{S})$ where

$$R^{\uparrow^d}(\mathbf{Q}) = \{s \mid (\forall Q)[sRQ \Rightarrow Q \in \mathbf{Q}]\}, \quad \text{for any } \mathbf{Q} \in \mathcal{P}(\mathcal{P}(\mathbf{S})).$$

Theorem 11. *Any monotone unary operator $g : B \to B$ can be written as a composition of a join-homomorphism $f : \mathcal{P}(\mathcal{P}(\mathcal{F}(B))) \to B$ and a meet-homomorphism $h : B \to \mathcal{P}(\mathcal{P}(\mathcal{F}(B)))$.*

Proof. Take any monotone unary operator $g : B \to B$. Then, for any $a \in B$,

$$g(a) \quad = \quad (h^{-1} \circ h)(g(a))$$
$$\text{since } h \text{ is a bijection and by definition of inverses}$$
$$= \quad h^{-1}(h(g(a)))$$
$$\text{by associativity of composition}$$
$$= \quad h^{-1}(g_{R_g}(h(a)))$$
$$\text{by Theorem 8}$$

$$
\begin{aligned}
&= & & h^{-1}(\{\mathsf{X} \mid \mathsf{X}R_g h(a)\}) \\
& & & \text{by definition of } g_{R_g} \text{ from } R_g \\
&= & & h^{-1}(\{\mathsf{X} \mid (\exists \mathsf{Z})[\mathsf{X}R_g \mathsf{Z} \text{ and } (\forall Z \in \mathsf{Z})[a \in Z]]\}) \\
& & & \text{by definition of } h \\
&= & & h^{-1}((R_g)^\uparrow \circ (\in^{\smile})^{\uparrow^d})(h(a)) \\
& & & \text{by definition of } (\cdot)\uparrow \text{ and } (\cdot)^{\uparrow^d} \\
&= & & h^{-1} \circ (R_g)^\uparrow \circ (\in^{\smile})^{\uparrow^d} \circ h(a) \\
& & & \text{by definition of composition.}
\end{aligned}
$$

Since h is an embedding, and $(R_g)^\uparrow$ is a join-homomorphism and $(\in^{\smile})^{\uparrow^d}$ is a meet-homomorphism, it follows that $h^{-1} \circ (R_g)^\uparrow$ is a join-homomorphism and $(\in^{\smile})^{\uparrow^d} \circ h$ is a meet-homomorphism. Therefore, $g = j \circ m$ where j is the join-homomorphism $h^{-1} \circ (R_g)^\uparrow$ and m is the meet-homomorphism $(\in^{\smile})^{\uparrow^d} \circ h$. □

This decomposition of a monotone unary operator is a generalisation of that in [GMM94] for monotone unary operators over a powerset Boolean algebra and used in [GaM91] for proving the completeness of Morgan's [MoR87] refinement laws.

5   Conclusion

It has long been known that a predicate transformer semantics for commands with both angelic- and demonic nondeterminism can be given in terms of monotone unary operators. But there is no conventional relational model. Binary multirelations have been proposed here as a relational representation for such commands. As a justification of the viability of this representation we have established a duality showing that binary multirelations and monotone unary operators are intertranslatable.

References

[BaW89]   Back, R.J.R. and J. von Wright. [1989]. Refinement Calculus, part 1: Sequential Programs. In: *Rex Workshop for Refinement of Distributed Systems, Nijmegen, The Netherlands.* Lecture Notes in Computer Science **430**.

[BaW92]   Back, R.J.R. and J. von Wright. [1992]. Combining angels, demons and miracles in program specifications. *Theoretical Computer Science* **100**. p 365–383.

[BaW95]   Back, R.J.R. and J. von Wright. [1995]. Games and winning strategies. *Information Processing Letters* **53** (3). p 165–172.

[BaW98]   Back, R.J.R. and J. von Wright. [1998]. *Refinement Calclulus: A Systematic Introduction.* Graduate Texts in Computer Science. New York: Springer-Verlag.

[BeS71]   Bell, J.L. and A.R. Slomson. [1971]. *Models and Ultraproducts.* 2nd Edition. Amsterdam: North-Holland.

[Bri93]   Brink, C. [1993]. Power structures. *Algebra Universalis* **30**. p 177–216.

[BrR01]   Brink, C. and I. Rewitzky. [2001] *A Paradigm for Program Semantics: Power Sructures and Duality.* Stanford: CSLI Publications.

[Dav79]   Davey, B.A. [1979]. On the lattice of subvarieties. *Houston Journal of Mathematics* **5** (2). p 183–192.

[DaP90]   Davey, B.A. and H.A. Priestley. [1990]. *Introduction to Lattices and Order.* Cambridge: Cambridge University Press.

[Dij75]   Dijkstra, E.W. [1975]. Guarded commands, nondeterminacy and formal derivation of programs. *Communications of the ACM* **18** (8). p 453–458.

[Dij76]   ——. [1976]. *A Discipline of Programming.* Englewood Cliffs, New Jersey: Prentice-Hall.

[GaM91]   Gardiner, P.H. and C.C. Morgan. [1991]. Data refinement of predicate transformers. *Theoretical Computer Science* **87** (1). p 143–162.

[GMM94]   Gardiner, P.H., C.E. Martin and O. de Moor. [1994]. An algebraic construction of predicate transformers. *Science of Computer Programming* **22** (1-2). p 21–44.

[Gau57]   Gautam, N.D. [1957]. The validity of equations of complex algebra. *Archiv für Mathematische Logik und Grundlagenforschung* **3**. p 117–124.

[GHK80]   Gierz, G., K.H. Hofman, K. Keimel, J.D. Lawson, M. Mislove and D.S. Scott. [1980]. *A Compendium of Continuous Lattices.* Berlin: Springer-Verlag.

[Hes94]   Hesselink, W.H. [1994]. Nondeterminism and recursion via stacks and games. *Theoretical Computer Science* **124**. p 273-295.

[Hoa96]   Hoare, C.A.R. [1996]. An algebra of games of choice. Unpublished manuscript, 4 pages.

[JoT51]   Jónsson, B. and A. Tarski. [1951]. Boolean algebras with operators I. *American Journal of Mathematics* **73**. p 891–939.

[JoT52]   ——. [1952]. Boolean algebras with operators II. *American Journal of Mathematics* **74**. p 127–167.

[Mor88]   Morgan, C.C. [1988]. The specification statement. *ACM Transactions of Programming Language Systems* **10** (3). p 403-491.

[MoR87]   Morgan, C.C. and K.A. Robertson. [1987]. Specification statements and refinement. *IBM Journal of Research and Development* **31** (5).

[Nel89]   Nelson, G. [1989]. A generalisation of Dijkstra's calculus. *ACM Transactions on Programming Languages and Systems* **11** (4). p 517–562.

[ReB95]   Rewitzky, I. and C. Brink. [1995]. Predicate transformers as power operations. *Formal Aspects of Computing* **7**. p 169–182.

[Sto37]   Stone, M.H. [1937]. Topological representations of distributive lattices and Brouwerian logics. *Casopis Pro Potování Mathematiky* **67**. p 1–25.

Author Index

Lecture Notes in Computer Science

For information about Vols. 1–2834
please contact your bookseller or Springer-Verlag

Vol. 2873: J. Lawry, J. Shanahan, A. Ralescu (Eds.), Modelling with Words. XIII, 229 pages. 2003. (Subseries LNAI)

Vol. 2874: C. Priami (Ed.), Global Computing. Proceedings, 2003. XIX, 255 pages. 2003.

Vol. 2875: E. Aarts, R. Collier, E. van Loenen, B. de Ruyter (Eds.), Ambient Intelligence. Proceedings, 2003. XI, 432 pages. 2003.

Vol. 2876: M. Schroeder, G. Wagner (Eds.), Rules and Rule Markup Languages for the Semantic Web. Proceedings, 2003. VII, 173 pages. 2003.

Vol. 2877: T. Böhme, G. Heyer, H. Unger (Eds.), Innovative Internet Community Systems. Proceedings, 2003. VIII, 263 pages. 2003.

Vol. 2878: R.E. Ellis, T.M. Peters (Eds.), Medical Image Computing and Computer-Assisted Intervention - MICCAI 2003. Part I. Proceedings, 2003. XXXIII, 819 pages. 2003.

Vol. 2879: R.E. Ellis, T.M. Peters (Eds.), Medical Image Computing and Computer-Assisted Intervention - MICCAI 2003. Part II. Proceedings, 2003. XXXIV, 1003 pages. 2003.

Vol. 2880: H.L. Bodlaender (Ed.), Graph-Theoretic Concepts in Computer Science. Proceedings, 2003. XI, 386 pages. 2003.

Vol. 2881: E. Horlait, T. Magedanz, R.H. Glitho (Eds.), Mobile Agents for Telecommunication Applications. Proceedings, 2003. IX, 297 pages. 2003.

Vol. 2882: D. Veit, Matchmaking in Electronic Markets. XV, 180 pages. 2003. (Subseries LNAI)

Vol. 2883: J. Schaeffer, M. Müller, Y. Björnsson (Eds.), Computers and Games. Proceedings, 2002. XI, 431 pages. 2003.

Vol. 2884: E. Najm, U. Nestmann, P. Stevens (Eds.), Formal Methods for Open Object-Based Distributed Systems. Proceedings, 2003. X, 293 pages. 2003.

Vol. 2885: J.S. Dong, J. Woodcock (Eds.), Formal Methods and Software Engineering. Proceedings, 2003. XI, 683 pages. 2003.

Vol. 2886: I. Nyström, G. Sanniti di Baja, S. Svensson (Eds.), Discrete Geometry for Computer Imagery. Proceedings, 2003. XII, 556 pages. 2003.

Vol. 2887: T. Johansson (Ed.), Fast Software Encryption. Proceedings, 2003. IX, 397 pages. 2003.

Vol. 2888: R. Meersman, Zahir Tari, D.C. Schmidt et al. (Eds.), On The Move to Meaningful Internet Systems 2003: CoopIS, DOA, and ODBASE. Proceedings, 2003. XXI, 1546 pages. 2003.

Vol. 2889: Robert Meersman, Zahir Tari et al. (Eds.), On The Move to Meaningful Internet Systems 2003: OTM 2003 Workshops. Proceedings, 2003. XXI, 1096 pages. 2003.

Vol. 2891: J. Lee, M. Barley (Eds.), Intelligent Agents and Multi-Agent Systems. Proceedings, 2003. X, 215 pages. 2003. (Subseries LNAI)

Vol. 2892: F. Dau, The Logic System of Concept Graphs with Negation. XI, 213 pages. 2003. (Subseries LNAI)

Vol. 2893: J.-B. Stefani, I. Demeure, D. Hagimont (Eds.), Distributed Applications and Interoperable Systems. Proceedings, 2003. XIII, 311 pages. 2003.

Vol. 2894: C.S. Laih (Ed.), Advances in Cryptology - ASIACRYPT 2003. Proceedings, 2003. XIII, 543 pages. 2003.

Vol. 2895: A. Ohori (Ed.), Programming Languages and Systems. Proceedings, 2003. XIII, 427 pages. 2003.

Vol. 2896: V.A. Saraswat (Ed.), Advances in Computing Science – ASIAN 2003. Proceedings, 2003. VIII, 305 pages. 2003.

Vol. 2897: O. Balet, G. Subsol, P. Torguet (Eds.), Virtual Storytelling. Proceedings, 2003. XI, 240 pages. 2003.

Vol. 2898: K.G. Paterson (Ed.), Cryptography and Coding. Proceedings, 2003. IX, 385 pages. 2003.

Vol. 2899: G. Ventre, R. Canonico (Eds.), Interactive Multimedia on Next Generation Networks. Proceedings, 2003. XIV, 420 pages. 2003.

Vol. 2901: F. Bry, N. Henze, J. Maluszyński (Eds.), Principles and Practice of Semantic Web Reasoning. Proceedings, 2003. X, 209 pages. 2003.

Vol. 2902: F. Moura Pires, S. Abreu (Eds.), Progress in Artificial Intelligence. Proceedings, 2003. XV, 504 pages. 2003. (Subseries LNAI).

Vol. 2903: T.D. Gedeon, L.C.C. Fung (Eds.), AI 2003: Advances in Artificial Intelligence. Proceedings, 2003. XVI, 1075 pages. 2003. (Subseries LNAI).

Vol. 2904: T. Johansson, S. Maitra (Eds.), Progress in Cryptology – INDOCRYPT 2003. Proceedings, 2003. XI, 431 pages. 2003.

Vol. 2905: A. Sanfeliu, J. Ruiz-Shulcloper (Eds.), Progress in Pattern Recognition, Speech and Image Analysis. Proceedings, 2003. XVII, 693 pages. 2003.

Vol. 2906: T. Ibaraki, N. Katoh, H. Ono (Eds.), Algorithms and Computation. Proceedings, 2003. XVII, 748 pages. 2003.

Vol. 2910: M.E. Orlowska, S. Weerawarana, M.P. Papazoglou, J. Yang (Eds.), Service-Oriented Computing – ICSOC 2003. Proceedings, 2003. XIV, 576 pages. 2003.

Vol. 2911: T.M.T. Sembok, H.B. Zaman, H. Chen, S.R. Urs, S.H.Myaeng (Eds.), Digital Libraries: Technology and Management of Indigenous Knowledge for Global Access. Proceedings, 2003. XX, 703 pages. 2003.

Vol. 2913: T.M. Pinkston, V.K. Prasanna (Eds.), High Performance Computing – HiPC 2003. Proceedings, 2003. XX, 512 pages. 2003.

Vol. 2914: P.K. Pandya, J. Radhakrishnan (Eds.), FST TCS 2003: Foundations of Software Technology and Theoretical Computer Science. Proceedings, 2003. XIII, 446 pages. 2003.

Vol. 2916: C. Palamidessi (Ed.), Logic Programming. Proceedings, 2003. XII, 520 pages. 2003.

Vol. 2918: S.R. Das, S.K. Das (Eds.), Distributed Computing – IWDC 2003. Proceedings, 2003. XIV, 394 pages. 2003.

Vol. 2923: V. Lifschitz, I. Niemelä (Eds.), Logic Programming and Nonmonotonic Reasoning. Proceedings, 2004. IX, 365 pages. 2004. (Subseries LNAI).

Vol. 2927: D. Hales, B. Edmonds, E. Norling, J. Rouchier (Eds.), Multi-Agent-Based Simulation III. Proceedings, 2003. X, 209 pages. 2003. (Subseries LNAI).

Vol. 2929: H. de Swart, E. Orlowska, G. Schmidt, M. Roubens (Eds.), Theory and Applications of Relational Structures as Knowledge Instruments. Proceedings. VII, 273 pages. 2003.